SIAM Studies in Applied Mathematics

This series of monographs focuses on mathematics and its applications to problems of current concern to industry, government, and society. These monographs will be of interest to applied mathematicians, numerical analysts, statisticians, engineers, and scientists who have an active need to learn useful methodology.

Series List

Vol. 1 *Lie–Bäcklund Transformations in Applications*
Robert L. Anderson and Nail H. Ibragimov

Vol. 2 *Methods and Applications of Interval Analysis*
Ramon E. Moore

Vol. 3 Ill-Posed Problems for Integrodifferential Equations in Mechanics and Electromagnetic Theory
Frederick Bloom

Vol. 4 *Solitons and the Inverse Scattering Transform*
Mark J. Ablowitz and Harvey Segur

Vol. 5 *Fourier Analysis of Numerical Approximations of Hyperbolic Equations*
Robert Vichnevetsky and John B. Bowles

Vol. 6 *Numerical Solution of Elliptic Problems*
Garrett Birkhoff and Robert E. Lynch

Vol. 7 *Analytical and Numerical Methods for Volterra Equations*
Peter Linz

Vol. 8 *Contact Problems in Elasticity: A Study of Variational Inequalities and Finite Element Methods*
N. Kikuchi and J. T. Oden

Vol. 9 *Augmented Lagrangian and Operator-Splitting Methods in Nonlinear Mechanics*
Roland Glowinski and P. Le Tallec

Vol. 10 *Boundary Stabilization of Thin Plate Splines*
John E. Lagnese

Vol. 11 *Electro-Diffusion of Ions*
Isaak Rubinstein

Vol. 12 *Mathematical Problems in Linear Viscoelasticity*
Mauro Fabrizio and Angelo Morro

Vol. 13 *Interior-Point Polynomial Algorithms in Convex Programming*
Yurii Nesterov and Arkadii Nemirovskii

Vol. 14 *The Boundary Function Method for Singular Perturbation Problems*
Adelaida B. Vasil'eva, Valentin F. Butuzov, and Leonid V. Kalachev

Vol. 15 *Linear Matrix Inequalities in System and Control Theory*
Stephen Boyd, Laurent El Ghaoui, Eric Feron, and Venkataramanan Balakrishnan

The Boundary Function Method for Singular Perturbation Problems

Adelaida B. Vasil'eva, Valentin F. Butuzov, and Leonid V. Kalachev

The Boundary Function Method for Singular Perturbation Problems

Society for Industrial and Applied Mathematics

siam *Philadelphia*

10 9 8 7 6 5 4 3 2 1

Library of Congress Cataloging-in-Publication

Vasil'eva, A. B. (Adelaida Borisovna), 1926–
The boundary function method for singular perturbation problems / Adelaida B. Vasil'eva, Valentin F. Butuzov, and Leonid V. Kalachev.
p. cm. — (SIAM studies in applied mathematics ; vol. 14)
Includes bibliographical references and index.
ISBN 0-89871-333-1
1. Boundary value problems—Numerical solutions. 2. Singular perturbations (Mathematics) I. Butuzov, V. F. (Valentin Fedorovich) II. Kalachev, Leonid V. III. Title. IV. Series: SIAM studies in applied mathematics ; 14.
QA379.V39 1995 94-42996
515'.35—dc20

Contents

Foreword

For many years, Adelaida Vasil'eva and her students and colleagues in a mathematics section of the Physics Department at Moscow State University have been doing substantial research on singular perturbation theory. This began with Vasil'eva's own thesis work, more than 40 years ago, under A. N. Tikhonov. It extensively developed his early, but somewhat incomplete, ideas on stability and boundary layer behavior for systems of ordinary differential equations. Under Vasil'eva's scientific leadership, the group typically carried out fairly detailed and elaborate asymptotic expansions for solutions of boundary layer problems, and they methodically justified the resulting approximations by carefully estimating the remainders.

The work was regularly reported in journal and survey articles and in the 1973 and 1990 textbooks *Asymptotic Expansions of Solutions to Singularly Perturbed Equations* and *Asymptotic Methods in the Theory of Singular Perturbations*, both published in Russian, coauthored jointly by Vasil'eva and her former student and colleague Valentin Butozov. The monographs' coverage of the group's innovative work is excellent (in addition to presenting the influential work of Vishil and Lyusternik), but significant asymptotics done elsewhere (at Moscow State University, in various institutes in Moscow, the Ukraine, and other parts of the former Soviet Union, as well as throughout the world) receive little direct attention. The importance of the group's effort became universally recognized, especially as its focus broaden to include partial differential equations of various types, as well as significant applications, such as combustion, chemical kinetics, and semiconductor device modeling. Among the most interesting mathematical issues clarified are (i) problems in the "critical" case, where the limiting equations have a manifold of solutions and nontraditional means must be employed to completely specify the limiting solution, and (ii) the construction of spike-like solutions and treatment of dissipative (or "contrast") structures. A short 1978 monograph on *Singularly Perturbed Equations in the Critical Case* was available in an English translation by F. A. Howes as a Mathematics Research Center report. The later publications largely employed the boundary function method of correction outer expansions in local regions of rapid change, rather than classical matching of inner and out expansions.

Leonid Kalachev came to the United States from Moscow, where he had been a student of Butuzov, about four years ago. He has a particularly broad understanding of applied asymptotics and related computational methods, and unusually good skills in languages and in communicating technical material. He is also very familiar with most relevant literature in English and Russian. We are, therefore, very fortunate that Leonid has been able to prepare this monograph for publication in English by SIAM. It is far from a literal translation. The text emphasizes mathematical motivations, but supresses proofs. It also nicely displays the importance and continuing promise of singular perturbations in applied contexts, as well as much outstanding mathematics that forms the best legacy of 50 years of this school of Tikhonov.

Robert E. O'Malley, Jr.
Seattle, February 1994

Preface

This book was written as a textbook on one of the effective asymptotic methods in the theory of *singular perturbations*, the *boundary function method.* Singularly perturbed equations are often used as mathematical models describing processes in physics, chemical kinetics, and mathematical biology, and they often arise during investigation of applied problems of technology and engineering. This explains the interest of applied scientists in asymptotic methods (both from the mathematical and practical viewpoints). Historical aspects of development of different asymptotic techniques can be found, e.g., in O'Malley [114]; a short review of the history of research on singularly perturbed problems in Russia (and other parts of the former Soviet Union) is given in Vasil'eva [147].

There are a number of introductory as well as advanced books on asymptotic methods that discuss some aspects of different perturbation techniques and their applications. Among them, let us mention, e.g., the monographs and textbooks by Chang and Howes [41], Cole [43], Bush [14], Erdélyi [48], Hinch [62], Hoppensteadt [64], Kevorkian [76], Murdock [97], Murray [98], Nayfeh [102]–[104], O'Malley [113], [114], Smith [128], Van Dyke [139], Vishik [162], and Wasow [166]. Many books are devoted to a description of some specific asymptotic method and present material starting from the basic underlying ideas and finishing with sophisticated calculations for concrete examples. This book can be considered both as an introduction to the basics of one of the asymptotic methods, the *boundary function method*, also known as the *method of boundary layer corrections* (cf. O'Malley [113], [114], Smith [128]), and as the survey of results, most of which (especially on the problems for partial differential equations) were presented only in the journal literature. The core of the book reflects the development of this asymptotic technique during the last forty years in the group of *small parameter* specialists in the Physics Department of Moscow State University (starting with the seminal results by Tikhonov [129]–[131], which were extended by Vasil'eva, Butuzov, and several generations of their students). Two monographs on the application of the boundary function method to singularly perturbed ordinary differential equations, integro-differential equations and equations with small delay as well as to singularly perturbed problems in the critical cases (*singular singularly perturbed* problems) were written by Vasil'eva and Butuzov [148], [149].

This book is based on the Russian edition of the textbook [151] by Vasil'eva and Butuzov. The American edition, which differs in many aspects from the Russian edition, was prepared by Kalachev. The text of the book has been revised substantially, some new material has been added in every chapter, and more examples, as well as the exercises and new references on asymptotic methods and their applications, were included.

The book consists of four chapters. In the first chapter, some basic ideas and definitions are given. The notions of *regular* and *singular* perturbations, asymptotic series, and asymptotic approximations with a small parameter are introduced. Some characteristic examples of asymptotic expansions for solutions of regularly and singularly perturbed problems, as well as the notion of a *boundary layer*, are discussed.

The second chapter starts with an analysis of the classical *theorem on the passage to the limit.* Then the algorithm for construction of the asymptotic approximation to the solution of initial value problems is described. The characteristic feature of the solution in this case is the presence of a boundary layer in the vicinity of the initial point. Next, discussion of singularly perturbed problems in the *critical cases* is presented, together with examples from chemical kinetics. A brief survey of results on boundary value problems with boundary and *interior* layers is given. A special section is devoted to *spike-type solutions* (contrast structures) along with the survey of results on their stability.

In the third chapter singularly perturbed partial differential equations are considered. In the beginning, the classical method of Vishik and Lyusternik is described on a simple example of an elliptic equation in a domain with *smooth boundary.* The next sections highlight more recent results that allow us to construct the asymptotic solution even in cases when the method of Vishik and Lyusternik, applied directly, does not work, as is the case, e.g., for problems in *domains with corner points.* Depending on the type of equation, we may use either the boundary function method, the *smoothing procedure*, or a combination of these two methods. Further, partial differential equations in *the critical cases* appearing, e.g., in the problems of chemical kinetics with diffusion, as well as asymptotic solutions of parabolic equations that are periodic in time, and hyperbolic systems are considered in this chapter.

We need to mention specifically that the main attention in Chapters 2 and 3 is devoted to algorithms to construct the asymptotic approximations of the solutions. To a much lesser extent, the questions of justification of asymptotic correctness of asymptotic solutions are discussed. The proofs are included only for problems where they are short. If the proof is lengthy, either the outline of the proof is presented or references are mentioned where the complete proof can be found.

The fourth chapter is entirely devoted to applied problems that are formulated in terms of singularly perturbed equations. The physical meaning of the small parameters for such problems is discussed. Asymptotic solutions for some problems of *chemical kinetics* and *combustion*, *semiconductor physics*, *heat conduction*, and *mathematical biology* are constructed using the boundary function method. Conclusions, derived on the basis of constructed approximations, are discussed together with possible extensions of the results. Due to the limited length of the book, some of the problems solved by the method are only briefly mentioned, and the corresponding references are presented, so the interested reader can easily find the detailed analysis in the journal literature.

We should mention that the boundary function method, although widely used, is not universal. It is applied most successfully to problems whose solutions exhibit boundary layer types of behavior. There exist problems where this method does not work. However, for those problems to which this method can be applied, it works in a most effective and easy way, which makes it advantageous as compared with the other methods. The boundary function method makes it possible to construct the asymptotic solution in a form that provides a uniform asymptotic approximation for the solution of the original problem over the domain of interest and allows one to estimate the remainder term, which justifies the algorithm. Conditions that allow one to verify for *each particular problem* whether the boundary function method can or cannot be used to obtain the asymptotic solution are discussed in this book for quite general classes of ordinary and partial differential equations. Let us name some *other asymptotic methods* described in the literature (including the monographs devoted exclusively to these methods): *matching* of asymptotic expansions (cf. Il'in [67], Kevorkian and Cole [77], Lagerstrom [83], Eckhaus [46], [47]), matching applied to *relaxation oscillations* (cf. Mischenko and Rosov [94], Grasman [56]), *WKB-method* and matching for *turning point problems* (cf. Fedoriuk [50], Maslov [90], Olver [112]), *multiple scales* techniques (cf. Cole [43], Kevorkian and Cole [77], Kapila [73]), *averaging* method (cf. Bogoliubov and Mitropolskii [12], Mitropolskii [95], Moiseev [96], Sanders and Verhulst [123]), *regularization* of singularly perturbed problems (cf. Lomov [87]).

This book is specifically written so it could be used as supplementary material in the courses on asymptotic methods and applied mathematics at the undergraduate or graduate levels, as well as for self-study by applied scientists who use asymptotic methods in their work. The authors intentionally omitted a lot of cumbersome details (especially in the proofs of theorems) and included a large number of examples and exercises that illustrate both the method and its numerous applications. We hope that this book will be a helpful addition to a number of already existing books devoted to singular perturbations.

The authors thank Professor R.E. O'Malley, Jr., who read the manuscript of the American edition and made numerous suggestions that considerably improved the final version of the text. The authors also thank Natalia Kalachev for help in the technical preparation of the manuscript. Leonid Kalachev would like to express his gratitude to all the faculty of the Department of Applied Mathematics of the University of Washington for moral support of his work on the American edition of the book.

Chapter 1

Basic Ideas

1.1 Regular and singular perturbations

Applied mathematics studies natural processes using mathematical models for them. Any mathematical model is somewhat approximate. It is not absolutely adequate for the process it describes. Deriving the mathematical model, one tries to capture all essential, dominant features of the process. On the other hand, the model should be "simple" enough to allow the analytical and/or numerical treatment leading to the information one wants to obtain about the process. A variety of models in physics, chemical kinetics, mathematical biology, and many other fields are quite naturally formulated in terms of differential equations. During the derivation of the model equations, some terms whose influence on the process is supposed to be negligible are often not taken into account. As a result, the model might be simplified considerably. Such simplifications often rely on physical intuition.

It is natural to pose the question about the role of these omitted terms. Is their influence on the process is really unimportant or, to the contrary, does including these terms in the model lead to a substantial change in the information we obtain from the mathematical model? To answer this question correctly, we must consider a more complicated extended model that includes the small terms that were omitted in the simplified version. Next, we study how close the solutions of the two models (extended and simplified) are.

If the model is formulated in terms of differential equations, taking into account the additional "small" terms will often lead to the appearance of some small factors (parameters) multiplying some of the terms in appropriately nondimensionalized equations. Then the question on the influence of the small parameters on the process reduces to the study of the dependence of solutions of such differential equations on small parameters. The terms containing small parameters are called *perturbations.* The extended model is then called perturbed, and the simplified model (that does not contain the small parameters) is called unperturbed.

Perturbations that occur in different problems can be formally divided into two classes: *regular perturbations* and *singular perturbations* . Before giving formal definitions, we mention that the primary qualitative difference between these two kinds of perturbations is that regular perturbations lead to small changes from the solution of the unperturbed problem. Unlike such regular perturbations, singular perturbations, thought to be small in some sense, cause considerable changes in the solution.

Let us now give some formal definitions. We consider two equations:

equation A_0: $L_0 u = f_0$;

equation A_ε: $L_0 u + \varepsilon L_1 u = f_0 + \varepsilon f_1$.

Here L_0 and L_1 are given operators, f_0 and f_1 are known functions, ε is a small scalar parameter (in what follows, we will consider $\varepsilon > 0$), u is the unknown function of the independent variable x (x might be either one- or multidimensional). Equation A_0 might be a simplified model of some process, and then equation A_ε corresponds to the extended model. The terms $\varepsilon L_1 u$ and εf_1 represent perturbations. If A_0 and A_ε are differential equations, we must add necessary initial (and/or boundary) conditions. These conditions might also contain a small parameter ε. Combining the equations and any corresponding additional conditions, we can consider the problems A_0 and A_ε. We denote the solution of A_0 by $u_0(x)$, and the solution of A_ε by $u_\varepsilon(x)$ for $x \in D$, where D is some domain.

The main question of perturbation theory might be posed as follows: does the difference $u_\varepsilon(x) - u_0(x)$ approach zero (in some norm space) as $\varepsilon \to 0$? In some sense the answer to this question depends on the choice of norm. In this book by the norm of a vector $u(x) = \mathrm{col}(u_1(x), \ldots, u_k(x))$ at a point x, we will understand the Euclidean norm $\| u(x) \| = \sqrt{u_1^2(x) + \ldots + u_k^2(x)}$. For example, if $u(x)$ is a scalar, then $\| u(x) \| = |u(x)|$. Hence, the norm of $u(x)$ depends on x.

Let us now formulate the definitions of regularly perturbed and singularly perturbed problems.

Definition 1.1 *The problem A_ε is called regularly perturbed in a domain D if*

$$\sup_D \| u_\varepsilon(x) - u_0(x) \| \to 0 \quad \text{when} \quad \varepsilon \to 0.$$

Otherwise, the problem A_ε is called singularly perturbed .

Here $\sup_D f(x)$ stands for the least upper bound of $f(x)$ in D. It follows from the definition that for a regularly perturbed problem the solution $u_0(x)$ of A_0 will be close to the solution u_ε of A_ε in the entire domain D for all sufficiently small ε. However, if the problem A_ε is singularly perturbed, then $u_0(x)$ will not be close to u_ε for all small ε at least in some part of domain D.

Let us illustrate the definition by several simple examples.

Example 1. Consider the initial value problem for scalar differential equation

$$A_\varepsilon : \quad \frac{du}{dx} = -u + \varepsilon x, \qquad 0 \le x \le 1, \quad u(0) = 1.$$

The exact solution is

$$u_\varepsilon(x) = (1 + \varepsilon)\exp(-x) + \varepsilon(x - 1).$$

Problem A_0 corresponding to A_ε has the form

$$A_0 : \quad \frac{du}{dx} = -u, \qquad 0 \le x \le 1, \quad u(0) = 1.$$

Hence,

$$u_0(x) = \exp(-x),$$

and

$$\sup_{[0,1]} \| u_\varepsilon(x) - u_0(x) \| = \varepsilon \max_{[0,1]} |\exp(-x) + x - 1| = \varepsilon \exp(-1) \to 0,$$

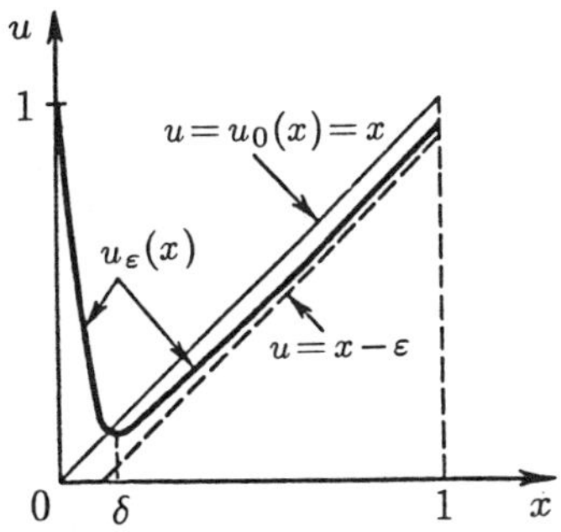

Figure 1.1: *The exact solution $u_\varepsilon(x)$ for small ε and the solution $u_0(x)$ of the reduced equation.*

as $\varepsilon \to 0$. According to our definition, the problem A_ε is regularly perturbed.

A generalization of this example is given by the initial value problem for the vector system of differential equations

$$A_\varepsilon : \quad \frac{du}{dx} = f(u, x, \varepsilon), \qquad 0 \le x \le x_0, \quad u(0) = u^0.$$

If $f(x)$ is a sufficiently smooth vector function, then A_ε also turns out to be regularly perturbed. This problem will be considered in § 1.3 in more detail.

It is worthwhile to emphasize the following important fact: we have considered the problem A_ε in the bounded (finite) interval $0 \le x \le x_0$. If A_ε were considered in the infinite interval $0 \le x \le \infty$, then it generally becomes singularly perturbed.

Example 2. Consider the initial value problem for the scalar differential equation, where small parameter $\varepsilon > 0$ enters as a factor in front of the derivative term:

$$A_\varepsilon : \quad \varepsilon \frac{du}{dx} = -u + x, \qquad 0 \le x \le 1, \quad u(0) = 1.$$

The solution of the problem is

$$u_\varepsilon(x) = (1 + \varepsilon)\exp(-x/\varepsilon) + x - \varepsilon.$$

In this case the equation A_0 that we obtain from A_ε for $\varepsilon = 0$ is not a differential, but an algebraic equation:

$$A_0 : \quad 0 = -u + x,$$

and therefore it does not need any initial or other condition to specify its solution. We have $u_0(x) = x$, and hence,

$$\sup_{[0,1]} \| u_\varepsilon(x) - u_0(x) \| = \max_{[0,1]} |(1 + \varepsilon)\exp(-x/\varepsilon) - \varepsilon| = 1.$$

Thus, $\sup_{[0,1]} \| u_\varepsilon(x) - u_0(x) \|$ does not approach zero as $\varepsilon \to 0$, and therefore, according to our definition, the problem A_ε is singularly perturbed.

The graphs of $u_0(x)$ and $u_\varepsilon(x)$ for small $\varepsilon > 0$ are presented in Fig. 1.1.

Note two features characteristic of this problem.

1. Equation A_0 is not a differential equation: for $\varepsilon = 0$, the differential equation A_ε reduces to an algebraic equation A_0. In this book we will call such A_0 the *reduced* (or *degenerate*) equation.

2. The solution $u_0(x)$ of A_0 fails to be close to the solution $u_\varepsilon(x)$ of the problem A_ε for small ε only in some small δ-vicinity of the initial point, but in the interval $[\delta, 1]$ (where δ is an arbitrarily small but fixed number) $u_0(x)$ is close to $u_\varepsilon(x)$ (see Fig. 1.1). The subinterval $[0, \delta]$, where the fast change of the solution $u_\varepsilon(x)$ from the initial value to values close to $u_0(x)$ takes place, is called a *boundary layer.*

Typical representatives of singularly perturbed problems are differential equations with small parameters multiplying the highest derivatives, for which the phenomenon of a boundary layer takes place. Such equations will be studied in this book. Our main goal will be the construction of approximations to the solution $u_\varepsilon(x)$ valid outside the boundary layer as well as within the boundary layer, i.e., so- called *uniform approximations* in the entire x-domain. This approximation will have an asymptotic character. The exact definition of an asymptotic approximation with respect to a small parameter will be introduced in § 1.2.

The following system generalizes Example 2:

$$A_\varepsilon \quad \begin{cases} \varepsilon \dfrac{dz}{dx} = F(z, y, x, \varepsilon), \\[2ex] \dfrac{dy}{dx} = f(z, y, x, \varepsilon), \qquad 0 \le x \le x_0, \\[2ex] z(0) = z^0, \qquad y(0) = y^0. \end{cases}$$

This system, together with the other initial and boundary value problems for singularly perturbed ordinary differential equations, will be studied in Chapter 2.

Exercise

For the following problems find the exact solutions u_ε and the solutions u_0 of the reduced problems with $\varepsilon = 0$. Check using Definition 1.1 whether these problems are regularly or singularly perturbed.

(a) $\dfrac{d^2u}{dx^2} + 2\dfrac{du}{dx} + \varepsilon u = 0, \quad 0 \le x \le 1, \quad \dfrac{du}{dx}(0) = 0, \quad u(0) = 1.$

(b) $\varepsilon^2\dfrac{d^2u}{dx^2} + 4\varepsilon\dfrac{du}{dx} + 3u = 0, \quad 0 \le x \le 1, \quad \dfrac{du}{dx}(0) = 0, \quad u(0) = 1.$

Example 3. Consider the following elliptic partial differential equation subject to Dirichlet boundary conditions:

$$A_\varepsilon \quad \begin{cases} \varepsilon \triangle u - k^2(x, y)u = f(x, y, \varepsilon), \qquad (x, y) \in \Omega, \\[2ex] u|_{\partial\Omega} = 0. \end{cases}$$

Here $\varepsilon > 0$ is a small parameter, $\triangle = \partial^2/\partial x^2 + \partial^2/\partial y^2$ is the Laplace operator, Ω is a bounded domain with a smooth boundary $\partial\Omega$. In this example, as well as in Example 2, the small parameter enters as a factor multiplying the derivatives. Problem A_0 obtained from A_ε by taking $\varepsilon = 0$ consists of one algebraic equation

$$A_0 : \quad -k^2(x, y)u = f(x, y, 0),$$

for which we need not impose any boundary conditions. Thus,

$$u_0 = -f(x, y, 0)/k^2(x, y).$$

Alhough the solution $u_\varepsilon(x, y)$ of problem A_ε is not yet known, it is a priori evident that $u_0(x, y)$ for small ε will not generally be close to $u_\varepsilon(x, y)$ at least in the vicinity of the boundary $\partial\Omega$ since $u_\varepsilon|_{\partial\Omega} = 0$, but $u_0(x, y)$ does not necessarily satisfy this condition. Therefore $\sup_\Omega \| u_\varepsilon(x, y) - u_0(x, y) \|$ does not approach zero as $\varepsilon \to 0$ and, according to our definition, problem A_ε is singularly perturbed. This problem will be discussed in Chapter 3. Applying the method of Vishik–Lyusternik, we will find that in the vicinity of $\partial\Omega$ a boundary layer is formed. Within the domain Ω (outside the boundary layer) the solution $u_0(x, y)$ of A_0 will be close to the solution $u_\varepsilon(x, y)$ of A_ε. The method of Vishik–Lyusternik will provide the uniform approximation to $u_\varepsilon(x, y)$ in $\overline{\Omega} = \Omega + \partial\Omega$. In Chapter 3 many other singularly perturbed problems for partial differential equations will also be considered.

1.2 Asymptotic approximations. Asymptotic and convergent series

Let $u_\varepsilon(x)$ be a solution of the problem A_ε defined in a domain D. Consider a function $U(x, \varepsilon)$ defined in a subdomain D_1 of D. In particular, D_1 might coincide with D. Recall that we assumed that $\varepsilon > 0$.

Definition 1.2 *The function $U(x, \varepsilon)$ is called an asymptotic approximation of the solution $u_\varepsilon(x)$ with respect to the parameter ε in the subdomain D_1 if*

$$\sup_{D_1} \| u_\varepsilon(x) - U(x, \varepsilon) \| \to 0 \quad \textit{as} \quad \varepsilon \to 0.$$

Moreover, if $\sup_{D_1} \| u_\varepsilon(x) - U(x, \varepsilon) \| = O(\varepsilon^k)$, then we say that $U(x, \varepsilon)$ is the asymptotic approximation of $u_\varepsilon(x)$ in D_1 to within accuracy of the order ε^k.

The symbolic notation $\alpha(\varepsilon) = O(\varepsilon^k)$ means that there exist the numbers $c > 0$ and $\varepsilon_0 > 0$ such that for $0 < \varepsilon \leq \varepsilon_0$ the inequality $\| \alpha(\varepsilon) \| \leq c\varepsilon^k$ holds. In what follows we will also use the symbolic notation $\alpha(\varepsilon) = o(\varepsilon^k)$, which means that

$$\lim_{\varepsilon \to 0} \| \alpha(\varepsilon) \| / \varepsilon^k = 0.$$

To illustrate this definition, let us return to Examples 1 and 2 of § 1.1. In Example 1 the function $U = u_0(x)$ is the asymptotic approximation of the solution $u_\varepsilon(x)$ in the whole interval $D = \{0 \leq x \leq 1\}$ with an accuracy of the order ε. In Example 2 the function $U = u_0(x)$ gives the asymptotic approximation of $u_\varepsilon(x)$ with accuracy of the order ε not in the entire interval $D = \{0 \leq x \leq 1\}$, but only in the subintervals $D_1 = \{\delta \leq x \leq 1\}$ for $\delta > 0$, i.e., outside the boundary layer.

By the *asymptotic method*, we understand some method of construction of an asymptotic approximation $U(x, \varepsilon)$ for the solution $u_\varepsilon(x)$ of problem A_ε. Often the construction of $U(x, \varepsilon)$ is reduced to solving a set of the problems that are simpler than A_ε. The practical value of an asymptotic method is determined by the possibility of effectively finding $U(x, \varepsilon)$ with the help of simpler problems.

Let us specifically note the similarities and differences between the asymptotic and numerical methods for solving singularly perturbed problems. Asymptotic approximation is an approximation to the solution of the original problem for some sufficiently small ε ($\varepsilon \to 0$). In real problems ε, although small, is fixed. Therefore it might happen that $U(x, \varepsilon)$ will be quite an inaccurate approximation to $u_\varepsilon(x)$. During a numerical solution, the differential equation A_ε is substituted by a finite difference equation. The interval of integration is divided into, say, n parts. We anticipate that the solution of the difference equation (for any fixed ε) approximates the solution $u_\varepsilon(x)$ of problem A_ε. Increasing n provides the required proximity of the solution of difference equation to $u_\varepsilon(x)$.

Standard numerical methods, however, often fail to work when ε is small. The difficulties arise, e.g., because of possible sharp changes of the solution $u_\varepsilon(x)$ in the boundary layer (and/or analogous internal layers, since such fast changes might also take place in the vicinities of interior points of the domain of interest). Possible approaches to constructing special numerical methods for singularly perturbed problems are presented in Ascher, Mattheij, and Russell [1], Bakhvalov [2], Yu. Boglaev [11], Doolan, Miller, and Schilders [44], Il'in [66], in proceedings and collections of papers [55], [60], [93], to name a few. Asymptotic approximation gives one the idea about the qualitative behavior of the solution. This should guide the construction of more sophisticated numerical methods. On the other hand, numerical methods are often used to obtain the successive terms of an asymptotic approximation (such problems do not normally contain small parameters), e.g., the generally nonlinear problem A_0 might not have an explicit analytical solution, and in such cases, it can be solved only numerically. Thus, asymptotic and numerical methods in some sense complement each other.

In this book we will consider methods for constructing uniform asymptotic approximations for the solutions $u_\varepsilon(x)$ of the singularly perturbed problems in a domain D. Usually we will construct the series in the powers of ε

$$\sum_{k=0}^{\infty} \varepsilon^k u_k(x,\varepsilon) \tag{1.2.1}$$

(where the $u_k(x,\varepsilon)$ are bounded functions), such that the nth partial sum (truncated series)

$$U_n(x,\varepsilon) = \sum_{k=0}^{n} \varepsilon^k u_k(x,\varepsilon)$$

will give the asymptotic approximation to the solution $u_\varepsilon(x)$ with an accuracy of order ε^{n+1} in domain D, i.e.,

$$\sup_D \| u_\varepsilon(x) - U_n(x,\varepsilon) \| = O(\varepsilon^{n+1}). \tag{1.2.2}$$

Definition 1.3 *The series* (1.2.1) *satisfying condition* (1.2.2) *is called the asymptotic series for the function* $u_\varepsilon(x)$ (*or the asymptotic expansion for* $u_\varepsilon(x)$) *in the domain* D *as* $\varepsilon \to 0$.

In what follows, instead of saying "we construct the asymptotic series" or "we construct the asymptotic expansion of function $u_\varepsilon(x)$ as $\varepsilon \to 0$," we will often use the shorter expressions: "we construct the asymptotics of solution" or "we construct the asymptotic solution."

In practice we need not construct the whole series (1.2.1) but can restrict our attention to the first several terms of the series. Often when referring to the asymptotic series we will mean the algorithm that allows us to obtain the terms of the asymptotic series up to any number n. (We will say that the series is constructed when such an algorithm is presented.)

We should mention the following important fact: the asymptotic series (1.2.1) might not converge to $u_\varepsilon(x)$ and might even diverge. The convergence of (1.2.1) to $u_\varepsilon(x)$ means that

$$\| u_\varepsilon(x) - U_n(x,\varepsilon) \| \to 0 \quad \text{as } n \to \infty. \tag{1.2.3}$$

According to the definition used, the asymptotic series (1.2.1) satisfies (1.2.2). This means that there exist numbers $c > 0$ and $\varepsilon_0 > 0$ (depending, generally, on n) such that for $0 < \varepsilon \le \varepsilon_0$ the following inequality holds:

$$\sup_D \| u_\varepsilon(x) - U_n(x,\varepsilon) \| \le c\varepsilon^{n+1}. \tag{1.2.4}$$

Although, if $0 < \varepsilon < 1$, the factor ε^{n+1} in the right-hand side of (1.2.4) approaches zero as $n \to \infty$, it does not follow that (1.2.3) holds since $c = c(n)$ depends on n, and this dependence might prevent $c(n)\varepsilon^{n+1}$ from tending to zero as $n \to \infty$.

Let us consider an example of a divergent asymptotic series of the form (1.2.1). We seek the solution to

$$\varepsilon \frac{du}{dx} = -\frac{u}{x^2} - \frac{1}{x} \tag{1.2.5}$$

in the form

$$u = \sum_{k=0}^{\infty} \varepsilon^k u_k(x).$$

Substitution into (1.2.5) implies

$$\varepsilon(u_0' + \varepsilon u_1' + \ldots + \varepsilon^{k-1} u_{k-1}' + \ldots) = -\frac{1}{x^2}(u_0 + \varepsilon u_1 + \ldots + \varepsilon^k u_k + \ldots) - \frac{1}{x}.$$

Equating coefficients by like powers of ε in both sides of the equation, we obtain

$$0 = -\frac{u_0}{x^2} - \frac{1}{x}, \quad u_0' = -\frac{u_1}{x^2}, \quad u_1' = -\frac{u_2}{x^2}, \ldots, \quad u_{k-1}' = -\frac{u_k}{x^2}, \ldots.$$

The coefficients of the series can be found successively:

$$u_0 = -x, \quad u_1 = x^2, \quad u_2 = -(2!)x^3, \ldots, \quad u_k = (-1)^{k+1}(k!)x^{k+1}, \ldots.$$

Thus, we constructed the series

$$\sum_{k=0}^{\infty} \varepsilon^k (-1)^{k+1} (k!) x^{k+1}, \tag{1.2.6}$$

which diverges for $\varepsilon > 0$ at all points x except $x = 0$.

Let us show that the divergent series (1.2.6) is, in fact, the asymptotic expansion for some solution of (1.2.5) in the interval $0 < x \le a$. The general solution of (1.2.5) has the form

$$u = c \exp\left(\frac{1}{\varepsilon x}\right) - \left(\int_0^x \frac{1}{\varepsilon t} \exp\left(-\frac{1}{\varepsilon t}\right) dt\right) \exp\left(\frac{1}{\varepsilon x}\right).$$

Consider the particular solution with $c = 0$:

$$u_\varepsilon(x) = -\left(\int_0^x \frac{1}{\varepsilon t}\exp\left(-\frac{1}{\varepsilon t}\right)dt\right)\exp\left(\frac{1}{\varepsilon x}\right).$$

Integrating by parts three times, we obtain

$$\begin{aligned}
u_\varepsilon(x) &= -\left(\int_0^x t\, d\exp\left(-\frac{1}{\varepsilon t}\right)\right)\exp\left(\frac{1}{\varepsilon x}\right)\\
&= -t\,\exp\left(-\frac{1}{\varepsilon t}\right)\Big|_0^x\cdot\exp\left(\frac{1}{\varepsilon x}\right) + \left(\int_0^x \exp\left(-\frac{1}{\varepsilon t}\right)dt\right)\exp\left(\frac{1}{\varepsilon x}\right)\\
&= -x + \left(\int_0^x \varepsilon t^2\, d\exp\left(-\frac{1}{\varepsilon t}\right)\right)\exp\left(\frac{1}{\varepsilon x}\right)\\
&= -x + \varepsilon x^2 - \left(\int_0^x 2\varepsilon t\exp\left(-\frac{1}{\varepsilon t}\right)dt\right)\exp\left(\frac{1}{\varepsilon x}\right)\\
&= -x + \varepsilon x^2 - 2\varepsilon^2 x^3 + \varepsilon^2\int_0^x 6t^2\exp\left(-\frac{1}{\varepsilon t} + \frac{1}{\varepsilon x}\right)dt.
\end{aligned}$$

Moreover, since $\exp(-1/\varepsilon t + 1/\varepsilon x) \leq 1$ for $0 < t \leq x$, the last term does not exceed

$$\varepsilon^2\int_0^x 6t^2 dt = 2\varepsilon^2 x^3 \leq 2a^3\varepsilon^2.$$

Therefore $u_\varepsilon(x) = -x + \varepsilon x^2 + O(\varepsilon^2)$.

Further integration by parts leads to the equality

$$\begin{aligned}
u_\varepsilon(x) &= -x + \varepsilon x^2 - 2\varepsilon^2 x^3 + \ldots + \varepsilon^n(-1)^{n+1}(n!)x^{n+1} + O(\varepsilon^{n+1})\\
&= \sum_{k=0}^{n} \varepsilon^k(-1)^{k+1}(k!)x^{k+1} + O(\varepsilon^{n+1}).
\end{aligned}$$

This relation shows that the divergent series (1.2.6) satisfies condition (1.2.2), and therefore this series is an asymptotic expansion for the solution $u_\varepsilon(x)$ of (1.2.5) as $\varepsilon \to 0$ in the interval $0 < x \leq a$.

1.3 Examples of asymptotic expansions for solutions of regularly and singularly perturbed problems

We introduce the following notation. If $x = \mathrm{col}(x_1, \ldots, x_k)$ is a k-dimensional vector, t is a scalar, and $f(x,t) = \mathrm{col}(f_1, \ldots, f_m)$ is an m-dimensional vector function, then

$$f_x(x,t) = \left(\frac{\partial f_i}{\partial x_j}\right) \quad \text{is an} \quad (m \times k)\text{–matrix, and} \quad f_t(x,t) = \mathrm{col}\left(\frac{\partial f_1}{\partial t}, \ldots, \frac{\partial f_m}{\partial t}\right)$$

is an m-dimensional vector.

1.3.1 The regularly perturbed problem

Consider an initial value problem for the differential equations system

$$A_\varepsilon: \quad \frac{dx}{dt} = f(x,t,\varepsilon), \qquad 0 \le t \le T, \quad x(0,\varepsilon) = x^0.$$

Here $x(t,\varepsilon)$ is an m-dimensional vector function. The corresponding problem A_0 can be written as

$$A_0: \quad \frac{dx}{dt} = f(x,t,0), \qquad 0 \le t \le T, \quad x(0) = x^0.$$

Let problem A_0 have a solution $x = \overline{x}(t)$, and let the function $f(x,t,\varepsilon)$ be infinitely differentiable in the domain

$$(1.3.1) \qquad \| x - \overline{x}(t) \| \le a, \qquad 0 \le t \le T, \quad 0 \le \varepsilon \le \varepsilon_0,$$

where a and ε_0 are positive constants.

We seek the solution of A_ε in the form of the series

$$(1.3.2) \qquad x(t,\varepsilon) = x_0(t) + \varepsilon x_1(t) + \cdots + \varepsilon^k x_k(t) + \cdots .$$

Substituting (1.3.2) into A_ε, formally expanding $f(x_0(t) + \varepsilon x_1(t) + \cdots + \varepsilon^k x_k(t) + \ldots, t, \varepsilon)$ into a Taylor series about the point $(x_0, t, 0)$, and collecting terms of like powers of ε, we arrive at the following equalities:

$$\begin{aligned}
&\frac{d}{dt}(x_0 + \varepsilon x_1 + \cdots + \varepsilon^k x_k + \cdots) \\
&= f(x_0,t,0) + \overline{f}_x\,(\varepsilon x_1 + \varepsilon^2 x_2 + \cdots + \varepsilon^k x_k + \cdots) + \overline{f}_\varepsilon\,\varepsilon + \cdots \\
(1.3.3) \qquad &= f(x_0,t,0) + \varepsilon(\overline{f}_x\, x_1 + \varphi_1) + \cdots + \varepsilon^k(\overline{f}_x\, x_k + \varphi_k) + \cdots \\
&\equiv \sum_{k=0}^{\infty} \varepsilon^k f_k,
\end{aligned}$$

$$(1.3.4) \qquad x_0(0) + \varepsilon x_1(0) + \cdots + \varepsilon^k x_k(0) + \cdots = x^0.$$

Here $\overline{f}_x = f_x(x_0,t,0)$, $\varphi_1 = \overline{f}_\varepsilon = f_\varepsilon(x_0,t,0)$, and the φ_k $(k > 1)$ are expressed recursively through x_i with indices $i < k$.

Equating coefficients of the same powers of ε in both sides of (1.3.3) and (1.3.4), we obtain the problems for the terms x_0, x_1, ... of the series (1.3.2):

$$\frac{dx_0}{dt} = f_0 \equiv f(x_0,t,0), \qquad x_0(0) = x^0,$$

$$\frac{dx_k}{dt} = f_k \equiv \overline{f}_x \cdot x_k + \varphi_k, \qquad x_k(0) = 0, \quad k = 1, 2, \ldots .$$

The problem for x_0 evidently coincides with A_0, so $x_0 = \overline{x}(t)$. The equations for x_k $(k = 1, 2, \ldots)$ (so-called *variational equations*) are linear differential equations, and thus have unique solutions with any given initial conditions. The terms of the series

(1.3.2) can be uniquely defined up to any order n. Note that the main term $x_0(t)$ of the series (1.3.2) is the solution of the problem A_0, which is, generally, simpler than A_ε. All the later terms of the series are solutions of linear initial value problems (the problem A_ε may be nonlinear).

Let us introduce the notation $X_n(t,\varepsilon)$ for the nth partial sum of the series (1.3.2):

$$X_n(t,\varepsilon) = \sum_{k=0}^{n} \varepsilon^k x_k(t).$$

Theorem 1.1 *For sufficiently small* ε $(0 < \varepsilon \leq \varepsilon_0$, *where* ε_0 *is some constant*), *problem* A_ε *has a unique solution* $x(t,\varepsilon)$, *and the series* (1.3.2) *is the asymptotic series for* $x(t,\varepsilon)$ *in the entire interval* $0 \leq t \leq T$, *i.e.,*

$$\sup_{[0,T]} \| x(t,\varepsilon) - X_n(t,\varepsilon) \| = O(\varepsilon^{n+1}). \tag{1.3.5}$$

We will present the outline of the proof of this classical theorem without much detail. Let us substitute the solution $x(t,\varepsilon) = X_n(t,\varepsilon) + v(t,\varepsilon)$ into the problem A_ε. For the remainder $v = x - X_n$, we obtain the problem

$$\frac{dv}{dt} = \overline{f}_x v + g(v,t,\varepsilon), \qquad v(0,\varepsilon) = 0, \tag{1.3.6}$$

where $g(v,t,\varepsilon) \equiv f(X_n(t,\varepsilon)+v,t,\varepsilon) - \overline{f}_x v - dX_n/dt$. Note that the function $g(v,t,\varepsilon)$ possesses the following two properties:

1*. $g(0,t,\varepsilon) = f(X_n(t,\varepsilon),t,\varepsilon) - \dfrac{dX_n}{dt} = \displaystyle\sum_{k=0}^{n} \varepsilon^k f_k + O(\varepsilon^{n+1}) - \frac{d}{dt}\sum_{k=0}^{n} \varepsilon^k x_k$

$$= \sum_{k=0}^{n} \varepsilon^k \left(f_k - \frac{dx_k}{dt} \right) + O(\varepsilon^{n+1}) = O(\varepsilon^{n+1}),$$

since $dx_k/dt = f_k$, i.e., $X_n(t,\varepsilon)$ satisfies the equation A_ε with an accuracy of the order ε^{n+1}. Another way of saying this is that the *discrepancy* (or *error*) that we obtain after substituting of X_n into A_ε is of the order $O(\varepsilon^{n+1})$.

2*. If $\| v_1(t,\varepsilon) \| \leq c_1\varepsilon$ and $\| v_2(t,\varepsilon) \| \leq c_1\varepsilon$, where c_1 is an arbitrary number, then a $c_2 > 0$ and an $\varepsilon_0 > 0$ can be found such that

$$\sup_{[0,T]} \| g(v_2,t,\varepsilon) - g(v_1,t,\varepsilon) \| \leq c_2\varepsilon \sup_{[0,T]} \| v_2(t,\varepsilon) - v_1(t,\varepsilon) \|$$

for $0 < \varepsilon \leq \varepsilon_0$.

This property follows from the formula of finite increments (mean value theorem) $g(v_2,t,\varepsilon) - g(v_1,t,\varepsilon) = g_v^*(v_2 - v_1)$ (here g_v^* means that the elements of the matrix g_v are calculated at some intermediate points), so the result follows from estimating the matrix g_v:

$$\begin{aligned} g_v &= f_x(X_n + v,t,\varepsilon) - \overline{f}_x = f_x(x_0 + O(\varepsilon) + v,t,\varepsilon) - f_x(x_0,t,0) \\ &= O(\varepsilon + \| v \|) = O(\varepsilon), \quad \text{when} \quad \| v \| = O(\varepsilon). \end{aligned}$$

Using the fundamental matrix $\Phi(t,\varepsilon)$ for the homogeneous equation $dv/dt = \overline{f}_x\, v$, we can pass on from the differential equation (1.3.6) for v to the equivalent integral equation

$$v(t,\varepsilon) = \int_0^t \Phi(t,\varepsilon) g(v(s,\varepsilon), s, \varepsilon) ds \equiv G(v,t,\varepsilon).$$

The integral operator $G(v,t,\varepsilon)$ inherits the two properties similar to the corresponding properties of the function $g(v,t,\varepsilon)$. The first property ($G(0,t,\varepsilon) = O(\varepsilon^{n+1})$) shows that $G(v,t,\varepsilon)$ is the quantity of the order ε^{n+1} for $v = 0$ (and thus for $v = O(\varepsilon^{n+1})$). From the second property it follows that $G(v,t,\varepsilon)$ is a contraction operator for sufficiently small $\| v \|$ and ε such that $c_2\varepsilon < 1$. Now we can apply the method of successive approximations to the integral equation, i.e.,

$$v_0 = 0 \quad \text{and} \quad v_i = G(v_{i-1}, t, \varepsilon), \qquad i = 1, 2, \dots .$$

Using the properties of the operator G mentioned above, it can be easily shown that for ε small enough the unique solution of the equation exists and satisfies the estimate

$$\max_{[0,T]} \| v(t,\varepsilon) \| = O(\varepsilon^{n+1}).$$

This completes the proof of the theorem.

Note that for the construction of $X_n(t,\varepsilon)$ and for the proof of the estimate (1.3.5) for any n, it is sufficient to assume that the function $f(x,t,\varepsilon)$ is $n+1$ times differentiable in the domain (1.3.1).

Let us emphasize once more two important features of the problem discussed above.

1. The solution $x_0 = \overline{x}(t)$ of problem A_0 is the asymptotic approximation for the solution $x(t,\varepsilon)$ of problem A_ε in the entire interval $0 \le t \le T$. (The accuracy of this approximation is of the order ε, as follows from (1.3.5).) Therefore A_ε is regularly perturbed.

2. The asymptotic solution (1.3.2) of the regularly perturbed problem A_ε is a power series in ε with coefficients depending on t and independent of ε.

Exercise

Consider the initial value problem

$$\frac{dx}{dt} = -x + \varepsilon x^2, \qquad 0 \le t \le 1, \quad x(0) = 1.$$

Find the asymptotic expansion of the solution in powers of ε using the regular perturbation technique. Show that the same approximation can be obtained from the exact solution by expanding it in the power series in ε for small ε.

1.3.2 Example of an asymptotic approximation for a singularly perturbed problem

Now consider the initial value problem for the scalar linear differential equation

$$A_\varepsilon: \quad \varepsilon \frac{dx}{dt} = -ax + f(t), \qquad 0 \le t \le T, \quad x(0,\varepsilon) = 0,$$

with a small parameter $\varepsilon > 0$ in front of the derivative. Here $a = \text{const} > 0$, and $f(t)$ is an infinitely differentiable function. The corresponding problem A_0 consists of the algebraic equation

$$0 = -ax + f(t),$$

and, hence $x = f(t)/a \equiv \overline{x}(t)$.

Let us try to construct an asymptotic expansion of the solution of problem A_ε in the form of the power series (1.3.2) in the powers of ε (cf. § 1.3.1):

$$x_0(t) + \varepsilon x_1(t) + \cdots + \varepsilon^k x_k(t) + \cdots . \tag{1.3.7}$$

Substituting this series into the equation A_ε, we obtain

$$\varepsilon \frac{d}{dt}(x_0 + \varepsilon x_1 + \cdots + \varepsilon^{k-1} x_{k-1} + \cdots) = -a(x_0 + \varepsilon x_1 + \cdots + \varepsilon^k x_k + \cdots) + f(t).$$

Equating coefficients of like powers of ε, we obtain

$$\begin{aligned} x_0 &= \frac{f(t)}{a}, \qquad x_1 = -\frac{1}{a}\frac{dx_0}{dt} = -\frac{f'(t)}{a^2}, \ldots, \\ x_k &= -\frac{1}{a}\frac{dx_{k-1}}{dt} = \frac{(-1)^k f^{(k)}(t)}{a^{k+1}}, \qquad k = 2, 3, \ldots . \end{aligned} \tag{1.3.8}$$

Note that the leading term $x_0(t)$ of the series (1.3.7) coincides with the solution $\overline{x}(t)$ of the problem A_0.

Series (1.3.7) satisfies formally the differential equation of A_ε, but it does not in general satisfy the initial condition since we defined the terms of the series without using the initial condition $x(0, \varepsilon) = x^0$. Therefore the series (1.3.7) cannot be expected to be the asymptotic expansion for the solution of the problem A_ε, at least in some vicinity of the initial point $t = 0$. Will the series (1.3.7) be the asymptotic expansion for $x(t, \varepsilon)$ outside a neighborhood of the initial point?

To answer this question, let us consider the exact solution of A_ε, which can be easily determined to be

$$x(t, \varepsilon) = \int_0^t \frac{1}{\varepsilon} \exp\left(-\frac{a}{\varepsilon}(t - s)\right) f(s) ds.$$

Integrating by parts, we obtain the asymptotic expansion of $x(t, \varepsilon)$; e.g., after integrating by parts twice, we will find

$$\begin{aligned} x(t, \varepsilon) &= \frac{1}{a} f(s) \exp\left(-\frac{a}{\varepsilon}(t - s)\right)\Big|_0^t - \frac{1}{a}\int_0^t f'(s) \exp\left(-\frac{a}{\varepsilon}(t - s)\right) ds \\ &= \frac{f(t)}{a} - \frac{f(0)}{a} \exp\left(-a\frac{t}{\varepsilon}\right) - \frac{\varepsilon}{a^2}\int_0^t f'(s) d \exp\left(-\frac{a}{\varepsilon}(t - s)\right) \\ &= \left[\frac{f(t)}{a} - \varepsilon\frac{f'(t)}{a^2}\right] + \left[-\frac{f(0)}{a} + \varepsilon\frac{f'(0)}{a^2}\right] \exp\left(-a\frac{t}{\varepsilon}\right) \\ &+ \frac{\varepsilon}{a^2}\int_0^t f''(s) \exp\left(-\frac{a}{\varepsilon}(t - s)\right) ds. \end{aligned}$$

Since

$$\int_0^t \exp\left(-\frac{a}{\varepsilon}(t-s)\right) ds = \frac{\varepsilon}{a}\left(1 - \exp\left(-a\frac{t}{\varepsilon}\right)\right) = O(\varepsilon),$$

the last term is of the order ε^2. Therefore, taking into account formulae (1.3.8), and introducing the new variable $\tau = t/\varepsilon$, we can rewrite this equality in the form

$$x(t,\varepsilon) = [x_0(t) + \varepsilon x_1(t)] + [-x_0(0)\exp(-a\tau) - \varepsilon x_1(0)\exp(-a\tau)] + O(\varepsilon^2).$$

Let us now introduce the notation $\Pi_0(\tau) + \varepsilon\Pi_1(\tau)$ for the terms depending on τ. Evidently, $x_0(0) + \Pi_0(0) = 0$ and $x_1(0) + \Pi_1(0) = 0$.

Continuing to integrate by parts, and taking into account expressions (1.3.8), we finally obtain

$$\begin{aligned} x(t,\varepsilon) &= [x_0(t) + \varepsilon x_1(t) + \cdots + \varepsilon^n x_n(t)] \\ &\quad + [\Pi_0(\tau) + \varepsilon\Pi_1(\tau) + \cdots + \varepsilon^n\Pi_n(\tau)] + O(\varepsilon^{n+1}) \\ &= \sum_{k=0}^{n} \varepsilon^k (x_k(t) + \Pi_k(\tau)) + O(\varepsilon^{n+1}) \end{aligned}$$

for an arbitrary n. Here each of the functions $\Pi_k(\tau)$ contains the factor $\exp(-a\tau)$ and satisfies $\Pi_k(0) = -x_k(0)$. Therefore the asymptotic expansion for $x(t,\varepsilon)$ has the form

$$x(t,\varepsilon) = \sum_{k=0}^{\infty} \varepsilon^k (x_k(t) + \Pi_k(\tau)). \tag{1.3.9}$$

Let us point out the characteristic features of this expansion.

1. The asymptotic expansion (1.3.9) consists of two series. One power series in ε is similar to the series in the regular case, with coefficients depending only on t. The other power series in ε has coefficients depending on the stretched time τ:

$$\Pi_0(\tau) + \varepsilon\Pi_1(\tau) + \cdots + \varepsilon^k\Pi_k(\tau) + \cdots . \tag{1.3.10}$$

2. Terms of the series (1.3.10) possess two important properties. First, since $x_k(0)+\Pi_k(0) = 0$, the series (1.3.9) satisfies the prescribed initial condition $x(0,\varepsilon) = 0$. Recall that series (1.3.7) does not in general satisfy this condition. We may say that the purpose of the series (1.3.10) is to satisfy the imposed initial condition (together with series (1.3.7)). Second, the functions $\Pi_k(\tau)$ decay exponentially as τ increases. Therefore they are important only in a small vicinity of the initial point $t = 0$ (in the boundary layer). For $t \geq \delta$, where δ is an arbitrarily small but fixed number, $\Pi_k(\tau) \to 0$ as $\varepsilon \to 0$ faster than any power of ε. Indeed, for $t \geq \delta$ we can write $\exp(-a\tau) = \exp(-at/\varepsilon) \leq \exp(-a\delta/\varepsilon) = O(\varepsilon^N)$ for any N.

We will call the terms $\Pi_k(\tau)$ *boundary functions.* The series (1.3.7) is the *regular series* or the *regular part* of the asymptotic solution. Series (1.3.10) is the *boundary layer series* or the *boundary layer part* of the asymptotic solution.

3. Unlike the regularly perturbed case of § 1.3.1, here the solution $x_0(t)$ of the problem A_0 (the main term of the regular series) is an asymptotic approximation to the solution $x(t,\varepsilon)$ of the problem A_ε not in the entire interval $0 \leq t \leq T$,

but only outside the boundary layer, i.e., in the subinterval $\delta \le t \le T$. To obtain the asymptotic approximation for $x(t,\varepsilon)$ in the whole interval $0 \le t \le T$, we must add the leading term $\Pi_0(\tau)$ of the boundary layer series to $x_0(t)$. Then $\sup_{[0,T]} |x(t,\varepsilon) - (x_0(t) + \Pi_0(\tau))| = O(\varepsilon)$, i.e., the sum $x_0(t) + \Pi_0(\tau)$ provides the uniform asymptotic approximation for $x(t,\varepsilon)$ with an accuracy of the order ε.

4. Outside the boundary layer where all the Π-functions are less than any power of ε, the regular series (1.3.7) is the asymptotic expansion for $x(t,\varepsilon)$.

5. It is worthwhile to point out that if we change the sign of a, then for any $t > 0$, the function $x(t,\varepsilon) \to \infty$ as $\varepsilon \to 0$. Therefore $x_0(t)$ will not be a valid asymptotic approximation (and series (1.3.7) will not be an asymptotic series) for $x(t,\varepsilon)$ in any subinterval of $0 \le t \le T$. This shows that, in the case of singularly perturbed differential equations, some special condition on the right-hand side of the equation should be imposed for the solution of A_0 to approximate to the solution of the problem A_ε, at least in some subdomain. In this concrete example such a condition is $a > 0$. For more general singularly perturbed problems we will introduce more general form of this condition in the next chapter. There we will construct the asymptotic expansions of the type (1.3.9) for solutions of a number of singularly perturbed systems of ordinary differential equations.

Exercises

1. Consider the problem

$$\varepsilon \frac{d^2u}{dx^2} + a\frac{du}{dx} = 0, \qquad a \ne 0, \qquad 0 \le x \le 1, \qquad \frac{du}{dx}(0) = -1, \quad u(0) = 1.$$

For which sign of a will the solution u_0 of the reduced problem

$$\frac{du_0}{dx} = 0, \qquad 0 \le x \le 1, \quad u_0(0) = 1$$

approximate the exact solution?

2. Find the exact solution $x(t,\varepsilon)$ of the problem

$$\varepsilon \frac{dx}{dt} = x - x^2, \qquad 0 \le t \le 1, \qquad x(0) = \frac{1}{2},$$

and represent it in the form $x = x_0 + \Pi_0(\tau)$ (similar to (1.3.9)), where $\Pi_0 \to 0$ as $\tau \to \infty$. What are x_0 and $\Pi_0(\tau)$?

Chapter 2

Singularly Perturbed Ordinary Differential Equations

2.1 Initial value problem

2.1.1 Theorem on passage to the limit

The presentation of this subsection is based on fundamental works by Tikhonov [129]–[131] on the dependence of solutions of differential equations on small parameters. We discuss conditions under which asymptotic solutions of certain initial value problems for singularly perturbed ordinary differential equations can be constructed using the boundary function method. A corresponding asymptotic algorithm is presented in the next subsection.

Consider the initial value problem (following the original papers, we will use the notation μ for the small parameter)

$$\mu\frac{dz}{dt} = F(z,y,t), \qquad \frac{dy}{dt} = f(z,y,t), \qquad 0 \le t \le T, \tag{2.1.1}$$

$$z(0,\mu) = z^0, \qquad y(0,\mu) = y^0. \tag{2.1.2}$$

Here $\mu > 0$, z and y are vector functions of arbitrary dimensions M and m, respectively. As we see, only part of (2.1.1) contains the small parameter multiplying the derivatives; the other part does not.

We assume that the *functions* $F(z,y,t)$ *and* $f(z,y,t)$ *are continuous together with their derivatives with respect to* z *and* y *in some domain* $G = \{\| z \| \le a,\ \| y \| \le a,\ 0 \le t \le T\}$.

We use the notation $z(t,\mu)$ and $y(t,\mu)$ for the solution of (2.1.1), (2.1.2). In general, it is not possible to find an exact solution of this problem. Our goal is to define an approximate solution using the fact that the parameter μ is small. Setting $\mu = 0$ in (2.1.1), we obtain

$$0 = F(\bar{z},\bar{y},t), \qquad \frac{d\bar{y}}{dt} = f(\bar{z},\bar{y},t). \tag{2.1.3}$$

The order of this system is lower than that of the original system since the first equation of (2.1.3) is no longer a differential equation. We will call (2.1.3) the *degenerate* or *reduced* system. For this system we should retain only the initial condition for $\overline{y}$, since no initial condition for $\overline{z}$ is needed:

$$\overline{y}(0) = y^0. \tag{2.1.4}$$

So, for $\mu = 0$, we obtain the problem (2.1.3), (2.1.4) instead of (2.1.1), (2.1.2). To solve it, we must start by expressing $\overline{z}$ from the first equation of (2.1.3) as a function of $\overline{y}$ and t. It is worth mentioning that this operation is not unique, i.e., the equation $F(\overline{z}, \overline{y}, t) = 0$ might have several solutions (roots) with respect to $\overline{z}$. In this case, the question of how to choose the appropriate root arises. Suppose we somehow choose one root $\overline{z} = \varphi(\overline{y}, t)$ (the theorem that will be formulated below selects the correct root). Substituting this root into the second equation of (2.1.3), we obtain the following equation:

$$\frac{d\overline{y}}{dt} = f(\varphi(\overline{y}, t), \overline{y}, t) \tag{2.1.5}$$

for $\overline{y}$. Solving this equation with the initial condition (2.1.4), we find $\overline{y}(t)$, and therefore also $\overline{z}(t) = \varphi(\overline{y}(t), t)$. It is seen that defining $\overline{z}(t)$ and $\overline{y}(t)$ happens to be an easier problem than obtaining the solution $z(t, \mu)$, $y(t, \mu)$ of the original problem (2.1.1), (2.1.2). Instead of solving the system of two differential equations, we must first solve the equation $F(\overline{z}, \overline{y}, t) = 0$ (not differential), and then the differential equation (2.1.5) with the initial condition (2.1.4) (note that (2.1.5) is an equation of the lower order than the original system (2.1.1) and that it does not depend on the small parameter).

It is natural to ask whether $\overline{z}(t)$ and $\overline{y}(t)$ are the asymptotic approximations for $z(t, \mu)$ and $y(t, \mu)$, respectively, as $\mu \to 0$. It should be noted that $\overline{z}(t)$ will not in general satisfy the initial condition (2.1.2), i.e., $\overline{z}(0) \neq z^0$, since the originally imposed initial value z^0 has not been used in the process of defining $\overline{z}(t)$. Therefore, at least in some vicinity of the initial point $t = 0$, the function $\overline{z}(t)$ will not be close to $z(t, \mu)$. But does $\overline{z}(t)$ approximate $z(t, \mu)$ outside this vicinity? In contrast with $\overline{z}$ and z, the values of $\overline{y}(t)$ and $y(t, \mu)$ coincide at $t = 0$: $\overline{y}(0) = y(0, \mu) = y^0$. Therefore, in some small neighborhood of $t = 0$, these functions will be close to each other. But is this true for the entire interval $0 \leq t \leq T$? The answer to these questions can be either yes or no depending on the conditions imposed on the systems (2.1.1) and (2.1.3). In particular, it depends on the choice of the root $\overline{z}(t) = \varphi(\overline{y}(t), t)$.

Let us formulate the conditions for the passage to the limit.

1^0. *Let the equation* $F(\overline{z}, \overline{y}, t) = 0$ *have an isolated root with respect to* $\overline{z}$: $\overline{z}(t) = \varphi(\overline{y}(t), t)$, $(\overline{y}, t) \in D = \{\| \overline{y} \| \leq a,\ 0 \leq t \leq T\}$, *and suppose the problem* (2.1.3), (2.1.4) *has a unique solution corresponding to this root in the interval* $0 \leq t \leq T$.

One can visualize the root $z = \varphi(y, t)$ as a surface in the space of variables (z, y, t) (see Fig. 2.1). The fact that the root is isolated means that there are no other roots in its vicinity. Elsewhere, other roots might exist.

The next two conditions show exactly which root one must choose. They are formulated for the so-called *associated system*:

$$\frac{d\tilde{z}}{d\tau} = F(\tilde{z}, y, t), \qquad \tau \geq 0. \tag{2.1.6}$$

The independent variable here is the new variable τ, and y and t enter the system (2.1.6) as parameters. This means that the system is considered for any values of

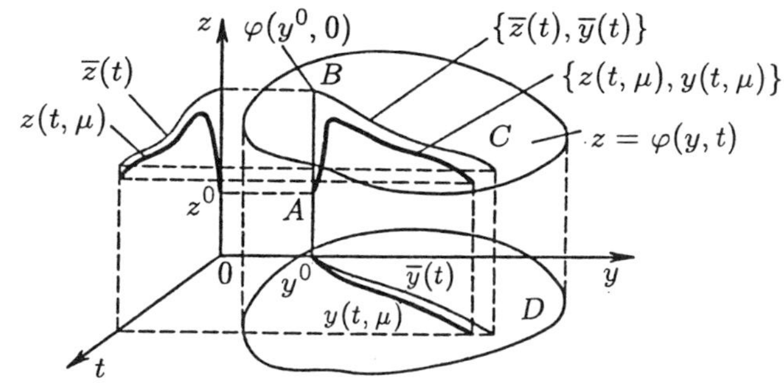

Figure 2.1: *The exact solution* $z(t,\mu)$, $y(t,\mu)$ *of* (2.1.1), (2.1.2) *and the solution* $\overline{z}(t)$, $\overline{y}(t)$ *of* (2.1.3), (2.1.4).

(y,t) in the domain D. By virtue of condition 1^0, $\tilde{z} = \varphi(y,t)$ is a rest point of system (2.1.6).

2^0. *Let the stationary point* $\tilde{z} = \varphi(y,t)$ *of the associated system be asymptotically stable in the sense of Lyapunov, uniformly in* $(y,t) \in D$ *as* $\tau \to \infty$.

This means that for any $\varepsilon > 0$, there exists $\delta = \delta(\varepsilon) > 0$ (one and the same for all $(y,t) \in D$) such that if $\| \tilde{z}(0) - \varphi(y,t) \| < \delta$, then $\| \tilde{z}(\tau) - \varphi(y,t) \| < \varepsilon$ for $\tau \geq 0$, and also $\tilde{z} \to \varphi(y,t)$ when $\tau \to \infty$.

There might exist several roots of the equation $F(z,y,t) = 0$ that satisfy condition 2^0. To make the final choice of the root, consider the associated system (2.1.6) for the initial parameters $y = y^0$ and $t = 0$:

$$\frac{d\tilde{z}}{d\tau} = F(\tilde{z}, y^0, 0), \tag{2.1.7}$$

with the initial condition

$$\tilde{z}(0) = z^0. \tag{2.1.8}$$

Here z^0 is the same vector that enters the initial condition (2.1.2). This vector, generally speaking, is not close to the rest point $\tilde{z} = \varphi(y^0, 0)$ of system (2.1.7) since the condition (2.1.2) has been imposed independently. Therefore the solution $\tilde{z}(\tau)$ of (2.1.7), (2.1.8) might not necessarily tend to the rest point $\varphi(y^0,0)$ as $\tau \to \infty$. We demand that $\tilde{z}(\tau)$ approach the rest point.

3^0. *Let the solution* $\tilde{z}(\tau)$ *of the problem* (2.1.7), (2.1.8) *exist for* $\tau \geq 0$ *and tend to the stationary point* $\varphi(y^0,0)$ *as* $\tau \to \infty$.

In such a case we will say that z^0 belongs to the *domain* (or *basin*) *of attraction* of the rest point $\varphi(y^0,0)$. The description of the domain of attraction when z is a scalar will be given below. Thus, conditions 2^0 and 3^0 allow us to choose the correct root of the equation $F(z,y,t) = 0$.

Theorem 2.1 (Tikhonov). *Under conditions* 1^0—3^0 *and for sufficiently small* μ, *the problem* (2.1.1), (2.1.2) *has a unique solution* $z(t,\mu)$, $y(t,\mu)$ *such that the following* *limiting equalities hold:*

$$\lim_{\mu \to 0} y(t, \mu) = \overline{y}(t) \quad \text{for } 0 \le t \le T,$$

$$\lim_{\mu \to 0} z(t, \mu) = \overline{z}(t) \quad \text{for } 0 < t \le T.$$

These equalities show that in the limit (as $\mu \to 0$) the solution $z(t, \mu)$, $y(t, \mu)$ of (2.1.1), (2.1.2) tends to the solution of the reduced problem (2.1.3), (2.1.4). That is why the above theorem is called the *theorem on passage to the limit* . Note that the transition to the limit for y takes place for all t in the interval $0 \le t \le T$. Moreover, this limiting process is uniform (as can be seen from the proof of the theorem). Transition to the limit for z takes place for any t except $t = 0$. This is quite clear since $z(0, \mu) = z^0 \ne \overline{z}(0)$. The limiting process for z will be uniform outside a small neighborhood of the initial point. We will call this neighborhood the *boundary* (*initial*) *layer.*

Thus, $\overline{y}$ will be the asymptotic approximation to $y(t, \mu)$ in entire interval $0 \le t \le T$, and $\overline{z}$ will be the asymptotic approximation to $z(t, \mu)$ for $\delta \le t \le T$, where δ is any arbitrarily small but fixed number. Possible behavior of the solution $z(t, \mu)$, $y(t, \mu)$ for small μ is shown in Fig. 2.1.

Here we will not prove Theorem 2.1 (see, e.g., Vasil'eva and Butuzov [148]). Let us only explain the role of conditions 2^0 and 3^0. If we make the change of variable $\tau = t/\mu$ in (2.1.1) (τ is a stretched time), we arrive at the system

$$\frac{dz}{d\tau} = F(z, y, \tau\mu), \qquad \frac{dy}{d\tau} = \mu f(z, y, \tau\mu) \tag{2.1.9}$$

with initial conditions

$$z|_{\tau=0} = z^0, \qquad y|_{\tau=0} = y^0.$$

Consider qualitatively the behavior of the solution when τ changes from zero to some value $\tau_0 > 0$. Since μ is small, the right-hand side of the second equation is small: $\mu f = O(\mu)$, i.e., the rate of change $dy/d\tau$ of the variable y with respect to the stretched time τ is of the order μ. Therefore the value of y, which equals y^0 at the initial moment of time, will change only slightly during the period $0 \le \tau \le \tau_0$ (no more than by order μ):

$$y = y^0 + O(\mu) \quad \text{for} \quad 0 \le \tau \le \tau_0.$$

Substituting this expression for y into the first equation of (2.1.9), we obtain

$$\frac{dz}{d\tau} = F(z, y^0 + O(\mu), \tau\mu). \tag{2.1.10}$$

In the bounded interval $0 \le \tau \le \tau_0$, this equation can be considered as a regular perturbation of the equation that we obtain from (2.1.10) when $\mu = 0$:

$$\frac{dz}{d\tau} = F(z, y^0, 0). \tag{2.1.11}$$

When μ is small, the solutions of (2.1.10) and (2.1.11) with the same initial condition $z|_{\tau=0} = z^0$ are close to each other in the interval $0 \le \tau \le \tau_0$. However, equation

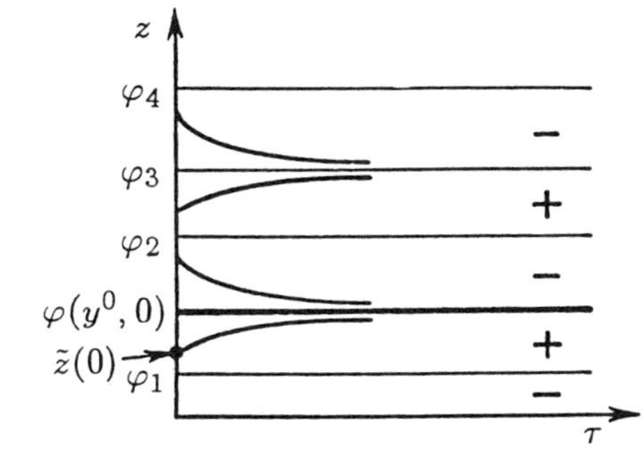

Figure 2.2: *The structure of the domains of attraction in scalar case.*

(2.1.11) is nothing but the associated equation (2.1.7). Therefore in $0 \leq \tau \leq \tau_0$, i.e., for $0 \leq t \leq \tau_0 \mu$, the behavior of the function $z(t,\mu)$ is approximately described by (2.1.7). By virtue of condition 3^0, the solution of (2.1.7) with initial condition (2.1.8) approaches $\varphi(y^0, 0)$ as τ grows (see Fig. 2.1). So, the condition 3^0 provides fast transition of $z(t,\mu)$ from the value z^0 at $t = 0$ to a value close to $\varphi(y,t)$ at $t = \tau_0\mu$. Condition 2^0 guarantees that $z(t,\mu)$ will remain near $\overline{z}(t)$ after that.

Let us now elucidate the structure of the *domain of attraction* of the asymptotically stable rest point $\tilde{z} = \varphi(y^0,0)$ of (2.1.7) in the case when z is a scalar function. Suppose that along with the root $\tilde{z} = \varphi(y^0,0)$, the equation $F(\tilde{z}, y^0, 0) = 0$ has also some other roots $\tilde{z} = \varphi_i$, $i = 1, 2, 3, \ldots$. These roots can be represented by straight lines in the plane $(\tilde{z}, \tau)$ (see Fig. 2.2). Let the function $F(\tilde{z}, y^0, 0)$ in the intervals between the roots have the signs shown in Fig. 2.2. Then if the initial condition $\tilde{z}(0)$ lies in the interval between φ_1 and $\varphi(y^0,0)$, where $F > 0$, the solution $\tilde{z}(\tau)$ for $\tau > 0$ will increase and monotonically approach the rest point $\varphi(y^0,0)$. If $\tilde{z}(0)$ lies in the interval between $\varphi(y^0,0)$ and φ_2, where $F < 0$, the solution $\tilde{z}(\tau)$ for $\tau > 0$ will decrease and again monotonically approach the rest point $\varphi(y^0,0)$. On the other hand, if we take an initial condition $\tilde{z}(0)$ out of the interval (φ_1, φ_2), the solution $\tilde{z}(\tau)$ will not tend to the rest point $\varphi(y^0,0)$.

Thus, in the case just considered, the rest point $\tilde{z} = \varphi(y^0,0)$ of the equation is asymptotically stable, and its domain of attraction is the interval (φ_1, φ_2).

It can be seen in Fig. 2.2 that the point $\tilde{z} = \varphi_3$ is also an asymptotically stable rest point, and its basin of attraction is the interval (φ_2, φ_4). Therefore the condition 2^0 of the theorem does not allow us to choose the right solution of the equation $F(z, y, t) = 0$ uniquely. Only condition 3^0 makes such a unique choice possible. If z^0 belongs to the interval (φ_1, φ_2), we must take the root $z = \varphi(y,t)$, but if z^0 is in (φ_2, φ_4), we take the root $z = \varphi_3(y,t)$. The situation when z^0 does not belong to the domain of attraction of any asymptotically stable rest point is also possible. Then the theorem cannot be applied.

Exercise

Find domains of attraction of the asymptotically stable rest points of the equation

$$\frac{dz}{d\tau} = z(1-z)(1+z).$$

It is worthwhile to mention that the domain of attraction has such a simple description only when z is a scalar function. For vector functions the structure of the domain of attraction becomes more complicated. In many aspects this question still remains open.

Exercise
The system

$$\frac{dz_1}{d\tau} = z_1(1 - z_1 - 2z_2),$$

$$\frac{dz_2}{d\tau} = z_2(1 - 2z_1 - z_2)$$

has asymptotically stable steady states $(1,0)$ and $(0,1)$. Show that the domain of attraction of $(1,0)$ is $D_1 = \{z_1 > 0, z_2 < z_1\}$, and the domain of attraction of $(0,1)$ is $D_2 = \{z_2 > 0, z_1 < z_2\}$. (Hint. Use phase plane analysis together with the fact that $z_1 = z_2$ is one of the separatrices for the saddle point $(\frac{1}{3}, \frac{1}{3})$, the point $(0,0)$ is an unstable node, and coordinate axes coincide with some of the phase trajectories of the system.)

2.1.2 Asymptotic algorithm

Theorem 2.1 is only the first step in studying the asymptotics of the solution of (2.1.1), (2.1.2). Since $\overline{z}(t)$ does not approximate $z(t,\mu)$ for small μ in the boundary layer, we must construct an asymptotic approximation for $z(t,\mu)$ that is valid in the entire interval $0 \le t \le T$ (a *uniform approximation*). Moreover, the theorem does not give the order of accuracy of the asymptotic approximation $\overline{y}(t)$ for the solution $y(t,\mu)$ in $0 \le t \le T$ and that of $\overline{z}(t)$ for $z(t,\mu)$ outside the boundary layer. Below we will show that this order is μ.

Our other goal is to construct an approximation with higher accuracy and, if possible, the complete asymptotic expansion for the solution of the problem (2.1.1), (2.1.2). To accomplish that, some additional conditions should be imposed on the functions F and f.

4^0. *We demand that the functions $F(z,y,t)$ and $f(z,y,t)$ are infinitely differentiable in domain G.*

Let $\lambda_i(t)$ $(i = 1,\dots,M)$ denote the eigenvalues of the matrix $\overline{F}_z(t) = F_z(\overline{z}(t), \overline{y}(t), t)$. We will replace condition 2^0 of Theorem 2.1 by the following condition (for which we will use the notation 2^0 as well).

2^0. *Let*

$$\text{Re}\lambda_i(t) < 0 \quad \textit{for} \quad 0 \le t \le T, \qquad i = 1,\dots,M.$$

Note that condition 2^0 of Theorem 2.1 follows from this condition.

We will seek the asymptotic expansion for the solution of (2.1.1), (2.1.2) in a form that is quite typical for singularly perturbed problems (this form is prompted by the form of the asymptotic solution (1.3.9) in § 1.3.2):

$$x(t,\mu) = \overline{x}(t,\mu) + \Pi x(\tau,\mu). \tag{2.1.12}$$

Here $\tau = t/\mu$, and x stands for both variables z and y, i.e. $x = \{z, y\}$,

$$\overline{x}(t,\mu) = \overline{x}_0(t) + \mu\overline{x}_1(t) + \cdots + \mu^k\overline{x}_k(t) + \cdots \tag{2.1.13}$$

is the regular part of the expansion,

$$\Pi x(\tau,\mu) = \Pi_0 x(\tau) + \mu\Pi_1 x(\tau) + \cdots + \mu^k \Pi_k x(\tau) + \cdots \tag{2.1.14}$$

is the boundary layer part. Substituting the series (2.1.12) into system (2.1.1), we obtain the equalities

$$\begin{aligned} \mu\frac{d\overline{z}}{dt} + \frac{d\Pi z}{d\tau} &= F(\overline{z}+\Pi z, \overline{y}+\Pi y, t), \\ \frac{d\overline{y}}{dt} + \frac{1}{\mu}\frac{d\Pi y}{d\tau} &= f(\overline{z}+\Pi z, \overline{y}+\Pi y, t). \end{aligned} \tag{2.1.15}$$

Next, we represent the right-hand sides of these equations in a form similar to (2.1.12): $F = \overline{F} + \Pi F$, $f = \overline{f} + \Pi f$. Let us illustrate how this can be done on the function F:

$$F(\overline{z}+\Pi z, \overline{y}+\Pi y, t) = F(\overline{z}(t,\mu), \overline{y}(t,\mu), t) + [F(\overline{z}(\tau\mu,\mu)$$

$$+\Pi z(\tau,\mu), \overline{y}(\tau\mu,\mu) + \Pi y(\tau,\mu), \tau\mu) - F(\overline{z}(\tau\mu,\mu), \overline{y}(\tau\mu,\mu), \tau\mu)] \equiv \overline{F} + \Pi F.$$

Equalities (2.1.15) can now be written as

$$\mu\frac{d\overline{z}}{dt} + \frac{d\Pi z}{d\tau} = \overline{F} + \Pi F, \qquad \frac{d\overline{y}}{dt} + \frac{1}{\mu}\frac{d\Pi y}{d\tau} = \overline{f} + \Pi f. \tag{2.1.16}$$

Let us substitute the series (2.1.13) and (2.1.14) for $\overline{x}(t,\mu)$ and $\Pi x(\tau,\mu)$ into (2.1.16) and represent $\overline{F}$, ΠF, $\overline{f}$, and Πf in the form of power series in μ:

$$\begin{aligned} \overline{F} &= F(\overline{z}_0(t) + \mu\overline{z}_1(t) + \cdots, \overline{y}_0(t) + \mu\overline{y}_1(t) + \cdots, t) \\ &= F(\overline{z}_0(t), \overline{y}_0(t), t) + \mu[\overline{F}_z(t)\overline{z}_1(t) + \overline{F}_y(t)\overline{y}_1(t)] + \cdots + \mu^k[\overline{F}_z(t)\overline{z}_k(t) \\ &\quad + \overline{F}_y(t)\overline{y}_k(t) + F_k(t)] + \cdots = \overline{F}_0 + \mu\overline{F}_1 + \cdots \mu^k\overline{F}_k + \cdots, \end{aligned}$$

where the elements of the matrices $\overline{F}_z(t)$ and $\overline{F}_y(t)$ are calculated at the point $(\overline{z}_0(t), \overline{y}_0(t), t)$, and the functions $F_k(t)$ are expressed recursively through $\overline{z}_i(t)$ and $\overline{y}_i(t)$ with $i < k$;

$$\begin{aligned} \Pi F = \ & F(\overline{z}_0(\tau\mu) + \mu\overline{z}_1(\tau\mu) + \cdots + \Pi_0 z(\tau) + \mu\Pi_1 z(\tau) + \cdots, \\ & \overline{y}_0(\tau\mu) + \mu\overline{y}_1(\tau\mu) + \cdots + \Pi_0 y(\tau) + \mu\Pi_1 y(\tau) + \cdots, \tau\mu) \\ & - F(\overline{z}_0(\tau\mu) + \mu\overline{z}_1(\tau\mu) + \cdots, \overline{y}_0(\tau\mu) + \mu\overline{y}_1(\tau\mu) + \cdots, \tau\mu) \\ = \ & [F(\overline{z}_0(0) + \Pi_0 z(\tau), \overline{y}_0(0) + \Pi_0 y(\tau), 0) - F(\overline{z}_0(0), \overline{y}_0(0), 0)] \\ & + \mu[F_z(\tau)\Pi_1 z(\tau) + F_y(\tau)\Pi_1 y(\tau) + G_1(\tau)] + \cdots \\ & + \mu^k[F_z(\tau)\Pi_k z(\tau) + F_y(\tau)\Pi_k y(\tau) + G_k(\tau)] + \cdots \\ = \ & \Pi_0 F + \mu\Pi_1 F + \cdots + \mu^k \Pi_k F + \cdots, \end{aligned}$$

where the elements of $F_z(\tau)$ and $F_y(\tau)$ are calculated at the point $(\bar{z}_0(0)+\Pi_0 z(\tau), \bar{y}_0(0)+\Pi_0 y(\tau), 0)$, and the functions $G_k(t)$ are expressed recursively through $\Pi_i z(\tau)$ and $\Pi_i y(\tau)$ with $i < k$. Similar expansions hold for $\bar{f}$ and Πf.

Let us now equate coefficients of like powers of μ on both sides of (2.1.16) (separately for the coefficients depending on t and on τ). This will provide the equations for the terms of the series (2.1.13) and (2.1.14).

For the leading term $\bar{x}_0(t) = \{\bar{z}_0(t), \bar{y}_0(t)\}$ of the regular part of the asymptotics, we obtain the system of equations

$$(2.1.17) \qquad 0 = \bar{F}_0 = F(\bar{z}_0(t), \bar{y}_0(t), t), \qquad \frac{d\bar{y}_0}{dt} = \bar{f}_0 \equiv f(\bar{z}_0(t), \bar{y}_0(t), t),$$

which evidently coincides with the reduced system (2.1.3).

For the leading term $\Pi_0 x(\tau) = \{\Pi_0 z(\tau), \Pi_0 y(\tau)\}$ of the boundary layer part of the asymptotics, we obtain

$$\begin{aligned} \frac{d\Pi_0 z}{d\tau} &= \Pi_0 F \equiv F(\bar{z}_0(0) + \Pi_0 z, \bar{y}_0(0) + \Pi_0 y, 0) - F(\bar{z}_0(0), \bar{y}_0(0), 0) \\ (2.1.18) \qquad &= F(\bar{z}_0(0) + \Pi_0 z, \bar{y}_0(0) + \Pi_0 y, 0), \\ \frac{d\Pi_0 y}{d\tau} &= 0. \end{aligned}$$

(Note that $F(\bar{z}_0(0), \bar{y}_0(0), 0) = 0$ by virtue of the first equation of (2.1.17).)

For the terms $\bar{x}_k(t)$ and $\Pi_k x(\tau)$ ($k \geq 1$) we have the equations (the so-called *variational equations*)

$$(2.1.19) \qquad \begin{aligned} \frac{d\bar{z}_{k-1}}{dt} &= \bar{F}_k \equiv \bar{F}_z(t)\bar{z}_k + \bar{F}_y(t)\bar{y}_k + F_k(t), \\ \frac{d\bar{y}_k}{dt} &= \bar{f}_k \equiv \bar{f}_z(t)\bar{z}_k + \bar{f}_y(t)\bar{y}_k + f_k(t); \end{aligned}$$

$$(2.1.20) \qquad \begin{aligned} \frac{d\Pi_k z}{d\tau} &= \Pi_k F \equiv F_z(\tau)\Pi_k z + F_y(\tau)\Pi_k y + G_k(\tau), \\ \frac{d\Pi_k y}{d\tau} &= \Pi_{k-1} f. \end{aligned}$$

Here the functions $f_k(t)$ and $F_k(t)$ are expressed recursively through the $\bar{z}_i(t)$ and $\bar{y}_i(t)$ with $i < k$. The functions $G_k(\tau)$ are expressed recursively through $\Pi_i z(\tau)$ and $\Pi_i y(\tau)$ with $i < k$, and $\Pi_{k-1} f$ is the coefficient of μ^{k-1} in the power series expansion for Πf with respect to μ (similar to the expansion for ΠF), e.g.,

$$(2.1.21) \qquad \begin{aligned} G_1(\tau) = \ & (F_z(\tau) - \bar{F}_z(0))(\bar{z}_0'(0)\tau + \bar{z}_1(0)) \\ & + (F_y(\tau) - \bar{F}_y(0))(\bar{y}_0'(0)\tau + \bar{y}_1(0)) + (F_t(\tau) - \bar{F}_t(0))\tau, \end{aligned}$$

$$(2.1.22) \qquad \Pi_0 f = f(\bar{z}_0(0) + \Pi_0 z(\tau), \bar{y}_0(0) + \Pi_0 y(\tau), 0) - f(\bar{z}_0(0), \bar{y}_0(0), 0).$$

To define the solutions of (2.1.19) and (2.1.20), we must impose initial conditions. Substituting expansion (2.1.12) into conditions (2.1.2), we obtain

$$\bar{z}_0(0) + \mu\bar{z}_1(0) + \cdots + \Pi_0 z(0) + \mu\Pi_1 z(0) + \cdots = z^0,$$

$$\bar{y}_0(0) + \mu\bar{y}_1(0) + \cdots + \Pi_0 y(0) + \mu\Pi_1 y(0) + \cdots = y^0.$$

Hence, equating the coefficients of like powers of μ in these relations implies

$$\bar{z}_0(0) + \Pi_0 z(0) = z^0, \qquad \bar{y}_0(0) + \Pi_0 y(0) = y^0, \tag{2.1.23}$$

$$\bar{z}_k(0) + \Pi_k z(0) = 0, \qquad \bar{y}_k(0) + \Pi_k y(0) = 0, \qquad k = 1, 2, \ldots. \tag{2.1.24}$$

Equations (2.1.23) contain four unknowns: $\bar{z}_0(0)$, $\Pi_0 z(0)$, $\bar{y}_0(0)$, and $\Pi_0 y(0)$ (a similar situation takes place in (2.1.24)). It is clear that by using only (2.1.23) it is impossible to define these four unknowns. We must take into consideration some additional ideas. First, it is important to mention that the first equation of (2.1.17) is not differential, and therefore we need not impose any initial condition on $\bar{z}_0(t)$. Thus, we need not treat $\bar{z}_0(0)$ in (2.1.23) as an unknown. The same is true for the quantities $\bar{z}_k(0)$ in the systems (2.1.24). Second, recall that the functions $\Pi_k x(\tau)$ should be boundary functions, i.e., they should approach zero as $\tau \to \infty$. It turns out that it is sufficient to impose this condition on only the functions $\Pi_k y(\tau)$, and as a result, it will also hold for $\Pi_k z(\tau)$. So, we introduce the condition

$$\Pi_k y(\infty) = 0, \qquad k = 0, 1, 2, \ldots \,. \tag{2.1.25}$$

Let us now show that (2.1.17)–(2.1.20) together with conditions (2.1.23)–(2.1.25) allow us to determine successively all the terms of the series (2.1.13), (2.1.14). From the second equation of (2.1.18), we obtain $\Pi_0 y(\tau) = \text{const}$, and since by virtue of (2.1.25) $\Pi_0 y(\infty) = 0$, we have

$$\Pi_0 y(\tau) \equiv 0.$$

This fact has quite clear interpretation: the function $y(t, \mu)$ does not have a boundary layer in the zeroth-order approximation.

Since $\Pi_0 y(0) = 0$, from the second equation of (2.1.23), we can find $\bar{y}_0(0) = y^0$. Thus, for $\bar{z}_0(t)$ and $\bar{y}_0(t)$ we have the system (2.1.17), which coincides with the reduced system (2.1.3), and the initial condition, which coincides with (2.1.4). For the solution $\bar{z}_0(t)$, $\bar{y}_0(t)$ of the problem (2.1.3), (2.1.4), let us take the solution that was mentioned in Theorem 2.1: $\bar{z}_0 = \bar{z}(t) = \varphi(\bar{y}, t)$, $\bar{y}_0 = \bar{y}(t)$. Then, the leading terms $\bar{z}_0(t)$ and $\bar{y}_0(t)$ of the regular part of the asymptotic expansion coincide with the limiting solution from Theorem 2.1.

Since $\bar{z}_0(t)$ is defined, the value $\bar{z}_0(0)$ is known, and therefore the first equation of (2.1.23) yields

$$\Pi_0 z(0) = z^0 - \bar{z}_0(0) = z^0 - \varphi(y^0, 0). \tag{2.1.26}$$

To find $\Pi_0 z(\tau)$ we must solve the first equation of (2.1.18) with this initial condition (also taking into account that $\Pi_0 y(\tau) = 0$). The change of variable $\tilde{z}(\tau) = \Pi_0 z(\tau) +$

$\varphi(y^0, 0)$ transforms the problem for $\Pi_0 z$ to the associated system (2.1.7) with the initial condition (2.1.8). By virtue of condition 3^0 of the theorem,

$$\Pi_0 z(\tau) \to 0 \quad \text{for } \tau \to \infty.$$

It can be proved (using condition 2^0) that $\Pi_0 z(\tau)$ possesses the exponential estimate (see Vasil'eva and Butuzov [148]):

$$\| \Pi_0 z(\tau) \| \leq c \exp(-\kappa\tau), \qquad \tau \geq 0. \tag{2.1.27}$$

Here and below we will use c and κ to represent appropriate positive numbers, which are, generally speaking, different in different inequalities.

When z is a scalar function, the estimate (2.1.18) can be obtained in a quite simple way. In this case condition 2^0 corresponds to the inequality

$$F_z(\bar{z}_0(t), \bar{y}_0(t), t) < 0 \quad \text{for } 0 \leq t \leq T. \tag{2.1.28}$$

The first equation of (2.1.18), according to the mean value theorem, can be written in the form

$$\frac{d\Pi_0 z}{d\tau} = F_z(\bar{z}_0(0) + \theta\Pi_0 z(\tau), y^0, 0)\, \Pi_0 z,$$

where $0 < \theta < 1$. Hence, taking into account (2.1.26), we obtain

$$\Pi_0 z(\tau) = (z^0 - \bar{z}_0(0)) \exp\left(\int_0^\tau F_z(\bar{z}_0(0) + \theta\Pi_0 z(s), y^0, 0) ds \right). \tag{2.1.29}$$

Since $\Pi_0 z(\tau) \to 0$ as $\tau \to \infty$, and $F_z(\bar{z}_0(0), y^0, 0) < 0$ by virtue of (2.1.28), there exist numbers $\kappa > 0$ and $\tau_0 > 0$ such that $F_z(\bar{z}_0(0) + \theta\Pi_0 z(t), y^0, 0) < -\kappa$ for $\tau \geq \tau_0$. Hence,

$$\int_{\tau_0}^{\tau} F_z(\bar{z}_0(0) + \theta\Pi_0 z(s), y^0, 0) ds < -\kappa(\tau - \tau_0)$$

for $\tau \geq \tau_0$, and therefore

$$\begin{aligned} |\Pi_0 z(\tau)| &= \left[|z^0 - \bar{z}_0(0)| \exp\left(\int_0^{\tau_0} F_z(\bar{z}_0(0) + \theta\Pi_0 z(s), y^0, 0) ds \right) \right] \\ &\quad \times \exp\left(\int_{\tau_0}^{\tau} F_z(\bar{z}_0(0) + \theta\Pi_0 z(s), y^0, 0) ds \right) \\ &\leq c \exp(-\kappa(\tau - \tau_0)) = c_1 \exp(-\kappa\tau) \end{aligned} \tag{2.1.30}$$

for $c_1 = c \exp(\kappa\tau_0)$. Denoting c_1 by c, we obtain the estimate (2.1.27) for $\Pi_0 z(\tau)$.

Thus, all the terms of the zeroth-order approximation are defined.

The higher-order terms of the asymptotic expansion can be constructed by the same scheme for any $k \geq 1$. Suppose we already defined all the terms up to order $k-1$, and the Π-functions (i.e., $\Pi_i z$ and $\Pi_i y$ for $i = 0, 1, \dots, k-1$) satisfy an exponential estimate of the type (2.1.27). Then the nonhomogeneous terms in the linear systems (2.1.19), (2.1.20), i.e., $d\bar{z}_{k-1}/dt$, $F_k(t)$, $f_k(t)$, $G_k(\tau)$, $\Pi_{k-1} f$, are known functions, and $G_k(\tau)$, $\Pi_{k-1} f$ possess the exponential estimate of the type (2.1.27). For $G_1(t)$ and

$\Pi_0 f$ such estimates follow directly from the formulae (2.1.21), (2.1.22) and from the exponential estimate for $\Pi_0 z(\tau)$.

From the second equation of (2.1.20) with condition (2.1.25), we obtain

(2.1.31) $$\Pi_k y(\tau) = \int_\infty^\tau \Pi_{k-1} f(s) ds.$$

Since $\| \Pi_{k-1} f(\tau) \| \le c \exp(-\kappa\tau)$, we have

$$\| \Pi_k y(\tau) \| \le \int_\tau^\infty c \exp(-\kappa s) ds = \frac{c}{\kappa} \exp(-\kappa\tau).$$

Again using the notation c for the constant c/κ, we arrive at the estimate of the type (2.1.27) for $\Pi_k y(\tau)$. From (2.1.31), we obtain

$$\Pi_k y(0) = \int_\infty^0 \Pi_{k-1} f(s) ds.$$

Therefore we can find $\overline{y}_k(0)$ from (2.1.24):

(2.1.32) $$\overline{y}_k(0) = -\Pi_k y(0) = \int_0^\infty \Pi_{k-1} f(s) ds.$$

To find $\overline{z}_k(t)$ and $\overline{y}_k(t)$ we now need to solve system (2.1.19) with initial condition (2.1.32). First, we express $\overline{z}_k$ through $\overline{y}_k$ from the first equation of (2.1.19):

(2.1.33) $$\overline{z}_k = \overline{F}_z^{-1}(t) \left[\frac{d\overline{z}_{k-1}}{dt} - \overline{F}_y(t) \overline{y}_k - F_k(t) \right]$$

(note that by virtue of condition 2^0, $\det \overline{F}_z(t) \neq 0$, and therefore $\overline{F}_z^{-1}(t)$ exists). Next, we substitute this expression into the second equation of (2.1.19). We obtain a linear differential equation for $\overline{y}_k(t)$. Solving it with the initial condition (2.1.32), we find $\overline{y}_k(t)$, and therefore also $\overline{z}_k(t)$ (by formula (2.1.33)).

The value of $\overline{z}_k(0)$ is now known. Thus, we can obtain the initial condition for $\Pi_k z$ from the first equation of (2.1.24): $\Pi_k z(0) = -\overline{z}_k(0)$. Finally, we must solve the first equation in (2.1.20) subject to the initial condition mentioned above. We should take into account that $\Pi_k y(\tau)$ is known, and since this function, as well as $G_k(\tau)$, has the exponential estimate, the function $\tilde{G}_k(\tau) = F_y(\tau) \Pi_k y(\tau) + G_k(\tau)$ satisfies

(2.1.34) $$\| \tilde{G}_k(\tau) \| \le c \exp(-\kappa\tau).$$

Moreover, $\Pi_k z(\tau)$ can be written in the form

(2.1.35) $$\Pi_k z(\tau) = -\Phi(\tau) \overline{z}_k(0) + \int_0^\tau \Phi(\tau) \Phi^{-1}(s) \tilde{G}_k(s) ds,$$

where $\Phi(\tau)$ is the fundamental matrix of the corresponding homogeneous equation

$$\frac{d\Phi}{d\tau} = F_z(\tau)\Phi, \quad \Phi(0) = I_M \text{ (an identity matrix of dimension } M).$$

It can be shown (see Vasil'eva and Butuzov [148]) that by virtue of condition 2^0 the following inequalities hold:

(2.1.36)
$$\| \Phi(\tau) \| \leq c\exp(-\kappa\tau), \qquad \tau \geq 0,$$
$$\| \Phi(\tau)\Phi^{-1}(s) \| \leq c\exp(-\kappa(\tau - s)), \qquad 0 \leq s \leq \tau.$$

In the case when z is a scalar function, we have

$$\Phi(\tau) = \exp\left(\int_0^\tau F_z(s)ds\right),$$

and the estimates (2.1.36) can be proved in the same way as the estimate (2.1.30).

Using the inequalities (2.1.34) and (2.1.36), from (2.1.35) we obtain

$$\begin{aligned}\| \Pi_k z(\tau) \| &\leq c\exp(-\kappa\tau) + \int_0^\tau c\exp(-\kappa(\tau - s))(c\exp(-\kappa s))ds \\ &\leq c\exp(-\kappa\tau) + c^2\tau\exp(-\kappa\tau).\end{aligned}$$

Let us take κ_1 such that $0 < \kappa_1 < \kappa$. Since $\tau\exp(-(\kappa - \kappa_1)\tau) \leq c$, we have $\tau\exp(-\kappa\tau) \leq c\exp(-\kappa_1\tau)$. Using again the notation κ for κ_1 and c for all constants of the same type, we finally obtain

$$\| \Pi_k z(\tau) \| \leq c\exp(-\kappa\tau), \qquad \tau \geq 0.$$

So, the algorithm described allows one to define the terms of the series (2.1.13) and (2.1.14) up to any order n inclusive. All the Π-functions decay exponentially.

2.1.3 Justification of the asymptotics

Let $X_n(t,\mu)$ denote the nth partial sum of the expansion (2.1.12):

$$X_n(t,\mu) = \sum_{k=0}^{n} \mu^k[\overline{x}_k(t) + \Pi_k x(\tau)].$$

Theorem 2.2 (Vasil'eva). *Under conditions* 1^0—4^0, *the series* (2.1.12) *is the asymptotic series as* $\mu \to 0$ *for the solution* $x(t,\mu) = \{z(t,\mu), y(t,\mu)\}$ *of the problem* (2.1.1), (2.1.2) *in the interval* $0 \leq t \leq T$, *i.e. the following estimate holds:*

(2.1.37)
$$\max_{0 \leq t \leq T} \| x(t,\mu) - X_n(t,\mu) \| = O(\mu^{n+1}).$$

This theorem can be proved by the same scheme as Theorem 1.1. Note that existence and uniqueness of the solution follow from Theorem 2.1. However, along with the proof of the estimate (2.1.37), one can also obtain uniqueness and existence.

For the remainder terms

$$u(t,\mu) = z(t,\mu) - Z_n(t,\mu), \qquad v(t,\mu) = y(t,\mu) - Y_n(t,\mu),$$

we obtain the equations

$$\mu\frac{du}{dt} = F_z u + F_y v + g_1(u,v,t,\mu),$$

$$\frac{dv}{dt} = f_z u + f_y v + g_2(u, v, t, \mu),$$

$$u(0, \mu) = 0, \qquad v(0, \mu) = 0,$$

where the elements of the matrices F_z, F_y, f_z, and f_y are calculated at the point $(\bar{z}_0(t) + \Pi_0 z(t/\mu), \bar{y}_0(t), t)$,

$$g_1(u, v, t, \mu) = F(Z_n + u, Y_n + v, t) - \mu \frac{dZ_n}{dt} - F_z u - F_y v,$$

and

$$g_2(u, v, t, \mu) = f(Z_n + u, Y_n + v, t) - \frac{dY_n}{dt} - f_z u - f_y v.$$

The functions g_1 and g_2 possess the following two properties, similar to those of function g from § 1.3.1.

1*. $g_1(0, 0, t, \mu) = O(\mu^{n+1})$, $\quad g_2(0, 0, t, \mu) = O(\mu^{n+1} + \mu^n \exp(-\kappa t/\mu))$.

2*. If $\| u_i(t, \mu) \| \leq c_1 \mu$, $\| v_i(t, \mu) \| \leq c_1 \mu$, $i = 1, 2$, then there exist numbers $c_2 > 0$ and $\mu_0 > 0$ such that for $0 < \mu \leq \mu_0$ the following inequalities hold:

$$\sup_{0 \leq t \leq T} \| g_i(u_2, v_2, t, \mu) - g_i(u_1, v_1, t, \mu) \|$$

$$\leq c_2 \mu \left[\sup_{0 \leq t \leq T} \| u_2 - u_1 \| + \sup_{0 \leq t \leq T} \| v_2 - v_1 \| \right].$$

Using the fundamental matrices $U(t, s, \mu)$ and $V(t, s, \mu)$ of the homogeneous systems

$$\mu \frac{du}{dt} = F_z \, u \quad \text{and} \quad \frac{dv}{dt} = f_y \, v,$$

we naturally consider the equivalent system of integral equations

$$\text{(2.1.38)} \qquad \begin{aligned} u(t, \mu) &= \int_0^t U(t, s, \mu) \frac{1}{\mu} \left[F_y \, v(s, \mu) + g_1(u, v, s, \mu) \right] ds, \\ v(t, \mu) &= \int_0^t V(t, s, \mu) \left[f_z \, u(s, \mu) + g_2(u, v, s, \mu) \right] ds. \end{aligned}$$

For the fundamental matrix $U(t, s, \mu)$ (see Vasil'eva and Butuzov [148]):

$$\| U(t, s, \mu) \| \leq c \exp\left(-\frac{\kappa}{\mu}(t - s) \right), \qquad 0 \leq s \leq t \leq T.$$

Substituting the expression for $v(t, \mu)$, defined by the second equation of (2.1.38), into the first equation, we obtain

$$\text{(2.1.39)} \qquad u(t, \mu) = \int_0^t K(t, s, \mu) u(s, \mu) ds + Q_1(u, v, t, \mu),$$

where K is a bounded kernel, and the operator Q_1 possesses the same two properties as function $g_1(u,v,t,\mu)$. The equation (2.1.39) can be replaced by the equivalent equation

$$(2.1.40)\quad u(t,\mu) = Q_1(u,v,t,\mu) + \int_0^t R(t,s,\mu)Q_1(u,v,s,\mu)ds \equiv G_1(u,v,t,\mu).$$

Here R is the resolvent of the kernel K. Next, let us substitute the expression (2.1.40) for $u(t,\mu)$ into the equation of (2.1.38):

$$(2.1.41)\quad \begin{aligned} v(t,\mu) &= \int_0^t V(t,s,\mu)\left[f_z(s,\mu)\,G_1(u,v,s,\mu) + g_2(u,v,s,\mu)\right] ds \\ &\equiv G_2(u,v,t,\mu). \end{aligned}$$

Integral operators G_1 and G_2 in (2.1.40) and (2.1.41) possess the same two properties as the function $g_1(u,v,t,\mu)$.

Applying the method of successive approximations to the system (2.1.40), (2.1.41), as in § 1.3.1, we find that a unique solution exists and satisfies the estimate

$$\max_{0\le t\le T} \| u(t,\mu) \| = O(\mu^{n+1}), \qquad \max_{0\le t\le T} \| v(t,\mu) \| = O(\mu^{n+1}),$$

i.e., the inequality (2.1.37) holds.

The detailed proof of the theorem can be found in Vasil'eva and Butuzov [148].

Exercise

Show that the property 2* holds for the functions g_1 and g_2.

Let us note in conclusion that to prove estimate (2.1.37) for any given n, it is sufficient to assume that the functions F and f are $n+2$ times continuously differentiable in some δ-tube about the curve $L_0 = L_1 \cup L_2$, where

$$L_1 = \{(z,y,t):\ z = \bar{z}_0(0) + \Pi_0 z(\tau),\ \tau \ge 0;\ y = y^0,\ t = 0\},$$

$$L_2 = \{(z,y,t):\ z = \bar{z}_0(t);\ y = \bar{y}_0(t),\ 0 \le t \le T\}.$$

The curve L_0 is the limiting curve (as $\mu \to 0$) of the integral curve corresponding to the solution $z(t,\mu)$, $y(t,\mu)$ of (2.1.1), (2.1.2) (see Fig. 2.1, where L_1 is the segment AB and L_2 is the curve BC).

Exercises

1. Find the zeroth- and the first-order approximations to the solution of

$$\begin{aligned} \mu\frac{dz}{dt} &= -az + by + f, \\ \frac{dy}{dt} &= cz + dy + g, \\ z(0) = 0,\quad & y(0) = 0, \qquad t \in [0,1], \end{aligned}$$

where a, b, c, d, f, and g are all positive constants, and $0 < \mu \ll 1$ is a small parameter.

2. Define the asymptotic solution up to the terms of the order $O(\mu)$ for

$$\mu\frac{dz}{dt} = -z + y + \mu x^2,$$
$$\frac{dy}{dt} = z + y,$$
$$z(0) = 0, \quad y(0) = 1, \quad t \in [0, 1].$$

3. Construct the zeroth-order terms of the asymptotic solution of

$$\mu\frac{dz}{dt} = z(1 - z) - y,$$
$$\frac{dy}{dt} = -z(1 - z) + y^2,$$
$$z(0) = 2, \quad y(0) = 0, \quad t \in [0, 1].$$

4. Consider the enzyme kinetics reaction during which a molecule of substrate S is converted irreversibly by a single molecule of enzyme E into a molecule of product P. Schematically, we can write

$$S + E \underset{k_1^-}{\overset{k_1^+}{\rightleftarrows}} C \overset{k_2}{\rightarrow} P + E,$$

where C stands for a molecule of an intermediate substrate-enzyme complex. (For a more detailed description of chemical kinetics problems see § 2.2.3.) When the initial concentration of enzyme is small, the nondimensionalized system of equations for concentrations s and c of substrate and complex, respectively, can be written in the form

$$\mu\frac{dc}{dt} = s - (s + k)c,$$
$$\frac{ds}{dt} = -s + (s + k - \lambda)c,$$
$$c(0) = 0, \quad s(0) = 1, \quad t \in [0, 1],$$

where k and λ are positive constants depending on $k_1^{\pm}$ and k_2, and $0 < \mu \ll 1$ is a small parameter.

Construct the zeroth-order terms of the asymptotic solution of this problem (cf., e.g., O'Malley [114] or Lin and Segel [86]). (Note that the regular function $\bar{s}_0$ will be defined implicitly by an algebraic relation, whereas $\bar{c}_0$ and the boundary functions will be defined explicitly.) We should mention that the regular part of the asymptotic solution of the zeroth order is usually referred to as the *Michaelis–Menten approximation* (see Murray [99] for biochemical details).

2.2 The critical case

2.2.1 The initial value problem for a system with a small nonlinearity

One of the main conditions in the theorem on passage to the limit is the condition of existence of an isolated root $\bar{z} = \varphi(\bar{y}, t)$ of the reduced equation $F(\bar{z}, \bar{y}, t) = 0$. In a variety of applied problems leading to singularly perturbed equations, in particular, in most problems of chemical kinetics, this condition is not satisfied because the reduced equation does not have an isolated root. It, instead, has a family of solutions depending on one or several parameters. Such a case will be called the *critical case* . (See also the related discussion of *singular singularly perturbed problems* in O'Malley [114].)

It turns out that under some quite general conditions, the asymptotics for the initial value problem in the critical case has the same form as in § 2.1.2. In particular, the solution of the initial value problem approaches one of the solutions of the reduced equation in the limit as $\mu \to 0$. However, the algorithm of constructing the asymptotic expansion undergoes some changes.

Let us consider this algorithm for a system of equations with a small nonlinearity:

$$\mu\frac{dx}{dt} = A(t)x + \mu f(x,t,\mu), \qquad 0 \le t \le T, \tag{2.2.1}$$

$$x(0,\mu) = x^0. \tag{2.2.2}$$

Here x and f are m-dimensional vector functions, $A(t)$ is an $(m \times m)$-matrix, and $\mu > 0$ is a small parameter. We assume that $A(t)$ and $f(x,t,\mu)$ are sufficiently smooth.

1^0. *Let the matrix* $A(t)$ *have zero as an eigenvalue of multiplicity* $0 < k < m$ *for each* $t \in [0,T]$:

(a) $\lambda_i(t) = 0,\ i = 1,2,\dots,k,$

while other eigenvalues of $A(t)$ *satisfy the inequality*

(b) $\mathrm{Re}\lambda_i(t) < 0,\ i = k+1,\dots,m.$

From condition 1^0(a) it follows that $\det A(t) = 0$, and therefore the reduced equation $A(t)\bar{x} = 0$ obtained from (2.2.1) when $\mu = 0$, has a family of solutions represented by a linear combination (with arbitrary coefficients) of the eigenvectors of the matrix $A(t)$ corresponding to $\lambda = 0$ (i.e., *null-vectors*).

2^0. *Let* k *null-vectors* $e_i(t)$, $i = 1,2,\dots,k$ *be linearly independent for each* $t \in [0,T]$.

Some other conditions will be imposed later during the construction of the asymptotic solution.

As in § 2.1.2, we will seek the asymptotic expansion of the solution of (2.2.1), (2.2.2) as a sum of regular and boundary function series:

$$x(t,\mu) = \bar{x}(t,\mu) + \Pi x(\tau,\mu) = \sum_{i=0}^{\infty} \mu^i(\bar{x}_i(t) + \Pi_i x(\tau)), \tag{2.2.3}$$

where $\tau = t/\mu$. Substituting (2.2.3) into (2.2.1), (2.2.2), and representing f in the form $f = \bar{f} + \Pi f$ (see § 2.1.2), we arrive at the equalities

$$\mu\frac{d\bar{x}}{dt} + \frac{d\Pi x}{d\tau} = A(t)\bar{x} + A(\tau\mu)\Pi x + \mu\bar{f} + \mu\Pi f, \tag{2.2.4}$$

$$\bar{x}(0,\mu) + \Pi x(0,\mu) = x^0. \tag{2.2.5}$$

Expanding $A(\tau\mu)$, $\bar{f}$ and Πf into power series in μ, and equating coefficients of like powers of μ on both sides of (2.2.4) and (2.2.5) (separately for coefficients depending on t and on τ), we obtain the problems for the terms of the series (2.2.3).

For $\bar{x}_0(t)$ we have the reduced equation

$$A(t)\bar{x}_0 = 0.$$

The general solution of this equation, by virtue of condition 2^0, can be written in the form

$$\bar{x}_0 = \sum_{i=1}^{k} \alpha_i(t) e_i(t), \tag{2.2.6}$$

where the $e_i(t)$ are the null vectors of the matrix $A(t)$, and the $\alpha_i(t)$ are some arbitrary scalar functions. Let us introduce the $(m \times k)$-matrix $e(t)$ whose columns are the eigenvectors $e_i(t)$ $(i = 1, \ldots, k)$, and the k-dimensional vector function $\alpha(t)$ with elements $\alpha_i(t)$. Then (2.2.6) can be written in the more compact form:

$$\bar{x}_0 = e(t)\alpha(t). \tag{2.2.7}$$

For $\Pi_0 x(\tau)$, we obtain the problem

$$\frac{d\Pi_0 x}{d\tau} = A(0)\Pi_0 x, \qquad \tau \geq 0, \tag{2.2.8}$$

$$\Pi_0 x(0) = x^0 - \bar{x}_0(0) = x^0 - \sum_{i=1}^{k} \alpha_i(0)e_i(0). \tag{2.2.9}$$

The general solution to (2.2.8) can be represented in the form

$$\Pi_0 x(\tau) = \sum_{i=1}^{k} c_i e_i(0) + \sum_{i=k+1}^{m} c_i w_i(\tau) \exp(\lambda_i(0)\tau), \tag{2.2.10}$$

where the c_i are arbitrary constants, the $e_i(0)$ are the null vectors of matrix $A(0)$, and $w_i(\tau)$ are known vector functions whose elements are polynomials in τ. In particular, if $\lambda_i(0)$ $(i = k+1, \ldots, m)$ are simple eigenvalues, then the w_i are the τ-independent eigenvectors of $A(0)$ corresponding to $\lambda_i(0)$.

By virtue of condition 1^0(b), the second sum in (2.2.10) approaches zero as $\tau \to \infty$. Let us require that the entire boundary function $\Pi_0 x(\tau)$ approach zero as $\tau \to \infty$, and let us impose a similar condition on all the Π-functions:

$$\Pi_i x(\infty) = 0.$$

For this condition to hold for $\Pi_0 x$, we must take $c_i = 0$, $i = 1, \ldots, k$. Thus,

$$\Pi_0 x(\tau) = \sum_{i=k+1}^{m} c_i w_i(\tau) \exp(\lambda_i(0)\tau). \tag{2.2.11}$$

Substituting this expression into the initial conditions (2.2.9), we obtain

$$\sum_{i=1}^{k} \alpha_i(0)e_i(0) + \sum_{i=k+1}^{m} c_i w_i(0) = x^0. \tag{2.2.12}$$

This is the system of m linear algebraic equations for the m unknowns $\alpha_i(0)$ $(i = 1, \ldots, k)$ and c_i $(i = k+1, \ldots, m)$. It is known (cf., e.g., Pontriagin [119]) that the vectors $e_i(0)$ $(i = 1, \ldots, k)$ and $w_i(0)$ $(i = k+1, \ldots, m)$ are linearly independent. Therefore the system (2.2.12) has a unique solution.

Thus, the function $\Pi_0 x(\tau)$ is completely determined by formula (2.2.11) and by virtue of the condition 1^0(b), it has the exponential estimate

$$\| \Pi_0 x(\tau) \| \leq c \exp(-\kappa\tau). \tag{2.2.13}$$

For the yet unknown functions $\alpha_i(t)$ in the expression (2.2.6) for $\bar{x}_0$, we obtained the initial values $\alpha_i(0)$. Let us introduce the notation $\alpha_i(0) = \alpha_i^0$, $\alpha(0) = \alpha^0$.

The functions $\alpha_i(t)$ are defined completely in the next step of the algorithm, i.e., during the solution of equation for $\bar{x}_1(t)$:

$$A(t)\bar{x}_1 = -f(\bar{x}_0(t), t, 0) + \frac{d\bar{x}_0}{dt} = -f(e(t)\alpha(t), t, 0) + \frac{d}{dt}(e(t)\alpha(t)). \tag{2.2.14}$$

Since $\det A(t) = 0$, this system is solvable if and only if the right-hand side of (2.2.14) is orthogonal to each of the eigenvectors $g_j(t)$ $(j = 1, \dots, k)$ of the adjoint matrix $A^*(t)$ corresponding to the eigenvalue $\lambda = 0$. For a real-valued matrix $A(t)$ the adjoint matrix $A^*(t)$ coincides with transpose of $A(t)$. Let us introduce the $(k \times m)$-matrix $g(t)$ whose rows are the vectors $g_j(t)$. The orthogonality condition can be written as

$$g(t)\left(-f(e(t)\alpha(t), t, 0) + e(t)\frac{d\alpha}{dt} + \frac{de}{dt}\alpha(t)\right) = 0. \tag{2.2.15}$$

The eigenvectors $g_j(t)$ can be chosen in such a way that $g(t)e(t) = I_k$, where I_k is the $(k \times k)$ identity matrix. Therefore, from (2.2.15), we obtain

$$\frac{d\alpha}{dt} = F(\alpha, t). \tag{2.2.16}$$

The explicit form of F is clear from comparison of (2.2.15) and (2.2.16). Thus, the solvability condition of (2.2.14) for $\bar{x}_1$ provides the differential equation for the unknown function $\alpha(t)$. Recall that the initial condition for $\alpha(t)$ has been found earlier.

3^0. *Let the equation* (2.2.16) *with the initial condition* $\alpha(0) = \alpha^0$ *have a solution* $\alpha(t)$ *for* $0 \le t \le T$.

Now $\bar{x}_0(t)$ is completely determined, and the solution of (2.2.14) can be written in the form

$$\bar{x}_1 = \tilde{x}_1(t) + \sum_{i=1}^{k} \beta_i(t)e_i(t) = \tilde{x}_1(t) + e(t)\beta(t),$$

where $\tilde{x}_1(t)$ is a known function (a particular solution of (2.2.14)), but $\beta(t)$ remains an arbitrary k-dimensional vector function.

For $\Pi_1 x(\tau)$ we obtain the problem

$$\frac{d\Pi_1 x}{d\tau} = A(0)\Pi_1 x + \varphi_1(\tau), \qquad \tau \ge 0, \tag{2.2.17}$$

$$\Pi_1 x(0) = -\bar{x}_1(0) = -\tilde{x}_1(0) - \sum_{i=1}^{k} \beta_i(0)e_i(0), \tag{2.2.18}$$

where $\varphi_1(\tau) = A'(0)\tau\Pi_0 x(\tau) + [f(\bar{x}_0(0) + \Pi_0 x(\tau), 0, 0) - f(\bar{x}_0(0), 0, 0)]$. Evidently, $\varphi_1(\tau)$ has an estimate of the type (2.2.13).

The general solution of (2.2.8) can be written in the form

$$\Pi_1 x = \tilde{\Pi}_1 x(\tau) + \sum_{i=1}^{k} d_i e_i(0) + \sum_{i=k+1}^{m} d_i w_i(\tau)\exp(\lambda_i(0)\tau).$$

Here $\tilde{\Pi}_1 x(\tau)$ is a known function (a particular solution of (2.2.17)), which can be chosen in such a way that it will have the estimate of the type (2.2.13), d_i are arbitrary constants, and $e_i(0)$ and $w_i(\tau)$ are the same vectors as in (2.2.10).

From condition $\Pi_1 x(\infty) = 0$, we obtain $d_i = 0 \ \ (i = 1, \ldots, k)$. Next, substituting the expression for $\Pi_1 x(\tau)$ into the initial condition (2.2.18), we arrive at the linear algebraic system

$$\sum_{i=1}^{k} \beta_i(0) e_i(0) + \sum_{i=k+1}^{m} d_i w_i(0) = -\tilde{x}_1(0) - \tilde{\Pi}_1 x(0).$$

We can uniquely define $\beta_i(0) \ \ (i = 1, \ldots, k)$ and $d_i \ \ (i = k+1, \ldots, m)$ from this system. Thus, $\Pi_1 x(\tau)$ will be completely determined (it will possess an exponential estimate of the type (2.2.13)), and for the unspecified $\beta(t)$ the initial condition will be defined. This function will be completely determined in the next step of construction of the asymptotics during the solution for $\overline{x}_2(t)$. The solvability condition for this equation will provide a linear differential equation for $\beta(t)$:

$$\frac{d\beta}{dt} = B(t)\beta + g_1(t), \tag{2.2.19}$$

where $B(t) = F_\alpha(\alpha(t), t)$ is a known matrix, and $g_1(t)$ is a known vector function. Solving this equation with the initial condition obtained above, we will determine $\beta(t)$, and thus, the function $\overline{x}_1(t)$ will be completely defined.

Construction of higher-order terms of the series (2.2.3) can be done analogously. In the ith step the solution of the equation for $\overline{x}_i(t)$ will contain an arbitrary k-dimensional vector function $\gamma(t)$. During the construction of $\Pi_i x(\tau)$ the initial condition $\gamma(0)$ is defined, and the solvability condition of the equation for $\overline{x}_{i+1}(t)$ provides a differential equation of the type (2.2.19):

$$\frac{d\gamma}{dt} = B(t)\gamma + g_i(t),$$

with $g_i(t)$ known, which allows one to determine $\gamma(t)$ uniquely.

The boundary functions $\Pi_i x(\tau)$ are constructed analogously to $\Pi_1 x(\tau)$ and also satisfy an estimate of the type (2.2.13).

Thus, construction of the series (2.2.3) is completed.

The main results about the series (2.2.3) can be formulated as a theorem.

Theorem 2.3 *Under conditions* 1^0–3^0, *for sufficiently small* μ, *the problem* (2.2.1), (2.2.2) *has a unique solution* $x(t, \mu)$, *and the series* (2.2.3) *is the asymptotic series for this solution in the interval* $0 \le t \le T$, *i.e., for any* n, *the following relation holds:*

$$\max_{[0,T]} \| x(t, \mu) - X_n(t, \mu) \| = O(\mu^{n+1}),$$

where $X_n(t, \mu)$ *is the* n*th partial sum of the series* (2.2.3).

The proof of this theorem is quite standard: the equation for the remainder is transformed into an equivalent integral equation with subsequent application of method of successive approximations (see Vasil'eva and Butuzov [149]).

Exercise

Construct asymptotic solutions in the interval $t \in [0,1]$ up to the terms of the order $O(\mu^2)$ for the equations

a). $\mu \frac{d}{dt}\begin{pmatrix} x \\ y \end{pmatrix} = \begin{pmatrix} -1 & 1 \\ 1 & -1 \end{pmatrix}\begin{pmatrix} x \\ y \end{pmatrix} - \mu \begin{pmatrix} x^2 \\ y^2 \end{pmatrix}$,

b). $\mu \frac{d}{dt}\begin{pmatrix} x \\ y \end{pmatrix} = \begin{pmatrix} -(1+t) & (1+t) \\ 1 & -1 \end{pmatrix}\begin{pmatrix} x \\ y \end{pmatrix} - \mu \begin{pmatrix} x \\ 0 \end{pmatrix}$,

c). $\mu \frac{d}{dt}\begin{pmatrix} x \\ y \end{pmatrix} = \begin{pmatrix} -1 & (1+t) \\ 1 & -(1+t) \end{pmatrix}\begin{pmatrix} x \\ y \end{pmatrix} + \mu \begin{pmatrix} 0 \\ y \end{pmatrix}$,

subject to initial conditions

$$\begin{pmatrix} x(0) \\ y(0) \end{pmatrix} = \begin{pmatrix} 1 \\ 0 \end{pmatrix}.$$

2.2.2 Other problems in the critical case

We considered (2.2.1), where the factor μ in front of the derivative was the same as the factor in front of the function f in its right-hand side. The case when the order of the small parameter before the derivative is higher than the order of the coefficient by f is also of interest. Let us discuss, e.g., the equation

$$\mu^2 \frac{dx}{dt} = A(t)x + \mu f(x,t,\mu), \tag{2.2.20}$$

where the matrix $A(t)$ satisfies the same conditions 1^0–3^0 as in § 2.2.1. The asymptotic solution of this equation with the initial condition (2.2.2) will contain, along with the regular terms $\overline{x}_i(t)$ and boundary functions $\Pi_i x(\tau)$ ($\tau = t/\mu$), one more boundary layer series in powers of μ with coefficients P_i, which are decaying functions of the stretched variable $s = t/\mu^2$.

As before, we obtain the expression (2.2.7) for $\overline{x}_0(t)$, where $\alpha(t)$ is an arbitrary k- dimensional vector function. This function is defined from the solvability condition on the equation for $\overline{x}_1(t)$, but now this condition provides an algebraic equation

$$g(t)f(e(t)\alpha(t), t, 0) = 0.$$

$3'$. *Let this equation have an isolated solution for $\alpha(t)$ on $0 \le t \le T$.*

For $\Pi_0 x(\tau)$ we also obtain not differential but algebraic equation $A(0)\Pi_0 x = 0$, and hence,

$$\Pi_0 x(\tau) = e(0)h(\tau).$$

Here $h(\tau)$ is an arbitrary k-dimensional vector function. The solvability condition for the nonhomogeneous algebraic system for $\Pi_1 x(\tau)$ now provides the differential equation for $h(\tau)$:

$$\frac{dh}{d\tau} = g(0)[f(\overline{x}_0(0) + e(0)h(\tau), 0, 0) - f(\overline{x}_0(0), 0, 0)]. \tag{2.2.21}$$

For the leading term $P_0 x(s)$ we obtain the equation

$$\frac{dP_0 x}{ds} = A(0)P_0 x$$

with initial condition

$$P_0x(0) = x^0 - \bar{x}_0(0) - e(0)h(0) = x^0 - \bar{x}_0(0) - \sum_{i=1}^{k} h_i(0)e_i(0).$$

The equation for P_0x has the same form as (2.2.8). Therefore

$$P_0x(s) = \sum_{i=1}^{k} c_i e_i(0) + \sum_{i=k+1}^{m} c_i w_i(s)\exp(\lambda_i(0)s).$$

From the condition that $P_0x(\infty) = 0$ it follows that $c_i = 0 \;\; (i = 1, \ldots, k)$. Substituting the expression for $P_0x(s)$ into the initial condition, we obtain

$$\sum_{i=1}^{k} h_i(0)e_i(0) + \sum_{i=k+1}^{m} c_i w_i(0) = x^0 - \bar{x}_0(0).$$

The values $h_i(0) \;\; (i = 1, \ldots, k)$ and $c_i \;\; (i = k+1, \ldots, m)$ can now be uniquely defined from this equation.

To complete the construction of the zeroth-order approximation, we must solve (2.2.21) with the known initial condition, which we will write as $h(0) = h^0$.

The essential requirements that guarantee the boundary layer structure of the solution of (2.2.20), (2.2.2) (in particular, the exponential estimate of the type (2.2.13) for the function $h(\tau)$, and thus, for the function $\Pi_0 x(\tau)$) are the following:

4^0. *Let the eigenvalues of the* $(k \times k)$*-matrix* $g(t)f_x(\bar{x}_0(t), t, 0)e(t)$ *have negative real parts for* $0 \le t \le T$.

Note that, by virtue of this condition, $h = 0$ is an asymptotically stable rest point of the equation (2.2.21).

5^0. *Let the initial condition* $h(0) = h^0$ *belong to the domain of attraction of the rest point* $h = 0$ *of* (2.2.21).

Higher-order terms of the asymptotic expansion can be constructed analogously. It can be shown that all the boundary functions $\Pi_i x(\tau)$ and $P_i x(s)$ will have the exponential decay estimates of the type (2.2.13). The detailed solution of (2.2.20), (2.2.2) is presented in Butuzov and Nefedov [26].

Exercise

Find the zeroth-order approximation to the solution of

$$\mu^2 \frac{d}{dt}\begin{pmatrix} x \\ y \end{pmatrix} = \begin{pmatrix} -1 & 1 \\ 1 & -1 \end{pmatrix}\begin{pmatrix} x \\ y \end{pmatrix} - \mu \begin{pmatrix} x-1 \\ y \end{pmatrix},$$

$$\begin{pmatrix} x(0) \\ y(0) \end{pmatrix} = \begin{pmatrix} 3 \\ 0 \end{pmatrix}, \qquad t \in [0, 1].$$

In Vasil'eva and Butuzov [149], along with the problem (2.2.1), (2.2.2), the initial value problem for the general nonlinear system

$$\mu \frac{dx}{dt} = F(x, t, \mu)$$

in the critical case is also considered. The critical case occurs when the reduced equation $F(x,t,0) = 0$ has a family of solutions $x = \varphi(\alpha(t),t)$ that depends on an arbitrary k-dimensional vector $\alpha(t)$. Some applied problems from chemical kinetics that lead to such singularly perturbed equations in critical cases are also discussed there.

Problems in the critical case also play an important role in regularizing initial and boundary value problems for differential-algebraic equations and constrained ordinary differential equations of the form

$$\frac{dx}{dt} = f(x,y,t), \qquad 0 = g(x,y,t),$$

where $f(x,y,t)$ and $g(x,y,t)$ are m- and n-dimensional vector functions, respectively, and the $n \times n$ matrix g_y is singular with constant positive rank $r < n$. For description of the regularization procedure, see O'Malley and Kalachev [116].

2.2.3 The equations of chemical kinetics

Let us illustrate the process of construction of the asymptotic approximation to the solution of a *nonlinear* singularly perturbed initial value problem in the *critical case* on the example of a system from chemical kinetics (such problems were mentioned in the previous subsection). First, let us discuss the statement of chemical kinetics problems in general. Consider n chemical reactions involving m substances, namely

$$a_{i1}X_1 + \cdots + a_{im}X_m \underset{k_i^-}{\overset{k_i^+}{\rightleftarrows}} b_{i1}X_1 + \cdots + b_{im}X_m, \qquad i = 1,\ldots,n.$$

Here X_i denotes the ith substance; a_{ik} and b_{ik} are integers denoting the number of molecules of the kth substance that participates in the ith forward and reverse reactions, respectively; k_i^+ and k_i^- are the rate constants of reactions.

If we denote the concentration of substance X_j by x_j, then the changes in x_j determined by ith reactions with rate constants k_i^+ and k_i^- (during the time dt) are given, respectively, by

$$dx_j = -k_i^+ x_1^{a_{i1}} \ldots x_m^{a_{im}} (a_{ij} - b_{ij}) dt$$

(for the forward reaction), and

$$dx_j = k_i^- x_1^{b_{i1}} \ldots x_m^{b_{im}} (a_{ij} - b_{ij}) dt$$

(for the reverse reaction). Consequently, the total change in x_j (as a result of all the reactions) is equal to

$$dx_j = \sum_{i=1}^{n} \gamma_{ij} w_i dt,$$

where

$$w_i = k_i^+ x_1^{a_{i1}} \ldots x_m^{a_{im}} - k_i^- x_1^{b_{i1}} \ldots x_m^{b_{im}},$$

and $\gamma_{ij} = b_{ij} - a_{ij}$. Thus, we obtain the system of differential equations

$$\text{(2.2.22)} \qquad \frac{dx_j}{dt} = \sum_{i=1}^{n} \gamma_{ij} w_i, \qquad j = 1,\ldots,m.$$

Under actual conditions, the rate constants of various reactions differ from each other. Large rate constants correspond to fast reactions. This property can be expressed by means of a small parameter μ. Suppose that first s reactions are fast, so

$$k_i^{\pm} = \frac{1}{\mu}\hat{k}_i^{\pm}, \qquad i = 1, \ldots, s,$$

with $\hat{k}_i^{\pm}$ moderate. Let us introduce the notation

$$\hat{w}_i = \mu w_i = \hat{k}_i^{+} x_1^{a_{i1}} \ldots x_m^{a_{im}} - \hat{k}_i^{-} x_1^{b_{i1}} \ldots x_m^{b_{im}}, \qquad i = 1, \ldots, s.$$

Now the system (2.2.22) can be rewritten as

$$\mu \frac{dx_j}{dt} = \sum_{i=1}^{s} \gamma_{ij}\hat{w}_i + \mu \sum_{i=s+1}^{n} \gamma_{ij} w_i, \qquad j = 1, \ldots, m. \tag{2.2.23}$$

Setting $\mu = 0$, we obtain the reduced system

$$0 = \sum_{i=1}^{s} \gamma_{ij}\hat{w}_i, \qquad j = 1, \ldots, m. \tag{2.2.24}$$

In practice, it often turns out that system (2.2.24) has a family of solutions that depends on one or more arbitrary parameters. Thus, the problem for singularly perturbed equation (2.2.23) is the *critical case*. One method for determining approximate solutions of the equations of chemical kinetics containing a small parameter is known in physical chemistry as the method of *quasi-stationary concentrations* of Semenov–Bodenstein. A number of works are devoted to its mathematical justification (i.e. justification of the passage to the limit as $\mu \to 0$); cf., e.g., Vasil'ev, Vol'pert, and Khudyaev [140], Vol'pert and Khudyaev [164], and Segel and Slemrod [125].

Let us now discuss an example of an actual chemical reaction. The system

$$\frac{dx_1}{dt} = -k_1^{+} x_1 + k_1^{-} x_2,$$

$$\frac{dx_2}{dt} = k_1^{+} x_1 - k_1^{-} x_2 - k_2^{+} x_2 x_3 - k_3^{+} x_2 x_4,$$

$$\frac{dx_3}{dt} = -k_2^{+} x_2 x_3,$$

$$\frac{dx_4}{dt} = -k_3^{+} x_2 x_4,$$

with conditions

$$x_i(0) = x_i^0,$$

where x_j are concentrations of reacting substancies, occurs in studying the reaction kinetics of organometallic compounds and was proposed by Kashin of Moscow State University. The rate constants have the orders of magnitude

$$k_1^{+} \sim 10, \quad k_1^{-} \sim 10^9, \quad k_2^{+} \sim 10^{10}, \quad k_3^{+} \sim 10^8, \quad k_2^{-} = k_3^{-} = 0.$$

Dividing each of the equations by k_3^+ and making substitutions

$$\mu = \frac{1}{k_3^+}, \qquad A = k_1^+, \qquad B = \frac{k_1^-}{k_3^+}, \qquad C = \frac{k_2^+}{k_3^+},$$

we obtain

$$\begin{aligned}
&\mu\frac{dx_1}{dt} = -\mu A x_1 + B x_2,\\
&\mu\frac{dx_2}{dt} = \mu A x_1 - B x_2 - C x_2 x_3 - x_2 x_4,\\
&\mu\frac{dx_3}{dt} = -C x_2 x_3,\\
&\mu\frac{dx_4}{dt} = -x_2 x_4.
\end{aligned} \tag{2.2.25}$$

Let us rewrite (2.2.25) with corresponding initial conditions as

$$\mu\frac{dx}{dt} = F(x, t, \mu), \qquad x(0) = x^0, \tag{2.2.26}$$

where $x = \mathrm{col}(x_1, x_2, x_3, x_4)$, and other notation is obvious from comparison with (2.2.25). In this example, $F(x, t, \mu) \equiv F(x)$ is an infinitely differentiable function of its arguments. In general, when an m-dimensional vector function F depends also on t and μ, we should demand that $F(x, t, \mu)$ be sufficienly smooth in some domain $D(x, t)$ of the (x, t)-space and for $\mu \in [0, \mu_0]$, where μ_0 is a constant. Let us seek the asymptotic solution of (2.2.26) in the form (2.2.3). The problems for asymptotic terms are obtained in the standard way by substituting (2.2.3) in (2.2.26) and rewriting nonlinear terms there in the form $F = \overline{F} + \Pi F$.

As we mentioned in § 2.2.2, the reduced equation in the critical case, in general, will have a family of solutions

$$x = \varphi(\alpha_1, \alpha_2, \ldots, \alpha_k, t) = \varphi(\alpha, t).$$

We assume (cf. Vasil'eva and Butuzov [149]) that $\varphi(\alpha, t)$ is sufficiently smooth and that the rank of the matrix $\varphi_\alpha(\alpha, t) = \partial\varphi/\partial\alpha(\alpha, t)$ is equal to k. Since we will discuss only the terms of the zeroth-order approximation, in what follows we will omit the subindex 0 to make notations simpler. In the case of (2.2.25), the reduced system

$$B\overline{x}_2 = 0, \qquad B\overline{x}_2 + C\overline{x}_2\overline{x}_3 + \overline{x}_2\overline{x}_4 = 0, \qquad C\overline{x}_2\overline{x}_3 = 0, \qquad \overline{x}_2\overline{x}_4 = 0$$

(where $\overline{x}_i$ are the regular terms of the zeroth-order approximation for x_i) has a family of solutions depending on three parameters (k=3), namely,

$$\overline{x}_1 = \alpha_1, \qquad \overline{x}_2 = 0, \qquad \overline{x}_3 = \alpha_2, \qquad \overline{x}_4 = \alpha_3.$$

Thus, $\varphi = \mathrm{col}(\alpha_1, 0, \alpha_2, \alpha_3)$ and, evidently, $\mathrm{rank}\,\partial\varphi/\partial\alpha = 3$.

Let us now check the conditions analogous to conditions 1^0 and 2^0 of § 2.2.1. Since, in general, $F(\varphi, t, 0) = 0$, differentiating this identity with respect to α, we obtain

$$F_x(\varphi(\alpha, t), t, 0)\varphi_\alpha = 0.$$

This implies that the matrix $F_x(\varphi(\alpha,t),t,0)$ has the eigenvalue $\lambda(\alpha,t)=0$ and that the columns of φ_α are eigenvectors corresponding to $\lambda=0$. Since rank$\varphi_\alpha=k$, the multiplicity of λ is not less than k. *We assume* (cf. Vasil'eva and Butuzov [149]) *that the multiplicity of* λ *is exactly* k *and that the remaining eigenvalues* $\lambda_i(\alpha,t)$ *of matrix* $F_x(\varphi(\alpha,t),t,0)$ *satisfy*

$$\mathrm{Re}\lambda_i(\alpha,t)<0.$$

In the case of (2.2.26),

$$F_x(\varphi(\alpha))=\begin{pmatrix} 0 & B & 0 & 0\\ 0 & -B-C\alpha_2-\alpha_3 & 0 & 0\\ 0 & -C\alpha_2 & 0 & 0\\ 0 & -\alpha_3 & 0 & 0\end{pmatrix},$$

and its eigenvalues are $\lambda_1=\lambda_2=\lambda_3=0$ and $\lambda_4=-B-C\alpha_2-\alpha_3$. Since rank$\varphi_\alpha=3$, it can be easily seen that a linearly independent eigenvector corresponds to each of $\lambda_i=0$, $i=1,2,3$. From $B>0$, $C>0$ and for α_2, α_3 nonnegative (recall that α_2, α_3 represent concentrations that are nonnegative), it follows that $\lambda_4<0$. So, corresponding conditions are satisfied.

For the Π-functions of the zeroth order in the general case, we obtain

$$\frac{d\Pi}{d\tau}=F(\varphi(\alpha(0),0)+\Pi,0,0)-F(\varphi(\alpha(0),0),0,0), \tag{2.2.27}$$

where $\tau=t/\mu$ and vector $\alpha(0)$ enters as a parameter. For any α this equation has a rest point $\Pi=0$. The characteristic equation corresponding to this rest point, $\det(F_x(\varphi(\alpha(0),0),0,0)-\lambda I_m)=0$, has the root $\lambda=0$ of multiplicity k and $(m-k)$ roots satisfying Re$\lambda_i<0$. Therefore the rest point $\Pi=0$ is not asymptotically stable in the sense of Lyapunov, that is, the solution with an initial value arbitrarily close to this rest point will not necessarily converge to it as $\tau\to\infty$. However, if we prescribe *special* initial conditions, then the solution will converge exponentially to the rest point $\Pi=0$ as $\tau\to\infty$. In other words, in a neighborhood of $\Pi=0$ there exists an $(m-k)$-dimensional *stable manifold* $\Omega(\alpha)$ such that if the initial values $\Pi(0)$ belong to $\Omega(\alpha)$, then $\Pi(\tau)$ will also belong to $\Omega(\alpha)$ for $\tau>0$ and (cf. Vasil'eva and Butuzov [149])

$$\|\Pi(\tau)\|\le c\exp(-\kappa\tau),\qquad \tau\ge 0,$$

for moderate constants c and $\kappa>0$.

For our problem the system (2.2.27) has the form (to simplify notation even more, we will write Π_i instead of Πx_i)

$$\frac{d\Pi_1}{d\tau}=B\Pi_2,$$

$$\frac{d\Pi_2}{d\tau}=-B\Pi_2-C\Pi_2(\alpha_2(0)+\Pi_3)-\Pi_2(\alpha_3(0)+\Pi_4),$$

$$\frac{d\Pi_3}{d\tau}=-C\Pi_2(\alpha_2(0)+\Pi_3),$$

$$\frac{d\Pi_4}{d\tau}=-\Pi_2(\alpha_3(0)+\Pi_4).$$

This system is sufficiently simple and can be integrated. Solving it with conditions $\Pi_i(\infty) = 0$, we obtain the exact representation

$$
\begin{aligned}
\Pi_2 = &\ -\Pi_1 + \left(\exp\left(-\frac{C}{B}\Pi_1\right) - 1\right)\alpha_2(0) \\
&+ \left(\exp\left(-\frac{\Pi_1}{B}\right) - 1\right)\alpha_3(0), \\
\Pi_3 = &\ \left(\exp\left(-\frac{C}{B}\Pi_1\right) - 1\right)\alpha_2(0), \\
\Pi_4 = &\ \left(\exp\left(-\frac{\Pi_1}{B}\right) - 1\right)\alpha_3(0).
\end{aligned}
\tag{2.2.28}
$$

for the *stable manifold* $\Omega(\alpha)$. The manifold $\Omega(\alpha)$ is one-dimensional since the three functions Π_i $(i = 2, 3, 4)$ are expressed through Π_1.

Substituting $\overline{x}_i + \Pi_i$ into the initial conditions in the zeroth order, we obtain the following system for $\alpha_1(0)$, $\Pi_1(0)$, $\alpha_2(0)$, and $\alpha_3(0)$:

$$
\begin{aligned}
x_1^0 = &\ \alpha_1(0) + \Pi_1(0), \\
x_2^0 = &\ -\Pi_1(0) + \left(\exp\left(-\frac{C}{B}\Pi_1(0)\right) - 1\right)\alpha_2(0) \\
&+ \left(\exp\left(-\frac{\Pi_1(0)}{B}\right) - 1\right)\alpha_3(0), \\
x_3^0 = &\ \alpha_2(0) + \left(\exp\left(-\frac{C}{B}\Pi_1(0)\right) - 1\right)\alpha_2(0), \\
x_4^0 = &\ \alpha_3(0) + \left(\exp\left(-\frac{\Pi_1(0)}{B}\right) - 1\right)\alpha_3(0).
\end{aligned}
\tag{2.2.29}
$$

By setting $X \equiv \Pi_1(0) = x_1^0 - \alpha_1(0)$, we can determine $\alpha_2(0)$ and $\alpha_3(0)$ in terms of X, namely,

$$\alpha_2(0) = x_3^0 \exp\left(\frac{C}{B}X\right),$$

$$\alpha_3(0) = x_4^0 \exp\left(\frac{1}{B}X\right).$$

Substituting into the first equation of (2.2.29), we obtain

$$X + x_3^0 \exp\left(\frac{C}{B}X\right) + x_4^0 \exp\left(\frac{1}{B}X\right) = x_3^0 + x_4^0 - x_2^0.$$

Elementary considerations show that this equation has a unique solution for $x_3^0 \geq 0$ and $x_4^0 \geq 0$. Thus, $\alpha_1(0)$, $\alpha_2(0)$, and $\alpha_3(0)$ are uniquely determined from (2.2.29).

One can also write equations for $\alpha_1(t)$, $\alpha_2(t)$, and $\alpha_3(t)$. As usual (for problems in the critical cases), these equations can be derived as solvability conditions for the regular terms of the first order. In the present case the solution of the equations for α_i is a matter of integrating by quadratures. Thus, one can determine $\overline{x}_1(t) = \alpha_1(t)$, $\overline{x}_2(t) = 0$, $\overline{x}_3(t) = \alpha_2(t)$ and $\overline{x}_4(t) = \alpha_3(t)$.

Exercise
Derive the equations for $\alpha_i(t)$ $(i = 1, 2, 3)$. Solve these equations with some nonzero initial conditions $\alpha_i(0)$ $(i = 1, 2, 3)$.

The determination of Π-functions in the zeroth-order approximation reduces to integration of the scalar equation

$$\frac{d\Pi_1}{d\tau} = B\Big\{-\Pi_1 + \Big(\exp\Big(-\frac{C}{B}\Pi_1\Big) - 1\Big)\alpha_2(0) + \Big(\exp\Big(-\frac{\Pi_1}{B}\Big) - 1\Big)\alpha_3(0)\Big\}$$

by quadratures subject to the initial condition $\Pi_1(0) = x_1^0 - \alpha_1(0)$. After determining $\Pi_1(\tau)$, the remaining $\Pi_i(\tau)$ $(i = 2, 3, 4)$ are found by means of the equations (2.2.28) for $\Omega(\alpha)$. Higher-order terms of the asymptotic expansion can be constructed as well (see Vasil'eva and Butuzov [149]).

Exercise
Consider three chemical reactions

$$A_1 \underset{k_1^-}{\overset{k_1^+}{\rightleftarrows}} 2A_2, \qquad A_2 \xrightarrow{k} A_3$$

involving three substances.

(a) Write the system of equations for concentrations x_1, x_2 and x_3 of the substances A_1, A_2, and A_3, respectively, describing these reactions. Note that differential equations for x_1 and x_2 do not contain x_3 and, therefore, could be considered independently of the equation for x_3.

(b) Assume that $k_1^+ = k_1^- = 1/\mu$, i.e., the forward and reverse reactions involving substance A_1 are fast, and $k << 1/\mu$, i.e. the last reaction is slow. Show that the system of equations for concentrations x_1 and x_2 has the form

$$\frac{dx_1}{dt} = -\frac{1}{\mu}x_1 + \frac{1}{\mu}x_2^2,$$

$$\frac{dx_2}{dt} = \frac{2}{\mu}x_1 - \frac{2}{\mu}x_2^2 - kx_2.$$

(c) Find the zeroth-order approximation of the solution subject to initial conditions $x_1(0) = a > 0$ and $x_2(0) = b > 0$.

d). Find the terms of the first-order approximation of the solution.

2.3 Boundary value problems

2.3.1 Conditionally stable case

Consider the system

(2.3.1) $$\mu\frac{dz}{dt} = F(z, y, t), \quad \frac{dy}{dt} = f(z, y, t), \qquad 0 \le t \le T,$$

in the case when y is an m-vector and z contains two scalar components z_1 and z_2. Let condition 1^0 of § 2.1.1 be satisfied, and let the eigenvalues $\lambda_{1,2}(t)$ of matrix $\overline{F}_z(t) = F_z(\overline{z}_0(t), \overline{y}_0(t), t)$ (here $\overline{z}_0(t)$, $\overline{y}_0(t)$ is the solution of the reduced problem (2.1.17) from § 2.1.2) satisfy instead of condition 2^0 (see § 2.1.2), condition

$2'$. $\lambda_1(t) < 0$, $\lambda_2(t) > 0$ *for* $0 \le t \le T$.

Simple examples show that the solution of such *initial value* problems are not bounded for $\mu \to 0$. Therefore the theorem on passage to the limit does not hold.

Example 1. Consider the problem

$$\mu \frac{dz_1}{dt} = z_2 - 1, \qquad \mu \frac{dz_2}{dt} = z_1 + 1,$$

$$z_1(0,\mu) = 0, \qquad z_2(0,\mu) = 1.$$

This system of two equations is independent of y. Matrix $\overline{F}_z$ has the eigenvalues $\lambda_1 = -1$ and $\lambda_2 = 1$. The exact solution of the problem is $z_1 = \cosh(t/\mu) - 1$ and $z_2 = \sinh(t/\mu) + 1$. Evidently, z_1 and z_2 are unbounded as $\mu \to 0$, and there is no passage to a limiting solution $\overline{z}_1 = -1$, $\overline{z}_2 = 1$ of the reduced system as $\mu \to 0$.

For the more general problem (2.3.1) in the case when condition $2'$ holds, we must replace initial conditions for z by boundary conditions (assume that $T = 1$):

$$z_1(0,\mu) = z_1^0, \qquad z_2(1,\mu) = z_2^0, \tag{2.3.2}$$

and retain an initial condition for y:

$$y(0,\mu) = y_0. \tag{2.3.3}$$

The study of this singularly perturbed problem shows that its solution tends to the solution of the reduced problem that retains only the initial condition for y. The asymptotic expansion of the solution now contains boundary functions that appear not only in the vicinity of the point $t = 0$, but also in the vicinity of $t = 1$.

Example 2. Consider the differential equations of example 1 with the boundary conditions $z_1(0,\mu) = 0$, $z_2(1,\mu) = 0$. The exact solution of this problem has the form

$$z_1 = -1 + \left(\frac{1+\exp(-1/\mu)}{1+\exp(-2/\mu)}\right) e^{-t/\mu} + \left(\frac{-1+\exp(-1/\mu)}{1+\exp(-2/\mu)}\right) e^{(t-1)/\mu},$$

$$z_2 = 1 - \left(\frac{1+\exp(-1/\mu)}{1+\exp(-2/\mu)}\right) e^{-t/\mu} + \left(\frac{-1+\exp(-1/\mu)}{1+\exp(-2/\mu)}\right) e^{(t-1)/\mu}.$$

These formulae clearly indicate the presence of the boundary layers in the vicinities of both ends of the interval $[0,1]$. Further note that z_1 and z_2 tend to the solution $\overline{z}_1 = -1$, $\overline{z}_2 = 1$ of the reduced system in the open interval $(0,1)$ as $\mu \to 0$.

Let us describe the algorithm to construct the asymptotic solution of the problem (2.3.1)– (2.3.3). We seek an asymptotic expansion of the solution in the form (where x stands for both variables z and y, i.e., $x = \{z, y\}$)

$$x(t,\mu) = \overline{x}(t,\mu) + \Pi x(\tau_0,\mu) + Qx(\tau_1,\mu). \tag{2.3.4}$$

Here $\overline{x}(t,\mu)$ is the regular power series

$$\overline{x}(t,\mu) = \overline{x}_0(t) + \mu\overline{x}_1(t) + \cdots + \mu^k\overline{x}_k(t) + \cdots;$$

$\Pi x(\tau_0,\mu)$ $(\tau_0 = t/\mu)$ is a boundary layer series in the vicinity of $t=0$:

$$\Pi x(\tau_0,\mu) = \Pi_0 x(\tau_0) + \mu\Pi_1 x(\tau_0) + \cdots + \mu^k\Pi_k x(\tau_0) + \cdots;$$

and $Qx(\tau_1,\mu)$ $(\tau_1 = (t-1)/\mu)$ is a boundary layer series in the vicinity of $t=1$:

$$Qx(\tau_1,\mu) = Q_0 x(\tau_1) + \mu Q_1 x(\tau_1) + \cdots + \mu^k Q_k x(\tau_1) + \cdots.$$

Substituting (2.3.4) into (2.3.1), we obtain

$$\begin{aligned} &\mu\frac{d\overline{z}}{dt} + \frac{d\Pi z}{d\tau_0} + \frac{dQz}{d\tau_1} = \overline{F} + \Pi F + QF, \\ &\frac{d\overline{y}}{dt} + \frac{1}{\mu}\frac{d\Pi y}{d\tau_0} + \frac{1}{\mu}\frac{dQy}{d\tau_1} = \overline{f} + \Pi f + Qf, \end{aligned} \tag{2.3.5}$$

where

$$\overline{F} = F(\overline{z}(t,\mu), \overline{y}(t,\mu), t),$$

$$\Pi F = F(\overline{z}(\tau_0\mu,\mu) + \Pi z(\tau_0,\mu), \overline{y}(\tau_0\mu,\mu) + \Pi y(\tau_0,\mu), \tau_0\mu)$$

$$-F(\overline{z}(\tau_0\mu,\mu), \overline{y}(\tau_0\mu,\mu), \tau_0\mu), \quad \text{and}$$

$$QF = F(\overline{z}(1+\tau_1\mu,\mu) + Qz(\tau_1,\mu), \overline{y}(1+\tau_1\mu,\mu) + Qy(\tau_1,\mu), 1+\tau_1\mu)$$

$$-F(\overline{z}(1+\tau_1\mu,\mu), \overline{y}(1+\tau_1\mu,\mu), 1+\tau_1\mu).$$

Similar expressions define $\overline{f}$, Πf, Qf.

We should note that there is some difference between (2.3.5) and (2.1.16) (see § 2.1.2). Earlier the expression $\overline{F} + \Pi F$ was obtained by the formal identical transformation

$$F(\overline{z} + \Pi z, \overline{y} + \Pi y, t) = \overline{F} + \Pi F, \tag{2.3.6}$$

but here the equality

$$F(\overline{z} + \Pi z + Qz, \overline{y} + \Pi y + Qy, t) = \overline{F} + \Pi F + QF \tag{2.3.7}$$

is formally incorrect, and the substitution of $\overline{F} + \Pi F + QF$ for F should be considered as one of the steps of the proposed algorithm to construct the series (2.3.4). This substitution can be justified as follows. Everywhere except in some neighborhood of the point $t=1$ the functions $Q_k x$ (and therefore also the function QF) will be smaller than any power of μ as $\mu \to 0$. Hence, the equality (2.3.7) will be satisfied with an accuracy of the order μ^N (here N is an arbitrary number), it will *almost* satisfy (2.3.6). On the other hand, everywhere except in some neighborhood of $t=0$ the Π-functions are less than any power of μ and equality (2.3.7) is also satisfied with an accuracy of order μ^N. Therefore, equality (2.3.7) holds with accuracy μ^N everywhere in $[0, 1]$.

For the coefficients of the expansion ΠF, a representation similar to that discussed on p. 21 (§ 2.1.2) holds (with τ replaced by τ_0); for QF, we have an analogous representation with $t = 0$ changed to $t = 1$ and τ changed to τ_1. Similar representations hold for Πf and Qf.

Substituting (2.3.4) into the boundary conditions (2.3.2), (2.3.3), and assuming a priori that $\Pi_k x(1/\mu)$ and $Q_k x(-1/\mu)$ are less than any power of μ, we obtain

$$\bar{z}_{10}(0) + \mu\bar{z}_{11}(0) + \cdots + \Pi_0 z_1(0) + \mu\Pi_1 z_1(0) + \cdots = z_1^0,$$

$$\bar{z}_{20}(1) + \mu\bar{z}_{21}(1) + \cdots + Q_0 z_2(0) + \mu Q_1 z_2(0) + \cdots = z_2^0, \tag{2.3.8}$$

$$\bar{y}_0(0) + \mu\bar{y}_1(0) + \cdots + \Pi_0 y(0) + \mu\Pi_1 y(0) + \cdots = y^0$$

(the first subindex of $\bar{z}_{1k}$ and $\bar{z}_{2k}$ denotes the component of vector $\bar{z}$).

In (2.3.5), let us equate separately the terms depending on t, τ_0, and τ_1, respectively. In the zeroth order, we have

$$0 = \bar{F}_0 = F(\bar{z}_0(t), \bar{y}_0(t), t), \qquad \frac{d\bar{y}_0}{dt} = \bar{f}_0 \equiv f(\bar{z}_0(t), \bar{y}_0(t), t) \tag{2.3.9}$$

(this is, evidently, the reduced system for the original system (2.3.1));

$$\frac{d\Pi_0 z}{d\tau_0} = \Pi_0 F \equiv F(\bar{z}_0(0) + \Pi_0 z, \bar{y}_0(0) + \Pi_0 y, 0),$$
$$\frac{d\Pi_0 y}{d\tau_0} = 0 \qquad (\tau_0 \geq 0); \tag{2.3.10}$$

$$\frac{dQ_0 z}{d\tau_1} = Q_0 F \equiv F(\bar{z}_0(1) + Q_0 z, \bar{y}_0(1) + Q_0 y, 1), \qquad \frac{dQ_0 y}{d\tau_1} = 0 \qquad (\tau_1 \leq 0).$$

Additional conditions are obtained from (2.3.8), equating the coefficients of μ^0:

$$\bar{z}_{10}(0) + \Pi_0 z_1(0) = z_1^0, \tag{2.3.11}$$

$$\bar{z}_{20}(1) + Q_0 z_1(0) = z_2^0,$$

$$\bar{y}_0(0) + \Pi_0 y(0) = y^0. \tag{2.3.12}$$

In addition, we impose on the boundary functions the decay conditions similar to those of § 2.1.2. For $\Pi_k y(\tau_0)$ and $Q_k y(\tau_1)$, e.g., we ask that

$$\Pi_k y(\infty) = 0, \qquad Q_k y(-\infty) = 0, \quad k = 0, 1, 2, \ldots.$$

Then $\Pi_0 y(\tau_0) \equiv 0$, $Q_0 y(\tau_1) \equiv 0$, and from (2.3.12) we obtain $\bar{y}_0(0) = y^0$. System (2.3.9) with this initial condition defines the reduced solution $\bar{z}_0(t)$, $\bar{y}_0(t)$. We assume that for this solution the condition $2'$ is satisfied.

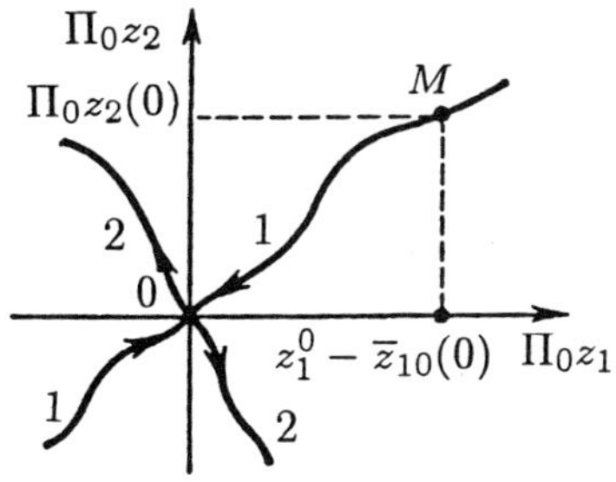

Figure 2.3: *Phase trajectory corresponding to the decaying to zero solution* $\Pi_0 z$ *of* (2.3.13), (2.3.14).

Let us now consider problem (2.3.10), (2.3.11). Taking into account the equality $\Pi_0 y(\tau_0) \equiv 0$, we have

$$\frac{d\Pi_0 z}{d\tau_0} = F(\bar{z}_0(0) + \Pi_0 z, y^0, 0), \tag{2.3.13}$$

$$\Pi_0 z_1(0) = z_1^0 - \bar{z}_{10}(0). \tag{2.3.14}$$

Thus, for the two-dimensional vector function $\Pi_0 z$ only one condition (2.3.14) is given. This condition cannot specify the solution uniquely. To find $\Pi_0 z$ and other Π-terms we demand that they decay to zero as $\tau_0 \to \infty$:

$$\Pi_k z(\infty) = 0, \qquad k = 0, 1, 2, \ldots.$$

By virtue of condition $2'$, the rest point $\Pi_0 z = 0$ of (2.3.13) is not asymptotically stable: it is a saddle point. This is illustrated in Fig. 2.3 (point O along with the two separatrices (1 and 2) passing through it). Arrows indicate the directions corresponding to increasing τ_0. For the solution $\Pi_0 z$ of (2.3.13), (2.3.14) to satisfy the condition $\Pi_0 z(\infty) = 0$, separatrix 1 should cross the straight line $\Pi_0 z_1 = z_1^0 - \bar{z}_{10}(0)$ (point M in Fig. 2.3). This will define the value $\Pi_0 z_2(0)$ (see Fig. 2.3) and thereby the entire solution $\Pi_0 z(\tau_0)$.

We impose the following condition analogous to condition 3^0 of § 2.1.1.

$3'$. *Let the straight line* $\Pi_0 z_1 = z_1^0 - \bar{z}_{10}(0)$ *intersect the separatrix which enters the rest point* (*saddle*) $\Pi_0 z = 0$ *of the system* (2.3.13)*as* $\tau_0 \to \infty$.

Remark 1. We call this case *conditionally stable* since the rest point $\Pi_0 z = 0$ is now not asymptotically stable, and "attracts" only those solutions whose initial conditions lie on the two branches of the separatrix 1.

Remark 2. It might happen that the separatrix 1 does not contain a point with abscissa $z_1^0 - \bar{z}_{10}(0)$ (see Fig. 2.4(a)) or, on the contrary, contains more than one such point (points M_1 and M_2 in Fig. 2.4(b)). In the first case, a solution with the assumed asymptotic behavior (2.3.4) does not exist. In the second case, two asymptotic solutions of type (2.3.4) exist. Depending on the choice of the point (M_1 or M_2) the asymptotic algorithm gives different results.

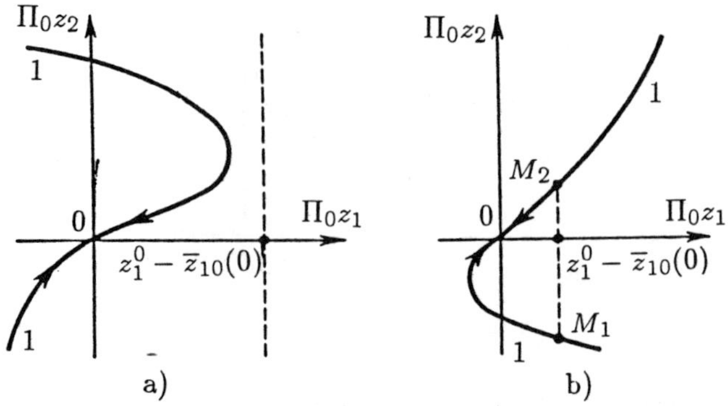

Figure 2.4: *The situation when* (2.3.13), (2.3.14) (a) *does not have a solution,* (b) *might have two solutions.*

It can be easily shown that $\Pi_0 z(\tau_0)$ satisfies an exponential estimate

(2.3.15) $$\| \Pi_0 z(\tau_0) \| \leq c \exp(-\kappa \tau_0), \qquad \tau_0 \geq 0.$$

Boundary function $Q_0 z(\tau_1)$ is defined analogously to $\Pi_0 z(\tau_0)$. In the case of $Q_0 z(\tau_1)$, the separatrix entering the saddle point as $\tau_1 \to -\infty$ is used. A condition similar to 3′ (we denote it 3″) should also be imposed. Function $Q_0 z(\tau_1)$ then has the exponential estimate of the type (2.3.15):

(2.3.16) $$\| Q_0 z(\tau_1) \| \leq c \exp(\kappa \tau_1), \qquad \tau_1 \leq 0.$$

This concludes the construction of the zeroth-order approximation.

Now let us consider the equations for $\overline{x}_1(t)$ and $\Pi_1 x(\tau_0)$:

(2.3.17) $$\frac{d\overline{z}_0}{dt} = \overline{F}_z(t)\overline{z}_1 + \overline{F}_y(t)\overline{y}_1, \qquad \frac{d\overline{y}_1}{dt} = \overline{f}_z(t)\overline{z}_1 + \overline{f}_y(t)\overline{y}_1,$$

where $\overline{F}_z(t) = F_z(\overline{z}_0(t), \overline{y}_0(t), t)$, and notation $\overline{F}_y(t)$, $\overline{f}_z(t)$, $\overline{f}_y(t)$ has analogous meaning,

(2.3.18) $$\frac{d\Pi_1 z}{d\tau_0} = \Pi_1 F \equiv F_z(\tau_0)\Pi_1 z + F_y(\tau_0)\Pi_1 y + G_1(\tau_0),$$
$$\frac{d\Pi_1 y}{d\tau_0} = \Pi_0 f,$$

where $F_z(\tau_0) = F_z(\overline{z}_0(0) + \Pi_0 z(\tau_0), y^0, 0)$, and notation $F_y(\tau_0)$ has similar meaning; $G_1(\tau_0)$ is given by formula (2.1.21) (see § 2.1.2). The additional conditions for (2.3.17), (2.3.18) are

(2.3.19) $$\overline{z}_{11}(0) + \Pi_1 z_1(0) = 0, \qquad \overline{y}_1(0) + \Pi_1 y(0) = 0,$$

(2.3.20) $$\Pi_1 z(\infty) = 0, \qquad \Pi_1 y(\infty) = 0.$$

The second equation of (2.3.18) with condition (2.3.20) yields

$$\Pi_1 y(\tau_0) = -\int_{\tau_0}^{\infty} \Pi_0 f(s) ds.$$

Hence,

$$\Pi_1 y(0) = -\int_{0}^{\infty} \Pi_0 f(s) ds,$$

and from the second equation of (2.3.19), we obtain

$$\overline{y}_1(0) = \int_{0}^{\infty} \Pi_0 f(s) ds.$$

This initial condition specifies the unique solution $\overline{z}_1(t)$, $\overline{y}_1(t)$ of the linear system (2.3.17). There now remains the first equation of (2.3.18) with boundary conditions $\Pi_1 z_1(0) = -\overline{z}_{11}(0)$, $\Pi_1 z(\infty) = 0$ to define $\Pi_1 z(\tau_0)$. It can be shown that these conditions specify $\Pi_1 z(\tau_0)$ uniquely (cf. Vasil'eva and Butuzov [148]).

The problem for $Q_1 x(\tau_1)$ has the form

$$\frac{dQ_1 z}{d\tau_1} = F_z(\tau_1) Q_1 z + F_y(\tau_1) Q_1 y + H_1(\tau_1), \qquad \frac{dQ_1 y}{d\tau_1} = Q_0 f, \tag{2.3.21}$$

$$\overline{z}_{21}(1) + Q_1 z_2(0) = 0, \tag{2.3.22}$$

$$Q_1 z(-\infty) = 0, \qquad Q_1 y(-\infty) = 0. \tag{2.3.23}$$

Here $F_z(\tau_1) = F_z(\overline{z}_0(1) + Q_0 z(\tau_1), \overline{y}_0(1), 1)$, and notation $F_y(\tau_1)$ has similar meaning; $H_1(\tau_1)$ has the same structure as $G_1(\tau_0)$ in (2.3.18).

The second equation of (2.3.21) with condition (2.3.23) implies

$$Q_1 y(\tau_1) = -\int_{\tau_1}^{-\infty} Q_0 f(s) ds.$$

Function $\overline{z}_{21}(t)$ is already determined, therefore $\overline{z}_{21}(1)$ is known, and for $Q_1 z(\tau_1)$ we have the first equation of (2.3.21) with the additional conditions $Q_1 z_2(0) = -\overline{z}_{21}(1)$, and $Q_1 z(-\infty) = 0$. This allows us to determine $Q_1 z(\tau_1)$ uniquely.

Analogously, $\overline{x}_k(t)$, $\Pi_k x(\tau_0)$, $Q_k x(\tau_1)$ can be obtained termwise for any arbitrary k. For Π- and Q-functions the exponential estimates of the types (2.3.15), (2.3.16) hold.

Example. Let us construct the zeroth-order terms of the asymptotic solution for

$$\mu \frac{dz_1}{dt} = y z_2 - 3, \qquad \mu \frac{dz_2}{dt} = y z_1, \qquad \frac{dy}{dt} = z_1 + 1, \qquad t \in (0,1),$$

$$z_1(0,\mu) = -1, \qquad z_2(1,\mu) = 1, \qquad y(0,\mu) = 1.$$

It is easily verified that the regular terms of the zeroth order in the expansion (2.3.4) are

$$\bar{y}_0(t) = 1 + t, \qquad \bar{z}_{10}(t) = 0, \qquad \bar{z}_{20}(t) = \frac{3}{1+t}.$$

The eigenvalues of matrix F_z are $\lambda_{1,2} = \pm\bar{y}_0(t)$ and, thus, condition 2′ is satisfied. Functions $\Pi_0 y$ and $Q_0 y$ are identically zero. The system (2.3.13) with condition (2.3.14) and decay conditions at infinity, has the form

$$\frac{d\Pi_0 z_1}{d\tau_0} = \Pi_0 z_2, \qquad \frac{d\Pi_0 z_2}{d\tau_0} = \Pi_0 z_1,$$

$$\Pi_0 z_1(0) = -1 - \bar{z}_{10}(0) = -1, \qquad \Pi_0 z_i(\infty) = 0,$$

where $\tau_0 = t/\mu$. From this system with conditions at $\tau_0 = \infty$, we obtain that $\Pi_0 z_1(\tau_0) = -\Pi_0 z_2(\tau_0)$. Solving now

$$\frac{d\Pi_0 z_1}{d\tau_0} = -\Pi_0 z_1, \qquad \Pi_0 z_1(0) = -1,$$

we obtain $\Pi_0 z_1(\tau_0) = -e^{-\tau_0}$, and $\Pi_0 z_2(\tau_0) = e^{-\tau_0}$.

Similarly, for $Q_0 z_1$ and $Q_0 z_2$, we have

$$\frac{dQ_0 z_1}{d\tau_1} = 2Q_0 z_2, \qquad \frac{dQ_0 z_2}{d\tau_1} = 2Q_0 z_1,$$

$$Q_0 z_1(0) = 1 - \bar{z}_{20}(1) = -\frac{1}{2}, \qquad Q_0 z_i(-\infty) = 0,$$

where $\tau_1 = (t-1)/\mu$. Analogously to the case of Π-functions, we obtain

$$Q_0 z_2 = Q_0 z_1 = -\frac{1}{2} e^{2\tau_1}.$$

All the zeroth-order terms have now been determined. From Theorem 2.4 below it will follow that

$$z_1(t,\mu) = -e^{-\frac{t}{\mu}} - \frac{1}{2} e^{2(t-1)/\mu} + O(\mu),$$

$$z_2(t,\mu) = \frac{3}{1+t} + e^{-\frac{t}{\mu}} - \frac{1}{2} e^{2(t-1)/\mu} + O(\mu),$$

$$y(t,\mu) = 1 + t + O(\mu).$$

Exercise

Find the zeroth-order terms of the asymptotic solution for

$$\mu\frac{dz_1}{dt} = yz_2 - 3, \qquad \mu\frac{dz_2}{dt} = yz_1, \qquad \frac{dy}{dt} = z_1 + 1, \qquad t \in (0,1),$$

$$z_1(1) = 1, \qquad z_2(0) = 1, \qquad y(0) = 1.$$

Remark. The algorithm described undergoes minor changes when (instead of (2.3.2)) boundary conditions $z_1(0) = z_1^0$, $z_1(1) = z_1^1$, or $z_2(0) = z_2^0$, $z_2(1) = z_2^1$ are given (the values of only one of the components of the vector z are prescribed on each endpoint of the interval $[0, 1]$).

Exercise

Find the zeroth- and the first-order approximations to the solution of

$$\mu\frac{dz_1}{dt} = yz_2 - 3, \qquad \mu\frac{dz_2}{dt} = 4yz_1, \qquad \frac{dy}{dt} = z_1 - 1, \qquad t \in (0, 1),$$

$$z_1(0) = 1, \qquad z_1(1) = 1, \qquad y(1) = 1.$$

Let $X_n(t, \mu)$ denote the nth partial sum of the expansion (2.3.4):

$$X_n(t, \mu) = \sum_{k=0}^{n} \mu^k[\overline{x}_k(t) + \Pi_k x(\tau_0) + Q_k x(\tau_1)].$$

Theorem 2.4 *Under conditions* 1^0 *and* 4^0 *of* § 2.1.2*and conditions* $2'$, $3'$ *and* $3''$, *for sufficiently small* μ *there exists a unique solution* $x(t, \mu)$ *of the problem* (2.3.1)–(2.3.3)*in a vicinity of the leading term of the asymptotic expansion*

$$\overline{x}_0(t) + \Pi_0 x\left(\frac{t}{\mu}\right) + Q_0 x\left(\frac{t-1}{\mu}\right).$$

The series (2.3.4)*is the asymptotic expansion of this solution in the interval* $0 \leq t \leq 1$ *for* $\mu \to 0$, *i.e.,*

$$\max_{0 \leq t \leq 1} \| x(t, \mu) - X_n(t, \mu) \| = O(\mu^{n+1}).$$

The proof of this theorem in the more general case, when z is a l-dimensional vector function, is presented in Vasil'eva and Butuzov [148].

Let us mention some generalizations of the problem discussed above.

1. The boundary conditions might be of the more complicated form:

$$R(z_1(0), z_2(0), y(0); z_1(1), z_2(1), y(1)) = 0, \tag{2.3.24}$$

where R is an $(m + 2)$-dimensional vector. In this case, a theorem on existence of a solution can also be proved, and an asymptotic expansion of the type (2.3.4) can be constructed using the results previously obtained in this subsection. For that purpose we must consider the problem (2.3.1)–(2.3.3) with z_1^0, z_2^0, y^0 as functions of μ, for which the following asymptotic representations hold:

$$z_i^0 = z_{i0}^0 + \mu z_{i1}^0 + \cdots, \qquad i = 1, 2; \quad y^0 = y_0^0 + \mu y_1^0 + \cdots. \tag{2.3.25}$$

The algorithm described earlier can be easily extended to this case. Let us now try to choose the z_{ik}^0 and y_k^0 such that the solution of the problem (2.3.1)–(2.3.3) with the boundary values (2.3.25) satisfies conditions (2.3.24). Substituting the asymptotic expansion of this solution into (2.3.24), we obtain

$$R(z_{10}^0 + \mu z_{11}^0 + \cdots, \bar{z}_{20}(0) + \Pi_0 z_2(0) + \cdots, y_0^0 + \mu y_1^0 + \cdots;$$

$$\bar{z}_{10}(1) + Q_0 z_1(0) + \cdots, z_{20}^0 + \mu z_{21}^0 + \cdots, \bar{y}_0(1) + \mu \bar{y}_1(1) + \cdots) = 0.$$

Let us expand the left-hand side in powers of μ:

$$R = R_0 + \mu R_1 + \cdots + \mu^k R_k + \cdots.$$

Equating coefficients of different powers of μ to zero, we obtain a sequence of equations $R_k = 0$ $(k = 0, 1, \ldots)$ for the unknowns z_{ik}^0 and y_k^0. For example, in the zeroth-order approximation we have

$$R_0 \equiv R(z_{10}^0, \bar{z}_{20}(0) + \Pi_0 z_2(0), y_0^0; \bar{z}_{10}(1) + Q_0 z_1(0), z_{20}^0, \bar{y}_0(1)) = 0.$$

This is the system of $m+2$ equations with $m+2$ unknowns z_{10}^0, z_{20}^0 and y_0^0 since according to the algorithm discussed previously in this subsection, $\Pi_0 z_2(0)$ depends on z_{10}^0 and y_0^0, $Q_0 z_1(0)$ depends on z_{20}^0 and y_0^0, and $\bar{y}_0(1)$ depends on y_0^0.

We assume that this system of equations is solvable and the corresponding Jacobian determinant is not equal to zero. It turns out that under this condition all the subsequent equations $R_k = 0$ will also be solvable, and therefore all coefficients in (2.3.25) will be determined. It can be shown that in the vicinity of $(z_{10}^0, z_{20}^0, y_0^0)$ there exists a unique $(z_1^0(\mu), z_2^0(\mu), y^0(\mu))$ such that the solution of the system (2.3.1) with boundary conditions $z_1(0,\mu) = z_1^0(\mu)$, $z_2(1,\mu) = z_2^0(\mu)$, $y(0,\mu) = y^0(\mu)$ will be the solution of (2.3.1), (2.3.24) as well. The asymptotic expansion of the solution of (2.3.1) – (2.3.3) with z_i^0 and y^0 given by (2.3.25) will be the *asymptotic expansion* as $\mu \to 0$ of the solution of the boundary value problem (2.3.1), (2.3.24) in the interval $0 \le t \le 1$.

Exercise

Find the zeroth-order approximation to the solution of

$$\mu\frac{dz_1}{dt} = \lambda_1\left(z_1 - \frac{1}{3}\right), \quad \mu\frac{dz_2}{dt} = \lambda_2\left(z_2 + \frac{2}{3}\right), \quad \frac{dy}{dt} = -\frac{3}{2}y - \frac{1}{2}, \quad t \in (0,1),$$

$$z_1(0) + z_2(0) - y(0) = s_1, \quad z_2(1) + y(1) = s_2, \quad z_1(1) + y(1) = s_3,$$

where s_i $(i = 1, 2, 3)$ are some constants, and

(a) $\lambda_1 = -1$, $\lambda_2 = 1$; (b) $\lambda_1 = 1$, $\lambda_2 = -1$.

Verify that in the case

(c) $\lambda_1 = -1$, $\lambda_2 = -1$,

the boundary function method cannot be applied.

Find the exact solutions in all three cases and verify that for (a) and (b) the constructed zeroth-order approximations are, in fact, the leading parts of the exact solutions. What is the behavior of the exact solution in case (c) as $\mu \to 0$?

2. The results of this subsection can be generalized to the case when the vector function z has an arbitrary dimension l. In this case instead of condition $2'$, the following condition must hold:

$$\text{Re}\lambda_i(t) < 0 \quad (i = 1, 2, \ldots, r),$$

(2.3.26)

$$\text{Re}\lambda_i(t) > 0 \quad (i = r+1, \ldots, l),$$

where the $\lambda_i(t)$ are simple eigenvalues of matrix $F_z(\overline{z}_0(t), \overline{y}_0(t), t)$; $\overline{z}_0(t)$ and $\overline{y}_0(t)$ is the solution of the corresponding reduced problem, and $t \in [0, 1]$.

The stationary solution of the equation, analogous to (2.3.13), will not be asymptotically stable. However, it can be shown (see Vasil'eva and Butuzov [148] and Coddington and Levinson [42]) that in the vicinity of $\Pi_0 z = 0$ there exists an r-dimensional *stable manifold* S^+ such that if the initial condition $\Pi_0 z(0)$ belongs to S^+, then $\Pi_0 z(\tau_0)$ will also belong to S^+ for $\tau_0 > 0$, and $\Pi_0 z(\tau_0) \to 0$ as $\tau_0 \to \infty$. (In the case of the two-dimensional system (2.3.13) such a condition corresponds to a point in the $(\Pi_0 z_1, \Pi_0 z_2)$-phase plane where the straight line given by (2.3.14) intersects the separatrix entering the rest point $\Pi_0 z = 0$ as $\tau_0 \to \infty$; it can be easily seen that the stable manifold S^+ in this case is the separatrix mentioned above.) Similarly, it can be shown that in the vicinity of the rest point $Q_0 z = 0$ of the equation for $Q_0 z(\tau_1)$ there exists an $(l - r)$-dimensional *unstable manifold* S^- such that if $Q_0 z(0)$ belongs to S^-, then $Q_0 z(\tau_1)$ also belongs to S^- for $\tau_1 < 0$, and $Q_0 z(\tau_1) \to 0$ as $\tau_1 \to -\infty$. Local structures of stable and unstable manifolds can be defined by linearization near the steady states $\Pi_0 z = 0$ and $Q_0 z = 0$, respectively. Linearized systems describe the behavior of the solutions belonging to S^+ and S^- in a sufficiently small neighborhood of corresponding stationary points. The problem of analytically describing the stable and unstable manifolds for general nonlinear systems is not solved. The explicit representations for S^+ and S^- can be obtained only in particular cases. Naturally, for the asymptotic algorithm to work, the number of boundary conditions prescribed at $t = 0$ and at $t = 1$ must correspond to the dimensions of the stable and unstable manifolds, respectively (i.e., the boundary conditions for z must be given as r components of vector function z prescribed at $t = 0$ and $l - r$ components of z prescribed at $t = 1$). This general case is discussed in detail in Vasil'eva and Butuzov [148].

Boundary conditions of type (2.3.24) can be treated by the method described in Esipova [49].

Example. Consider a system, analogous to (2.3.13), in the case when F is a vector function of dimension n depending on $\Pi_0 z$ linearly, i.e. $F(\overline{z}_0(0 + \Pi_0 z, y^0, 0) \equiv A\Pi_0 z$, where A is a constant $(l \times l)$-matrix:

$$\frac{d\Pi_0 z}{d\tau_0} = A\Pi_0 z.$$

If the eigenvalues of A satisfy the conditions

$$\mathrm{Re}\lambda_i < 0 \qquad (i = 1, 2, \ldots, r),$$

$$\mathrm{Re}\lambda_i > 0 \qquad (i = r + 1, \ldots, l),$$

the r-dimensional stable manifold can be constructed explicitly (in this case it will be an r- dimensional *stable subspace*).

Let us make the change of variables

$$\Pi_0 z = T \begin{pmatrix} \Pi^+ \\ \Pi^- \end{pmatrix},$$

where T is a matrix that transforms A to a Jordan canonical form:

$$T^{-1}AT = \begin{pmatrix} C^+ & 0 \\ 0 & C^- \end{pmatrix}.$$

Here the $(r \times r)$-matrix C^+ has eigenvalues with negative real parts and the $(n-r) \times (n-r)$-matrix C^- has eigenvalues with positive real parts. We then arrive at the system

$$\frac{d\Pi^+}{d\tau_0} = C^+\Pi^+, \tag{2.3.27}$$

$$\frac{d\Pi^-}{d\tau_0} = C^-\Pi^-, \tag{2.3.28}$$

By virtue of conditions on the eigenvalues of C^+, the rest point $\Pi^+ \equiv 0$ of (2.3.27) will be asymptotically stable as $\tau_0 \to \infty$ (the rest point $\Pi^- \equiv 0$ of (2.3.28) will be asymptotically stable as $\tau_0 \to -\infty$). Consider a set of trajectories of the system (2.3.27), (2.3.28) satisfying the initial condition $\Pi^-(0) = 0$ with arbitrary $\Pi^+(0)$. It can be easily seen that, since $\Pi^-(\tau_0) \equiv 0$, this set of trajectories will correspond to the stable manifold S^+. The explicit representation for it can be obtained if we return to the original variables

$$\Pi_0 z = \begin{pmatrix} \Pi_0^1 z \\ \Pi_0^2 z \end{pmatrix} = \begin{pmatrix} T_{11} & T_{21} \\ T_{12} & T_{22} \end{pmatrix} \begin{pmatrix} \Pi^+ \\ 0 \end{pmatrix}.$$

Hence,

$$\Pi_0^1 z = T_{11}\Pi^+, \qquad \Pi_0^2 z = T_{12}\Pi^+.$$

Assume that $\det T_{11} \neq 0$. Then, excluding Π^+ from these expressions, we obtain the equation describing S^+ in variables that are the components of $\Pi_0 z$:

$$\Pi_0^2 z = T_{12}T_{11}^{-1}\Pi_0^1 z.$$

It can be observed that on the manifold S^+ the original system for $\Pi_0 z$ reduces to an r- dimensional system

$$\frac{d\Pi_0^1 z}{d\tau_0} = T_{11}C^+T_{11}^{-1}\Pi_0^1 z.$$

Similarly, the explicit representation for the unstable manifold S^- can be obtained.

Remark 1. There are many boundary value problems where solutions with boundary layer asymptotics exist when $\lambda_i(t)$ satisfy condition 2^0 of § 2.1.2, but not conditions $2'$ or (2.3.26). In that case the boundary layer appears only in the vicinity of the point $t = 0$ (see Vasil'eva and Butuzov [148, §13] for details).

Exercise

Solve the problem from the exercise on page 50 with $\lambda_1 = 1$ and $\lambda_2 = 1$.

Remark 2. The boundary value problems in which some of $\lambda_i(t)$ are equal to zero in the entire interval $0 \le t \le T$, and therefore the reduced equation has a family of solutions (the critical case), are discussed in Vasil'eva and Butuzov [149]. An example of such problem from the theory of semiconductor devices is considered in § 4.3.

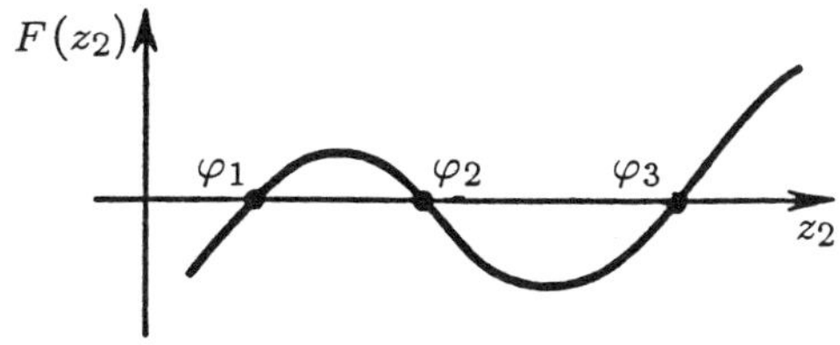

Figure 2.5: *Three roots of* $F(z_2) = 0$.

2.3.2 Internal transition layer

Let us now return to system (2.3.1). Suppose that equation $F(\overline{z}, \overline{y}, t) = 0$ has several solutions (roots) $\overline{z} = \varphi_i(\overline{y}, t)$. Before, we considered the case when only one of such solutions was used in the construction of the asymptotic expansion. Some other types of asymptotic behavior, when several such solutions are used in the asymptotic algorithm, are also possible. In these cases we obtain solutions with transitions from one root to another or solutions with so-called *internal transition layers.*

Let us study this phenomenon on an example of the system equivalent to the scalar second-order equation $\mu^2\, d^2z/dt^2 = F(z, t)$:

$$\mu\frac{dz_1}{dt} = F(z_2, t), \quad \mu\frac{dz_2}{dt} = z_1, \qquad 0 \le t \le 1, \tag{2.3.29}$$

with the boundary conditions

$$z_2(0, \mu) = 0, \qquad z_2(1, \mu) = 0. \tag{2.3.30}$$

First, we consider the autonomous system that can be obtained from (2.3.29) by setting $t = T = \text{const}$ in the right-hand side of the first equation, and then making the change of variable $t = \tau\mu$:

$$\frac{dz_1}{d\tau} = F(z_2, T) \equiv F(z_2), \qquad \frac{dz_2}{d\tau} = z_1. \tag{2.3.31}$$

(This is the *associated system* (see § 2.1.1) corresponding to (2.3.29) for $t = T$.) Assume that the function $F(z_2)$ has three simple roots $z_2 = \varphi_i$, $i = 1, 2, 3$, and moreover, that $F_{z_2}(\varphi_i) > 0$, $i = 1, 3$; $F_{z_2}(\varphi_2) < 0$ (see Fig. 2.5). Let us introduce the (z_1, z_2) phase plane. Since $z_1 dz_1 = F(z_2)dz_2$ (by virtue of (2.3.31)),

$$\frac{z_1^2}{2} = \int_{\varphi_2}^{z_2} F(z)dz + c \equiv \Phi(z_2) + c,$$

and hence,

$$z_1 = \pm\sqrt{2}\sqrt{\Phi(z_2) + c}, \tag{2.3.32}$$

where c is a constant. Depending on the function $\Phi(z_2)$, the family of the phase curves (2.3.32) looks differently. One can distinguish three cases. The graphs of $\Phi(z_2)$ and

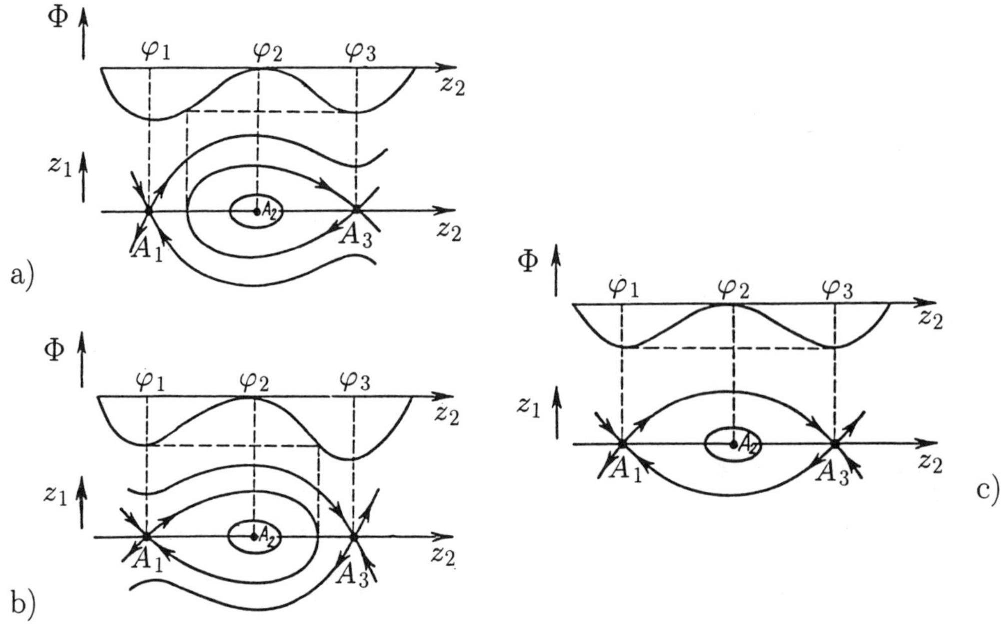

Figure 2.6: *Different types of* $\Phi(z_2)$ *and corresponding phase trajectories.*

the corresponding families of phase trajectories for each of these cases are presented in Fig. 2.6. These cases are as follows:

(a)

$$\int_{\varphi_1}^{\varphi_2} F(z)dz > \int_{\varphi_3}^{\varphi_2} F(z)dz, \quad \textit{i.e.,} \quad \Phi(\varphi_1) < \Phi(\varphi_3);$$

(b)

$$\int_{\varphi_1}^{\varphi_2} F(z)dz < \int_{\varphi_3}^{\varphi_2} F(z)dz, \quad \textit{i.e.,} \quad \Phi(\varphi_1) > \Phi(\varphi_3);$$

(c)

$$\int_{\varphi_1}^{\varphi_2} F(z)dz = \int_{\varphi_3}^{\varphi_2} F(z)dz, \tag{2.3.33}$$

i.e., $\Phi(\varphi_1) = \Phi(\varphi_3)$. In this last case the saddle points A_1 and A_2 (see Fig. 2.6(c)) are connected by separatrices, i.e., a homoclinic orbit consisting of two heteroclinic orbits (a so-called *cell*) occurs.

Now consider the system (2.3.29) (with variable t).

1^0. *Let the function* $F(z_2, t)$ *have continuous partial derivatives up to the second order inclusively in some domain* D *of the* (z_2, t) *plane.*

2^0. *Let the equation* $F(z_2, t) = 0$ *have three simple roots* $z_2 = \varphi_i$, $i = 1, 2, 3$, *such that*

(a) $\varphi_1(t) < \varphi_2(t) < \varphi_3(t)$ for $0 \le t \le 1$;

(b)*the region* $\{(z_2, t) : \varphi_1(t) \le z_2 \le \varphi_3(t),\ 0 \le t \le 1\}$ *entirely belongs to the domain* D;

(c)$F_{z_2}(\varphi_i(t), t) > 0$ *for* $i = 1, 3,\ 0 \le t \le 1$; *and* $F_{z_2}(\varphi_2(t), t) < 0$ *for* $0 \le t \le 1$.

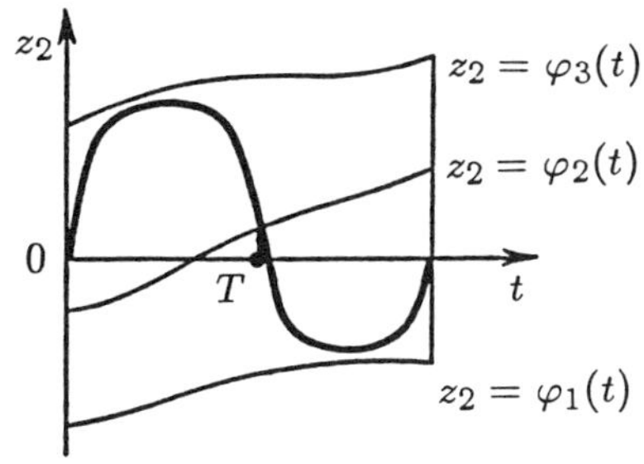

Figure 2.7: *The solution of* (2.3.29), (2.3.30)*with internal transition layer.*

It follows from condition 2^0 that for each fixed $t \in [0,1]$ the graph of the function $F(z_2, t)$ has the form shown in Fig. 2.5. As t changes, this graph may deform. It turns out that the phenomenon of a transition from root $\varphi_1(t)$ to root $\varphi_3(t)$ (or from $\varphi_3(t)$ to $\varphi_1(t)$; both of these two roots are conditionally stable since the corresponding matrix has eigenvalues $\lambda_{1,2}^{(i)} = \pm\sqrt{F_{z_2}(\varphi_i(t), t)}, \quad i = 1, 3$) for the solution of (2.3.29), (2.3.30) can take place if for some value of $t = T$ a homoclinic orbit consisting of two heteroclinic orbits exists in the corresponding phase plane for the associated system.

One can come to this conclusion using the asymptotic formulae of § 2.3.1. Let us show that there exists a solution of the type scetched in Fig. 2.7. We will call the value of $t = T$, for which $z_2 = \varphi_2$, the *point of transition.*

We approach this problem as follows. Assuming that T is some yet unknown value between $t = 0$ and $t = 1$, we will construct the asymptotic expansion of the solution of (2.3.29) in the interval $[0, T]$ satisfying the conditions $z_2(0, \mu) = 0$, $z_2(T, \mu) = \varphi_2(T)$ and approaching $\{0, \varphi_3(t)\}$ as $\mu \to 0$. In other words, we will construct the asymptotic solution defined in subinterval $[0, T]$, or the asymptotics of the left part of the solution (see Fig. 2.7). The algorithm of the § 2.3.1 will be used for that purpose. We note that at the end points $t = 0$ and $t = T$ the same component z_2 is prescribed (see the remark before Theorem 2.4).

Similarly, we construct the asymptotic solution of (2.3.29) in the interval $[T, 1]$, satisfying the conditions $z_2(T, \mu) = \varphi_2(T)$, $z_2(1, \mu) = 0$ and approaching $\{0, \varphi_1(t)\}$ as $\mu \to 0$, i.e., we construct the asymptotics of the right part of the solution (see Fig. 2.7).

Equating the values of expressions for z_1 corresponding to the left and right parts of the asymptotic solution at the point T determines T. Note that T is, generally speaking, a function of μ, and it should be sought in the form $T = T_0 + \mu T_1 + \mu^2 T_2 + \cdots$. Thus, the equality for the zeroth-order approximation of z_1 provides us the value of T_0. In what follows we will restrict ourselves to the construction of only the zeroth-order approximation. Therefore we may omit the subindex 0 in T_0 (below $T_0 = T$).

The asymptotic solution in the interval $[0, T]$ is expressed through the following formulae with accuracy of the order μ (the index 1 above the boundary functions shows that they are defined on the left subinterval; the index 2 will be used for the boundary functions defined on the right subinterval):

$$
\text{(2.3.34)}\qquad
\begin{aligned}
z_1(t,\mu) &= \overset{(1)}{\Pi}_0 z_1(\tau_0) + \overset{(1)}{Q}_0 z_1(\tau_T) + O(\mu),\\
z_2(t,\mu) &= \varphi_3(t) + \overset{(1)}{\Pi}_0 z_2(\tau_0) + \overset{(1)}{Q}_0 z_2(\tau_T) + O(\mu),
\end{aligned}
$$

where $\tau_0 = t/\mu \geq 0$, $\tau_T = (t-T)/\mu \leq 0$, and $\overset{(1)}{\Pi}_0\, z_i$ and $\overset{(1)}{Q}_0\, z_i$ are defined as solutions of the problems

$$
\frac{d\overset{(1)}{\Pi}_0 z_1}{d\tau_0} = F\left(\varphi_3(0) + \overset{(1)}{\Pi}_0 z_2, 0\right), \qquad \frac{d\overset{(1)}{\Pi}_0 z_2}{d\tau_0} = \overset{(1)}{\Pi}_0 z_1,
$$

$$
\overset{(1)}{\Pi}_0 z_2(0) = -\varphi_3(0), \quad \overset{(1)}{\Pi}_0 z_i(\infty) = 0, \qquad i = 1, 2;
$$

$$
\text{(2.3.35)}\qquad \frac{d\overset{(1)}{Q}_0 z_1}{d\tau_T} = F\left(\varphi_3(T) + \overset{(1)}{Q}_0 z_2, T\right), \qquad \frac{d\overset{(1)}{Q}_0 z_2}{d\tau_T} = \overset{(1)}{Q}_0 z_1,
$$

$$
\text{(2.3.36)}\qquad \overset{(1)}{Q}_0 z_2(0) = \varphi_2(T) - \varphi_3(T), \quad \overset{(1)}{Q}_0 z_i(-\infty) = 0, \qquad i = 1, 2.
$$

Analogously, the asymptotic solution of (2.3.29) in the subinterval $[T, 1]$ has the form

$$
\text{(2.3.37)}\qquad
\begin{aligned}
z_1(t,\mu) &= \overset{(2)}{\Pi}_0 z_1(\tau_T) + \overset{(2)}{Q}_0 z_1(\tau_1) + O(\mu),\\
z_2(t,\mu) &= \varphi_1(t) + \overset{(2)}{\Pi}_0 z_2(\tau_T) + \overset{(2)}{Q}_0 z_2(\tau_1) + O(\mu),
\end{aligned}
$$

where $\tau_T = (t-T)/\mu \geq 0$, $\tau_1 = (t-1)/\mu \leq 0$, and $\overset{(2)}{\Pi}_0\, z_i$ and $\overset{(2)}{Q}_0\, z_1$ are defined as the solutions of the problems

$$
\text{(2.3.38)}\qquad \frac{d\overset{(2)}{\Pi}_0 z_1}{d\tau_T} = F\left(\varphi_1(T) + \overset{(2)}{\Pi}_0 z_2, T\right), \qquad \frac{d\overset{(2)}{\Pi}_0 z_2}{d\tau_T} = \overset{(2)}{\Pi}_0 z_1,
$$

$$
\text{(2.3.39)}\qquad \overset{(2)}{\Pi}_0 z_2(0) = \varphi_2(T) - \varphi_1(T), \quad \overset{(2)}{\Pi}_0 z_i(\infty) = 0, \qquad i = 1, 2;
$$

$$
\frac{d\overset{(2)}{Q}_0 z_1}{d\tau_1} = F\left(\varphi_1(1) + \overset{(2)}{Q}_0 z_2, 1\right), \qquad \frac{d\overset{(2)}{Q}_0 z_2}{d\tau_1} = \overset{(2)}{Q}_0 z_1,
$$

$$
\overset{(2)}{Q}_0 z_2(0) = -\varphi_1(1), \quad \overset{(2)}{Q}_0 z_i(-\infty) = 0, \qquad i = 1, 2.
$$

Equating the expressions (2.3.34) and (2.3.37) for z_1 at $t = T$, and taking into account that at this point $\overset{(1)}{\Pi}_0\, z_i$ and $\overset{(2)}{Q}_0\, z_i$ are less than any power of μ, we obtain (to the zeroth order)

$$\overset{(1)}{Q}_0\, z_1(0) = \overset{(2)}{\Pi}_0\, z_1(0). \tag{2.3.40}$$

The relation above is the equation for T. We note that conditions (2.3.36), (2.3.39) provide (to the zeroth order) the equality of the expressions (2.3.34) and (2.3.37) for z_2 at the point $t = T$.

Let us rewrite (2.3.40) in a different form. We make the change of variables $\tilde{z}_1 = \overset{(1)}{Q}_0\, z_1$, $\tilde{z}_2 = \varphi_3(T) + \overset{(1)}{Q}_0\, z_2$. In these new variables the problem (2.3.35), (2.3.36) becomes

$$\frac{d\tilde{z}_1}{d\tau_T} = F(\tilde{z}_2, T), \quad \frac{d\tilde{z}_2}{d\tau_T} = \tilde{z}_1, \qquad \tau_T \le 0, \tag{2.3.41}$$

$$\tilde{z}_2(0) = \varphi_2(T), \quad \tilde{z}_1(-\infty) = 0, \qquad \tilde{z}_2(-\infty) = \varphi_3(T).$$

Analogously, if we make the change of variables $\hat{z}_1 = \overset{(2)}{\Pi}_0\, z_1$, $\hat{z}_2 = \varphi_1(T) + \overset{(2)}{\Pi}_0\, z_2$ in (2.3.38), (2.3.39), we obtain

$$\frac{d\hat{z}_1}{d\tau_T} = F(\hat{z}_2, T), \quad \frac{d\hat{z}_2}{d\tau_T} = \hat{z}_1, \quad \tau_T \ge 0, \tag{2.3.42}$$

$$\hat{z}_2(0) = \varphi_2(T), \quad \hat{z}_1(\infty) = 0, \quad \hat{z}_2(\infty) = \varphi_1(T).$$

In the new variables (2.3.40) can be written as

$$\tilde{z}_1(0) = \hat{z}_1(0). \tag{2.3.43}$$

Evidently, both (2.3.41) and (2.3.42) coincide (up to notation) with the system (2.3.31). Assume that corresponding phase portrait of the system is the one shown in Fig. 2.6(a). Then the point A on the separatrix entering the saddle $(0, \varphi_3(T))$ as $\tau_T \to -\infty$ corresponds to the value $\tilde{z}_2(0) = \varphi_2(T)$, and the point B (which does not coincide with A) on the separatrix entering the saddle $(0, \varphi_1(T))$ as $\tau_T \to \infty$ corresponds to the value $\hat{z}_2(0) = \varphi_2(T)$ (see Fig. 2.8). Therefore condition (2.3.43) is not satisfied. Similarly, condition (2.3.43) will not be satisfied in the phase portrait of Fig. 2.6(b). Only the phase portrait of Fig. 2.6(c) provides the necessary equality. So T must be such a value of t that the homoclinic orbit consisting of two heteroclinic orbits is formed in the phase plane of the associated system. The condition guaranteeing existence of such T will be formulated below (condition 4^0) after the transformation of (2.3.43) to a special form.

Further, for the existence of the asymptotic solution in each of the subintervals $[0, T]$ and $[T, 1]$, the phase portraits of the *associated* systems must satisfy for $t = 0$ and $t = 1$ other conditions corresponding to conditions $3'$ and $3''$ of § 2.3.1:

$3'$. *In the phase plane of the associated system* $d\tilde{z}_1/d\tau_0 = F(\tilde{z}_2, 0)$, $d\tilde{z}_2/d\tau_0 = \tilde{z}_1$, *let the straight line* $\tilde{z}_2 = 0$ *intersect the separatrix entering the saddle* $(0, \varphi_3(0))$ *as* $\tau_0 \to \infty$.

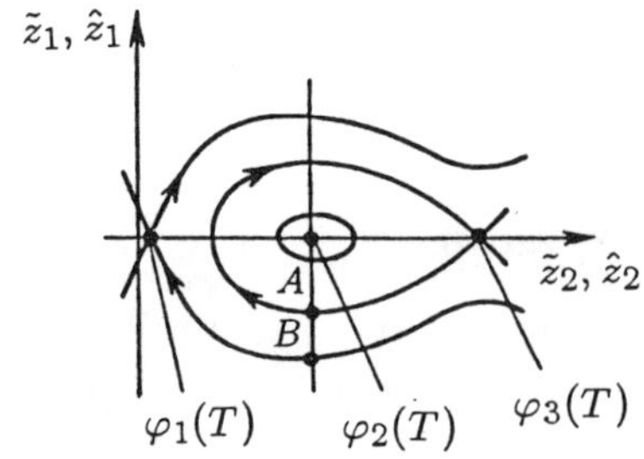

Figure 2.8: *The values of* $\tilde{z}_1(0)$ *and* $\hat{z}_1(0)$ *corresponding to* $\tilde{z}_2(0) = \hat{z}_2(0) = \varphi_2(T)$ *do not coincide, so transition is not possible.*

$3''$. *In the phase plane of the associated system* $d\hat{z}_1/d\tau_1 = F(\hat{z}_2, 1)$, $d\hat{z}_2/d\tau_1 = \hat{z}_1$, *let the straight line* $\hat{z}_2 = 0$ *intersect the separatrix entering the saddle* $(0, \varphi_1(1))$ *as* $\tau_1 \to -\infty$.

Let us now rewrite (2.3.43) in a different form. The separatrix of the system (2.3.41) that enters the saddle point $(0, \varphi_3(T))$ as $\tau_T \to -\infty$ is described by the equation

$$\tilde{z}_1 = -\sqrt{2}\sqrt{\int_{\varphi_3(T)}^{\tilde{z}_2} F(z,T)dz}, \tag{2.3.44}$$

and the separatrix of the system (2.3.42) that enters the saddle point $(0, \varphi_1(T))$ as $\tau_T \to \infty$ is described by

$$\hat{z}_1 = -\sqrt{2}\sqrt{\int_{\varphi_1(T)}^{\hat{z}_2} F(z,T)dz}, \tag{2.3.45}$$

Substituting these expressions into (2.3.43), we obtain an equation for T as

$$\int_{\varphi_3(T)}^{\varphi_2(T)} F(z,T)dz = \int_{\varphi_1(T)}^{\varphi_2(T)} F(z,T)dz \quad \text{or} \quad \int_{\varphi_3(T)}^{\varphi_1(T)} F(z,T)dz = 0. \tag{2.3.46}$$

Note that the relation above coincides with condition (2.3.33).

4^0. *Let* (2.3.46)*have a solution* $T = T_0$ *with* $0 < T_0 < 1$.

For the construction of the higher-order terms of the asymptotics (as well as for the proof of existence of such solution, as shown in Fig. 2.7), we will need one more condition (see Vasil'eva and Butuzov [148] for details).

5^0.

$$\frac{d}{dT}\int_{\varphi_3(T)}^{\varphi_1(T)} F(z,T)dz\bigg|_{T=T_0} \neq 0,$$

(so, the root T of (2.3.46) is simple).

Theorem 2.5 *Under conditions* 1^0–5^0, *for sufficiently small* μ, *there exists a solution of the boundary value problem* (2.3.29), (2.3.30)*for which*

$$\lim_{\mu \to 0} z_2(t, \mu) = \begin{cases} \varphi_3(t) & \text{for } 0 < t < T_0, \\ \varphi_1(t) & \text{for } T_0 < t < 1, \end{cases}$$

$$\lim_{\mu \to 0} z_1(t, \mu) = 0 \quad \text{for } 0 < t < T_0, \quad T_0 < t < 1.$$

Remark 1. Along with the constructed solution, there might also exist a solution for which $z_2(t, \mu)$ approaches $\varphi_1(t)$ in $(0, T_0)$ and $\varphi_3(t)$ in $(T_0, 0)$ as $\mu \to 0$.

Remark 2. If the homoclinic orbit consisting of two heteroclinic orbits (a *cell*) in the associated system (2.3.31) occurs for several values of T in the interval $(0, 1)$, then solutions with multiple transition layers might exist.

Remark 3. The solutions mentioned above might coexist with each other as well as with solutions without transition layers (such solutions were constructed in § 2.3.1). Thus, the solution of (2.3.29), (2.3.30) is not, in general, unique.

Remark 4. The results obtained in this section can be easily generalized to the case of systems containing both *fast* and *slow* variables (functions z and y, respectively, from § 2.3.1).

Exercise

Consider the problem

$$\mu \frac{dz_1}{dt} = z_2(z_2 - (1 + t))(z_2 - 3),$$

$$\mu \frac{dz_2}{dt} = z_1,$$

$$z_2(0) = 3, \qquad z_2(1) = 0.$$

(a) Find the solutions of the reduced equations and check that condition 4^0 is satisfied for $T = 0.5$.

(b) Show that condition 5^0 holds at $T = 0.5$.

(c) Find the zeroth-order approximation to the solution. Note that the transition layer near $t = T = 0.5$ (i.e., the solution of equation (2.3.31)) can be found explicitly (see Murray [99, p. 305]).

Remark 5. For some problems of type (2.3.31), a phase plane *cell* occurs for any T. In this case, under certain conditions, a solution as shown in Fig. 2.7 also exists, and the asymptotics for this solution can be constructed. The equation defining T will be more complicated ((2.3.46) will be satisfied identically). Let us write the equation for T in the special choice of $\varphi_2(t) \equiv 0$:

$$\text{(2.3.47)} \qquad \left. \int_0^{\varphi_i(T)} \frac{\displaystyle\int_{\varphi_i(T)}^{z} F_t(z, T)\,dz}{\left[2\displaystyle\int_{\varphi_i(T)}^{z} F(z, T)\,dz\right]^{1/2}} dz \right|_{i=1}^{i=3} = 0.$$

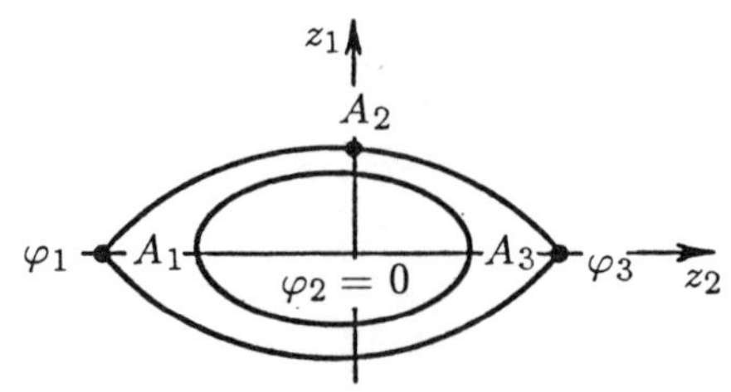

Figure 2.9: *The* cell *(homoclinic orbit consisiting of two heteroclinic orbits)in the phase plane of the autonomous system.*

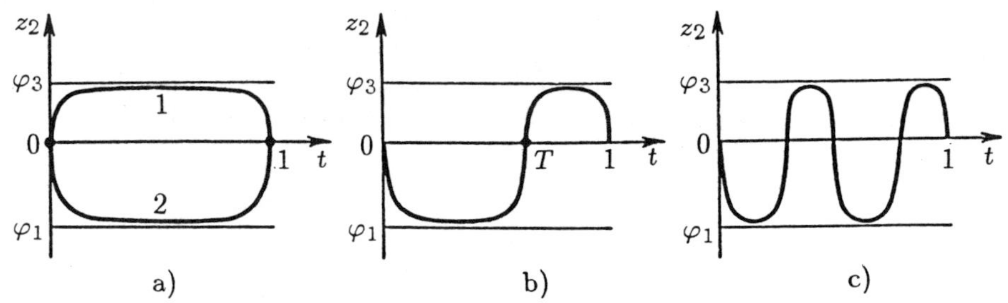

Figure 2.10: *Possible solutions for the autonomous case.*

This equation was obtained in Vasil'eva [141].

Equation (2.3.47) becomes an identity if the system (2.3.47) is autonomous, i.e., if $F(z_2, t) = F(z_2)$ (does not depend on t). In this case only the velocity of motion along the trajectories in the phase plane changes with μ, but the trajectories themselves remain the same.

Suppose that the *cell* (homoclinic orbit consisting of two heteroclinic orbits) exists in the phase plane of the autonomous system (see Fig. 2.9). Consider a trajectory corresponding to a closed cycle inside this homoclinic orbit. If we fix the time of revolution (i.e., to complete one cycle), then as $\mu \to 0$ the cycle will approach the boundary of the *cell*, and in the limit will consist of two heteroclinic orbits (the separatrices).

If the boundary conditions (2.3.30) are imposed, there might exist a solution corresponding to the part of the cycle lying to the right of the z_1-axis (the curve 1 in Fig. 2.10(a) shows the dependence of z_2 on t for this solution). The solution corresponding to the part of the cycle lying to the left of the z_1-axis might exist as well (the curve 2 in Fig. 2.10(a) shows the dependence of z_2 on t for this case). The theory of § 2.3.1 can be applied to construct such asymptotic solutions.

However, along with these solutions, there might exist a solution with a transition from φ_1 to φ_3 to which a complete revolution along the cycle corresponds (see

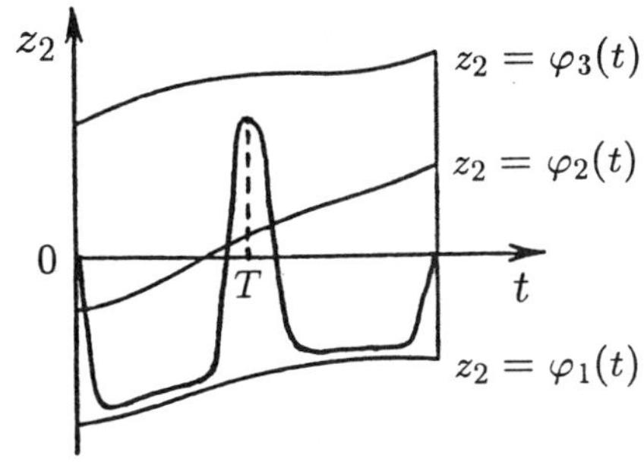

Figure 2.11: *A spike-type solution in the case when a homoclinic orbit exists in the phase-portrait of the second-order equation equivalent to a system of two first-order equations.*

Fig. 2.10b). To define the point of transition T with accuracy of the order μ, one can perform an approximate calculation, assuming that the motion takes place almost along the separatrix (recall that μ is small). The motion along the separatrix is described by the equation (here we use formulae (2.3.44), (2.3.45))

$$\frac{dz_2}{d\tau} = \pm\sqrt{2}\sqrt{\Phi(z_2) - \Phi(\varphi_3)},$$

where $\Phi(z) = \int_0^z F(z)dz$ and $\Phi(\varphi_3) = \Phi(\varphi_1)$ (the condition for existence of the *cell* in the phase plane). The time to move along the part of the trajectory close to the segment A_1A_2 of the separatrix (see Fig. 2.9), and the time of motion along the part of the trajectory near A_2A_3 are defined by the formulae

$$\tau_1 = -\int_0^{\varphi_1+\delta} \frac{dz}{\sqrt{\Phi(z) - \Phi(\varphi_1)}}, \qquad \tau_2 = \int_0^{\varphi_3-\delta} \frac{dz}{\sqrt{\Phi(z) - \Phi(\varphi_3)}}.$$

Using these expressions, $\lim_{\delta\to 0}\tau_1/\tau_2 = \lambda_3/\lambda_1$ where $\lambda_i = \sqrt{F_z(\varphi_i)}$ $(i = 1, 3)$. This implies that the ratio of lengths of the subintervals $[0, T]$ and $[T, 1]$ (see Fig. 2.10(b) is $T/(1 - T) = \lambda_3/\lambda_1$, and hence, $T = \lambda_3/(\lambda_3 + \lambda_1)$. This relation was obtained in Vasil'eva and Tupchiev [158] for an arbitrary system of two equations.

Evidently, cases corresponding to several revolutions along the closed cycle are also possible (any fixed number of cycles are allowed). The case of two cycles is shown in Fig. 2.10(c).

2.4 Spike-type solutions and other contrast (dissipative) structures

2.4.1 The main types of contrast structures

Solutions, which have internal transition layers, are often referred to as *contrast* or *dissipative* structures. One of such solutions, described in § 2.3.2, is shown in Fig. 2.7. It is called a *threshold-type* contrast structure (the term *step-type* is also used). In the case when a homoclinic orbit exists in the phase plane (as shown in Figures 2.6(a),(b), the solution with an internal transition layer of *spike-type* might exist (cf. Butuzov and Vasil'eva [38], [39] and Vasil'eva, Tupchiev, and Yarkin [159]); see Fig. 2.11. The value of T now is defined from a more complicated equation than (2.3.46). We will not present here the discussion of this case (detailed analysis of which is given in Butuzov and Vasil'eva [38]) since in the following subsections we will investigate spike-type solutions ("bursting" phenomenon) for a problem more complicated than (2.3.29). If (2.3.46) has several solutions, then the contrast structures with several *thresholds* might exist. Similarly, solutions with several spikes are possible, as well as solutions containing thresholds and spikes simultaneously.

Equation

$$\mu^2 \frac{d^2 z}{dx^2} = F(z, x) \tag{2.4.1}$$

is equivalent to system (2.3.29) (from now on we will denote the independent variable by x). It describes the stationary solution of the diffusion equation with a small diffusion coefficient μ^2:

$$\mu^2 u_{xx} - u_t = F(u, x). \tag{2.4.2}$$

This equation is often used as a mathematical model for a variety of problems of chemical and biological kinetics, nonlinear optics, semiconductor physics, and other applications. Stationary solutions of (2.4.1), $z(x, \mu)$, and, in particular, the contrast structures, represent the stationary regimes approached by the solution $u(x, t, \mu)$ of (2.4.2) for sufficiently large time t. The attention of many researchers in different areas of the natural sciences has been attracted lately to the phenomena of the spatial and temporal structures. An extensive bibliography on this topic is presented, e.g., in Haken [57], Murray [99], Nicolis and Prigogine [110], Polak and Mikhailov [117], and Romanovskii, Stepanov, and Chernavskii [122].

2.4.2 Spike-type solutions for the system of two equations: statement of the problem and asymptotic algorithm

Very often a system of two reaction-diffusion equations is considered as a model for describing such phenomena. The system in the spatially one-dimensional case has the form

$$u_t = D_1 u_{xx} - f(u, v), \qquad v_t = D_2 v_{xx} - g(u, v).$$

For $D_1 \ll D_2$, *spikes* (*contrast spatial structures*) might be formed in this system. These structures are characterized by the presence of subintervals of slow and fast change of the variable u with respect to x (see Fig. 2.12). They are described by the corresponding stationary system, which can be rewritten in the form (here we set $D_1 = \mu^2$, where $0 < \mu \ll 1$ is a small parameter; and $D_2 = 1$)

$$\mu^2 u'' = f(u, v), \qquad v'' = g(u, v). \tag{2.4.3}$$

Let us consider (2.4.3) on the interval $0 \le x \le l$, with the boundary conditions

$$u'(0) = u'(l) = v'(0) = v'(l) = 0 \tag{2.4.4}$$

(other boundary conditions can be taken as well).

Below, the asymptotic expansion for the spike-type solution of (2.4.3), (2.4.4) will be constructed. In the process of obtaining the asymptotic solution, we will find that there might exist *three types* of points x_* (*"bursting" points*), where the fast change (spike) of variable u takes place. The first type of such points is considered in this

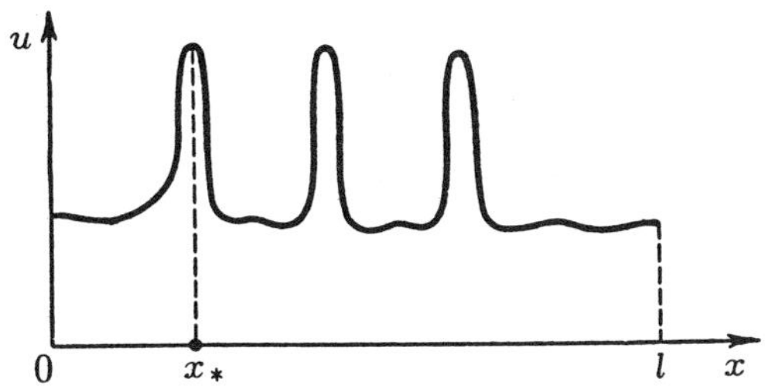

Figure 2.12: *Spike-type solution (contrast spatial structure).*

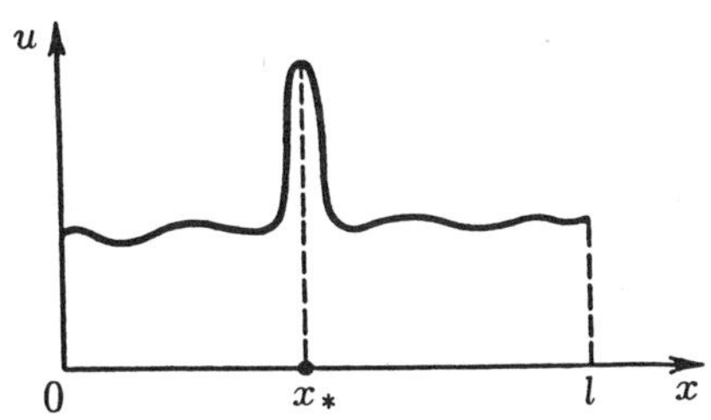

Figure 2.13: *The solution with one spike.*

subsection and in § 2.4.3; the second type is considered in § 2.4.4; and the third type in §§ 2.4.5 and 2.4.6. Here we will present only the asymptotic algorithm. The questions of existence (and uniqueness) of the solution to (2.4.3), (2.4.4) in the vicinity of the constructed asymptotics will not be discussed. In what follows we assume that the functions f and g are sufficiently smooth.

To simplify the presentation, let us consider the case when the solution of (2.4.3), (2.4.4) has only one point x_* ($0 < x_* < l$) in the vicinity of which there exists a spike of the function $u(x, \mu)$ (see Fig. 2.13). The case of several such points can be studied analogously.

Let us assume that x_* is a point of extreme value for u. Then

(2.4.5) $$u'(x_*, \mu) = 0.$$

We represent the value of x_* in the form

(2.4.6) $$x_* = x_0 + \mu x_1 + \cdots + \mu^k x_k + \cdots .$$

One of the problems is defining x_* (i.e., of determining the coefficients x_k).

Using the method described in § 2.3.1, we will construct the asymptotics for the solution of (2.4.3), (2.4.4) separately in the intervals $0 \leq x \leq x_*$ and $x_* \leq x \leq l$, performing matching at the point x_*. The asymptotic expansion on $0 \le x \le x_*$ is sought in the form

$$
\begin{aligned}
u &= \bar{u}_0(x) + \mu\bar{u}_1(x) + \cdots + \mu\Pi_1 u(\tau_0) + \cdots + Q_0 u(\tau) + \mu Q_1 u(\tau) + \cdots, \\
v &= \bar{v}_0(x) + \mu\bar{v}_1(x) + \cdots + \mu^3\Pi_3 v(\tau_0) + \cdots + \mu^2 Q_2 v(\tau) + \cdots,
\end{aligned} \tag{2.4.7}
$$

where $\tau_0 = x/\mu$ and $\tau = (x - x_*)/\mu$. The functions $\bar{u}_i(x)$ and $\bar{v}_i(x)$ represent the regular part of the asymptotics; while the functions Π_i describe the boundary layer in the vicinity of the point $x = 0$ (since the boundary conditions (2.4.4) prescribe not the values of the functions u and v but their derivatives, the Π-functions for the fast variable u appear starting from the term of the order μ, and for the slow variable v they start to appear at the order μ^3). The functions Q_i (and mainly the zeroth-order term $Q_0 u$) describe the fast change in the solution to the left of the point x_*. As in § 2.3.1, we impose the decay conditions on the boundary functions, i.e., $\Pi_i(+\infty) = 0$ and $Q_i(-\infty) = 0$.

The asymptotic expansion of the solution on $x_* \le x \le l$ has the analogous form:

$$
\begin{aligned}
u &= \bar{u}_0(x) + \mu\bar{u}_1(x) + \cdots + P_0 u(\tau) + \mu P_1 u(\tau) + \cdots + \mu R_1 u(\tau_1) + \cdots, \\
v &= \bar{v}_0(x) + \mu\bar{v}_1(x) + \cdots + \mu^2 P_2 v(\tau) + \cdots + \mu^3 R_3 v(\tau_1) + \cdots,
\end{aligned} \tag{2.4.8}
$$

where $\tau_1 = (x - l)/\mu$.

Substituting (2.4.7) into (2.4.3), the boundary conditions (2.4.4) at $x = 0$, and (2.4.5), and representing the functions f and g in a form similar to (2.4.7) (e.g., $f = \bar{f} + \Pi f + Qf$; see (2.3.5)), we obtain in the standard way equations for the terms of the asymptotic expansion (2.4.7). Analogously, equations for the terms of the series (2.4.8) can be written out.

For $\bar{u}_0$ and $\bar{v}_0$ we obtain the system of equations

$$
f(\bar{u}_0, \bar{v}_0) = 0, \qquad \bar{v}''_0 = g(\bar{u}_0, \bar{v}_0). \tag{2.4.9}
$$

1^0. *Let the equation $f(u, v) = 0$ have two solutions with respect to u for each $v \in I$ (here I is some interval)*:

$$
u = \varphi(v), \qquad u = \chi(v)
$$

(let, e.g., $\varphi(v) < \chi(v)$), and suppose

$$
f_u(\varphi(v), v) > 0, \qquad \text{while } f_u(\chi(v), v) < 0. \tag{2.4.10}
$$

(We should note that other solutions are also possible, but only these two will be used in the construction of the asymptotic solution.) *In addition, we assume that there exists a function $\psi(v)$ (see Fig. 2.14) such that*

$$
\int_{\varphi(v)}^{\psi(v)} f(u, v)\,du = 0. \tag{2.4.11}
$$

Setting $\bar{u}_0 = \varphi(\bar{v}_0)$, we obtain the equation

$$
\bar{v}''_0 = g(\varphi(\bar{v}_0), \bar{v}_0) \tag{2.4.12}
$$

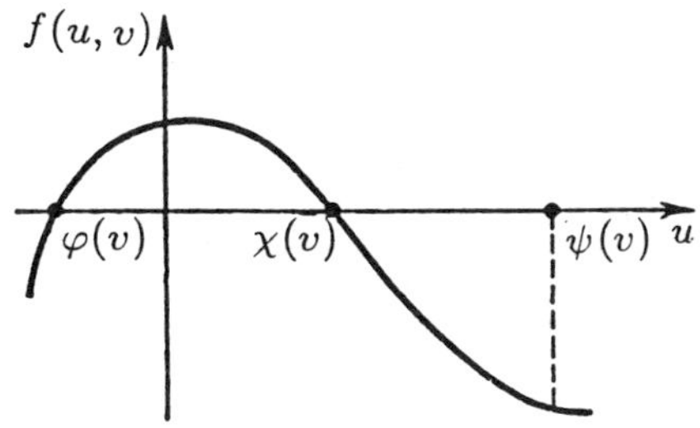

Figure 2.14: *Value of function $\psi(v)$ for some v.*

for $\overline{v}_0$.

Taking into account the form of (2.4.7), (2.4.8), we obtain the boundary conditions

(2.4.13) $$\overline{v}'_0(0) = \overline{v}'_0(l) = 0.$$

The condition that $v(x,\mu)$ and $v'(x,\mu)$ should be continuous at the point x_* leads to the equalities $\overline{v}_0(x_0-0) = \overline{v}_0(x_0+0)$, $\overline{v}'_0(x_0-0) = \overline{v}'_0(x_0+0)$. Therefore $\overline{v}_0(x)$ can be defined in the entire interval $[0,l]$ as the solution of the problem (2.4.12), (2.4.13). The equalities for $\overline{v}_0(x,\mu)$ and $\overline{v}'_0(x,\mu)$ at x_0 will then be satisfied.

The following cases will be considered separately:

2^0. *The problem* (2.4.12), (2.4.13)*has a nonconstant solution* $\overline{v}_0 = \overline{v}_0(x)$.

$2'$. *The problem* (2.4.12), (2.4.13)*has a constant solution* $\overline{v}_0 = \text{const}$.

The first case will be discussed in this subsection and in §§ 2.4.3, 2.4.4, the second case will be considered in §§ 2.4.5, 2.4.6.

Let us first assume that condition 2^0 is satisfied, so $\overline{v}_0 = \overline{v}_0(x)$ and $\overline{u}_0 = \varphi(\overline{v}_0(x))$.

For Q_0u we obtain the equation

(2.4.14) $$\frac{d^2Q_0u}{d\tau^2} = f(\overline{u}_0(x_0) + Q_0u, \overline{v}_0(x_0)), \qquad -\infty < \tau \le 0,$$

and the boundary conditions

(2.4.15) $$Q'_0u(0) = 0, \qquad Q_0u(-\infty) = 0$$

(the first condition follows from (2.4.5)).

Note that although $\overline{u}_0(x)$ and $\overline{v}_0(x)$ are known functions, the point x_0 is yet unknown, and therefore the unknown quantities $\overline{v}_0(x_0)$ (for which we will now use the simpler notation v_0) and $\overline{u}_0(x_0) = \varphi(v_0) \equiv \varphi_0$ enter (2.4.14). By virtue of (2.4.10) the rest point $Q_0u = 0$ of (2.4.14) is a saddle. In the phase plane, the separatrix that emerges from the saddle point passes around the other stationary point $(\chi_0 - \varphi_0)$, which is the center, and then returns to the saddle (as $\tau \to \infty$). Thus, the corresponding trajectory in the phase plane is a homoclinic orbit (the arrows in Fig. 2.15 show the direction of motion along the trajectory for growing τ). The upper part of this separatrix is described by the equation

(2.4.16) $$Q'_0u = \sqrt{2\int_{\varphi_0}^{\varphi_0+Q_0u} f(\xi, v_0)d\xi} \equiv \Phi(Q_0u).$$

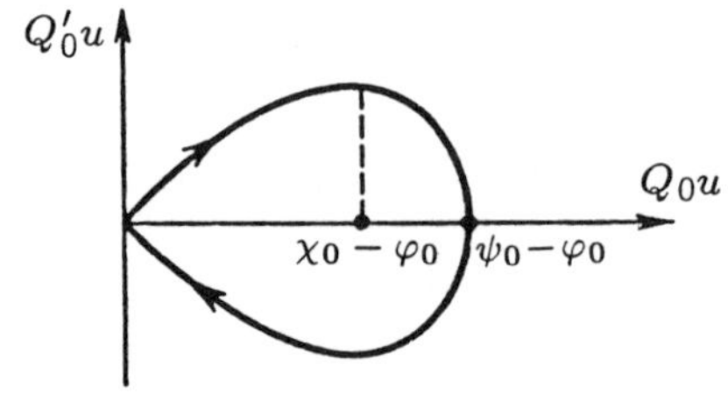

Figure 2.15: *The upper part of the homoclinic orbit corresponds to the solution* Q_0u *of* (2.4.14), (2.4.15);*the lower part corresponds to the solution* P_0u *of* (2.4.18).

The solution of this differential equation for $\tau \le 0$ with the initial condition $Q_0u(0) = \psi_0 - \varphi_0$ (for $\psi_0 = \psi(v_0)$, where $\psi(v)$ is the function determined by relation (2.4.11)) provides the nontrivial solution of (2.4.14), (2.4.15) as well. This solution depends on v_0 as a parameter.

The stationary point $Q_0u = 0$ of (2.4.16) is asymptotically stable for $\tau \to -\infty$. Indeed, $\Phi'(Q_0u) = f(\varphi_0 + Q_0u, v_0)/\Phi(Q_0u)$, and by virtue of (2.4.10)

$$\Phi'(0) = \sqrt{f_u(\varphi_0, v_0)} > 0.$$

Hence, for Q_0u, the usual for boundary function estimate holds:

$$|Q_0u(\tau)| \le c\exp(\kappa\tau), \qquad -\infty < \tau \le 0. \tag{2.4.17}$$

Analogously, for P_0u we obtain the problem

$$\frac{d^2P_0u}{d\tau^2} = f(\varphi_0 + P_0u, v_0), \qquad 0 \le \tau < \infty,$$
$$P_0'u(0) = 0, \qquad P_0u(\infty) = 0. \tag{2.4.18}$$

Its nontrivial solution is represented in the phase plane by the lower part of the separatrix (see Fig. 2.15), which is described by the equation

$$P_0'u = -\Phi(P_0u).$$

The solution of this equation for $\tau \ge 0$ with the initial condition $P_0u(0) = \psi_0 - \varphi_0$ is just the function $P_0u(\tau)$ sought, for which the estimate

$$|P_0u(\tau)| \le c\exp(-\kappa\tau), \qquad 0 \le \tau < \infty \tag{2.4.19}$$

holds.

So, the terms of the asymptotic solution to the zeroth order are constructed, but the point x_0 is yet not defined. We note that the condition that the function $u(x,\mu)$ be continuous at the point x_* at the zeroth order is satisfied: $\overline{u}_0(x_0 - 0) + Q_0u(0) =$

$\overline{u}_0(x_0+0)+P_0u(0)$. This follows from the continuity of the function $\overline{u}_0(x)$ at the point x_0, and from the relation $Q_0u(0)=P_0u(0)$.

For $\overline{u}_1$ and $\overline{v}_1$ we have the linear system

$$\overline{f}_u\overline{u}_1+\overline{f}_v\overline{v}_1=0, \qquad \overline{v}''_1=\overline{g}_u\overline{u}_1+\overline{g}_v\overline{v}_1$$

(where $\overline{f}_u=f_u(\overline{u}_0(x),\overline{v}_0(x))$, and the notation $\overline{f}_v$, $\overline{g}_u$, $\overline{g}_v$ has similar meaning), which implies

$$\overline{u}_1=-\overline{f}_u^{-1}\overline{f}_v\overline{v}_1=\varphi'(\overline{v}_0(x))\overline{v}_1,$$

$$\overline{v}''_1=A(x)\overline{v}_1, \tag{2.4.20}$$

for $A(x)=\overline{g}_v+\overline{g}_u\varphi'(\overline{v}_0)$. The boundary conditions at the ends of the interval are

$$\overline{v}'_1(0)=\overline{v}'_1(l)=0. \tag{2.4.21}$$

Since v and v' must be continuous at the point x_*, we obtain

$$\overline{v}_1(x_0-0)=\overline{v}_1(x_0+0), \qquad \overline{v}'_1(x_0-0)+Q'_2v(0)=\overline{v}_1(x_0+0)+P'_2v(0).$$

Thus, to define $\overline{v}_1(x)$ we must first determine Q_2v and P_2v. For Q_2v we have

$$\frac{d^2Q_2v}{d\tau^2}=Q_0g\equiv g(\varphi_0+Q_0u(\tau),v_0)-g(\varphi_0,v_0), \qquad Q_2v(-\infty)=0.$$

This yields

$$Q_2v(\tau)=\int_{-\infty}^{\tau}ds\int_{-\infty}^{s}Q_0g(\sigma)d\sigma, \qquad Q'_2v(\tau)=\int_{-\infty}^{\tau}Q_0g(\sigma)d\sigma.$$

From these formulae and the exponential estimate (2.4.17) for $Q_0u(\tau)$ and, consequently, the analogous estimate for $Q_0g(\tau)$, one can easily obtain the estimate

$$|Q_2v(\tau)|\le c\exp(\kappa\tau), \qquad -\infty<\tau\le 0.$$

Since, according to (2.4.16), $d\sigma=dQ_0(\sigma)/\Phi(Q_0u(\sigma))$,

$$Q'_2v(\tau)=\int_{\varphi_0}^{\varphi_0+Q_0u(\tau)}\frac{g(u,v_0)-g(\varphi_0,v_0)}{\Phi(u-\varphi_0)}du, \quad \text{and hence,}$$

$$Q'_2v(0)=\int_{\varphi_0}^{\psi_0}\frac{g(u,v_0)-g(\varphi_0,v_0)}{\Phi(u-\varphi_0)}du.$$

In a similar way, we obtain

$$Q_2v(0)=\int_{\varphi_0}^{\psi_0}\left[\int_{\varphi_0}^{s}\frac{g(u,v_0)-g(\varphi_0,v_0)}{\Phi(u-\varphi_0)}du\right]\frac{ds}{\Phi(s-\varphi_0)}.$$

Analogous calculations for P_2v provide the equalities

$$P_2v(0)=Q_2v(0), \qquad P'_2v(0)=Q'_2v(0).$$

Hence $\overline{v}_1$ must satisfy the following conditions:

(2.4.22) $\overline{v}_1(x_0+0)-\overline{v}_1(x_0-0)=0, \qquad \overline{v}_1'(x_0+0)-\overline{v}_1'(x_0-0)=2Q_2'v(0).$

To find $\overline{v}_1(x)$, we must solve the linear equation (2.4.20) with the additional conditions (2.4.21), (2.4.22).

3^0. *Let the problem* (2.4.20), (2.4.21)*have only the trivial solution.*

Then, as can be easily seen, the problem (2.4.20)—(2.4.22) for $\overline{v}_1(x)$ has a unique solution, and therefore $\overline{v}_1(x)$ and $\overline{u}_1(x)$ are defined (they still depend on the yet unknown value x_0). It is important to mention that $\overline{v}_1(x)$ and $\overline{u}_1(x)$ have discontinuous derivatives at x_0.

For Q_1u we obtain the problem

$$\frac{d^2Q_1u}{d\tau^2}=f_u\,Q_1u+f_1(\tau), \qquad -\infty<\tau\le 0, \tag{2.4.23}$$

$$Q_1'u(0)=-\overline{u}_0'(x_0)=-\varphi'(v_0)\overline{v}_0'(x_0), \qquad Q_1u(-\infty)=0. \tag{2.4.24}$$

Here $f_1(\tau)=(f_v+f_u\varphi'(v_0))[\overline{v}_0'(x_0)(x_1+\tau)+\overline{v}_1(x_0)]$, where f_u and f_v are calculated at the point $(\varphi_0+Q_0u(\tau),v_0)$; and x_1 is the coefficient of μ in expansion (2.4.6).

Note that the homogeneous equation corresponding to (2.4.23) has the nontrivial solution $Q_1u=Q_0'u(\tau)$ satisfying the conditions $Q_1u(0)=0$ and $Q_1u(-\infty)=0$. This can be verified by differentiating (2.4.14), which leads to $(Q_0'u(\tau))''=f_u\,Q_0'u(\tau)$.

Multiplying (2.4.23) by $Q_0'u(\tau)$ and integrating from $-\infty$ to 0, we obtain

$$Q_1u(0)=-\frac{1}{f(\psi_0,v_0)}\int_{-\infty}^0 f_1(\tau)Q_0'u(\tau)d\tau.$$

Let us use the expression for $f_1(\tau)$ and the equalities

$$\int_{-\infty}^0 f_u(\varphi_0+Q_0u,v_0)Q_0'u(\tau)d\tau=\int_{-\infty}^0 Q_0'''u(\tau)d\tau=Q_0''u(0)=f(\psi_0,v_0),$$

$$\int_{-\infty}^0 f_u(\varphi_0+Q_0u,v_0)Q_0'u(\tau)\tau d\tau=\int_{-\infty}^0 \tau Q_0'''u(\tau)d\tau$$

$$=\tau Q_0''u(\tau)|_{-\infty}^0-\int_{-\infty}^0 Q_0''u(\tau)d\tau=-Q_0'u(\tau)|_{-\infty}^0=0$$

to write $Q_1u(0)$ as

$$Q_1u(0)=-\frac{1}{f(\psi_0,v_0)}\int_{-\infty}^0 f_v(\varphi_0+Q_0u,v_0)[\overline{v}_0'(x_0)(x_1+\tau)+\overline{v}_1(x_0)]Q_0'u(\tau)d\tau$$

$$-\varphi'(v_0)[\overline{v}_0'(x_0)(x_1+\tau)+\overline{v}_1(x_0)]$$

$$=-\frac{1}{f(\psi_0,v_0)}\int_{\varphi_0}^{\psi_0} f_v(u,v_0)\left[\overline{v}_0'(x_0)\left(x_1+\int_{\psi_0}^{u}\frac{ds}{\Phi(s-\varphi_0)}\right)+\overline{v}_1(x_0)\right]du \tag{2.4.25}$$

$$-\varphi'(v_0)[\overline{v}_0'(x_0)(x_1+\tau)+\overline{v}_1(x_0)].$$

Thus, if the solution of (2.4.23), (2.4.24) exists, its value at $\tau = 0$ is given by the formula above.

To prove the existence of $Q_1u(\tau)$ and to determine its exponential decay as $\tau \to -\infty$, we will use the fundamental system of solutions for the homogeneous equation (2.4.23). Let us take $Y_1 = Q_0'u(\tau) = \Phi(Q_0u(\tau))$ as one of the nontrivial solutions. A solution Y_2 linearly independent from Y_1 can be defined to satisfy the following conditions: $Y_2(-1) = 0$, $Y_2'(-1) = 1/Y_1(-1)$. Then the Wronskian $Y_1Y_2' - Y_2Y_1'$ is 1, and Y_2 can be represented by the well-known formula $Y_2 = Y_1(\tau)\int_{-1}^{\tau} ds/Y_1^2(s)$. We note that this solution is bounded as $\tau \to 0$ together with its derivative and that $Y_2(0) = -1/f(\psi_0, v_0)$. However, for $\tau \to -\infty$ this solution will grow exponentially. It can be shown that

$$\lim_{\tau \to -\infty} Y_1(\tau)Y_2(\tau) = -\frac{1}{2\sqrt{f_u(\varphi_0, v_0)}}.$$

Since $Y_1 = \sqrt{f_u(\varphi_0, v_0)}(Q_0u + o(Q_0u))$ as $\tau \to -\infty$,

$$Y_2 = -\frac{1}{2f_u(\varphi_0, v_0)(Q_0u + o(Q_0u))}$$

as $\tau \to -\infty$. Consider the particular solution

$$\tilde{Q}_1u = -Y_1(\tau)\int_0^{\tau} Y_2(\xi)f_1(\xi)d\xi + Y_2(\tau)\int_{-\infty}^{\tau} Y_1(\xi)f_1(\xi)d\xi.$$

of the nonhomogeneous equation (2.4.23). Note that the condition (2.4.25) is evidently satisfied.

Let us study the behavior of $\tilde{Q}_1u$ as $\tau \to -\infty$. Since

$$f_1(\tau) = (Q_0f_v + \varphi'(v_0)\,Q_0f_u)[\overline{v}_0'(x_0)(x_1 + \tau) + \overline{v}_1(x_0)],$$

$$|f_1(\tau)| \le (c_1 + c_2|\tau|)Q_0u(\tau),$$

where c_1 and c_2 are constants. Since $|Y_2(\tau)| \le c/Q_0u(\tau)$, the integral in the first term in the expression for $\tilde{Q}_1u$ grows no faster than some power of τ. Since $Y_1(\tau)$ exponentially approaches zero as $\tau \to -\infty$, the whole first term in $\tilde{Q}_1u$ tends exponentially to zero as $\tau \to -\infty$.

For the second term in the expression for $\tilde{Q}_1u$, we have

$$\left|\int_{-\infty}^{\tau} Y_1(\xi)f_1(\xi)d\xi\right| \le \int_0^{Q_0u} (c_1 + c_2|\xi|)Q_0u\,dQ_0u$$

$$\le c\int_0^{Q_0u} (Q_0u)^{1-\gamma}dQ_0u \le c(Q_0u)^{2-\gamma},$$

provided $2 - \gamma > 1$. The product of this quantity and $Y_2(\tau)$ (recall that $|Y_2(\tau)| \le c/Q_0u(\tau)$) is also exponentially small.

The derivative of $\tilde{Q}_1u$ has the form

$$\tilde{Q}_1'u = -Y_1'(\tau)\int_0^{\tau} Y_2(\xi)f_1(\xi)d\xi + Y_2'(\tau)\int_{-\infty}^{\tau} Y_1(\xi)f_1(\xi)d\xi,$$

and therefore $\tilde{Q}_1'u(0) = Y_2'(0)\int_{-\infty}^0 Y_1(\xi)f_1(\xi)d\xi \equiv p$.

Thus, the constructed solution $\tilde{Q}_1 u$ of (2.4.23) satisfies the conditions

$$\tilde{Q}_1'u(0) = p, \qquad \tilde{Q}_1 u(-\infty) = 0.$$

To obtain a solution satisfying conditions (2.4.24), it is sufficient to add to $\tilde{Q}_1 u$ the solution $Y_1 = Q_0'u(\tau)$ of the homogeneous equation with multiplicative factor $c = -(\varphi'(v_0)\overline{v}_0'(x_0) + p)/f(\psi_0, v_0)$. So,

$$Q_1 u(\tau) = \tilde{Q}_1 u(\tau) + cQ_0'u(\tau).$$

The problem for $P_1 u$ is studied analogously, and we obtain the equality

$$P_1 u(0) = -\frac{1}{f(\psi_0, v_0)}\int_{\varphi_0}^{\psi_0} f_v(u, v_0)\left[\overline{v}_0'(x_0)\left(x_1 - \int_{\psi_0}^{u}\frac{ds}{\Phi(s-\varphi_0)}\right) + \overline{v}_1(x_0)\right]du$$
$$-\varphi'(v_0)[\overline{v}_0'(x_0)(x_1+\tau) + \overline{v}_1(x_0)]$$

The condition that $u(x,\mu)$ be continuous at the point x_* to the first order produces the equation $\overline{u}_1(x_0 - 0) + Q_1 u(0) = \overline{u}_1(x_0 + 0) + P_1 u(0)$. Since $\overline{u}_1(x)$ is continuous at x_0, $Q_1 u(0) = P_1 u(0)$. Using the expressions for $Q_1 u(0)$ and $P_1 u(0)$, we obtain the equation

(2.4.26) $$F(v_0)\cdot \overline{v}_0'(x_0) = 0,$$

where

(2.4.27) $$F(v_0) = \int_{\varphi(v_0)}^{\psi(v_0)} f_v(u, v_0)\left(\int_{\psi(v_0)}^{u}\frac{ds}{\Phi(s-\varphi(v_0))}\right)du.$$

Equation (2.4.26) can be split into two equations: $F(v_0) = 0$ and $\overline{v}_0'(x_0) = 0$. In connection with that, we will distinguish between two types of points x_*.

Consider the first type of the "bursting" points x_*.

4^0. *Let the equation* $F(v_0) = 0$ *have a solution* $v_0 = v_0^0$ *with* $F'(v_0^0) \neq 0$.

To define x_0, we have now obtained the equation

$$\overline{v}_0(x_0) = v_0^0.$$

5^0. *Let this equation have a solution* $x_0 \in (0, l)$ *with* $\overline{v}_0'(x_0) \neq 0$.

Knowing x_0, we can finally find $Q_0 u$, $P_0 u$, $Q_2 v$, $P_2 v$, $\overline{v}_1(x)$, $\overline{u}_1(x)$. The yet unknown value x_1 enters linearly the expressions for $Q_1 u$, $P_1 u$.

In the next step, in quite a similar way, we can determine successively the functions $Q_1 u$, $P_1 u$, $Q_3 v$, $P_3 v$, $\overline{v}_2(x)$, $\overline{u}_2(x)$ (which depend on x_1), and $Q_2 u$, $P_2 u$ (which depend on x_1 and x_2). From the condition that the function $u(x,\mu)$ is continuous at the point x_*, we obtain the equation $Q_2 u(0) = P_2 u(0)$. Simple but quite cumbersome calculations show that this equation can be transformed to

(2.4.28) $$F'(v_0^0)(\overline{v}_0'(x_0))^2 x_1 = k_1,$$

where k_1 is known. Since $F'(v_0^0)\overline{v}_0'(x_0) \neq 0$, we can uniquely define x_1, and this will finally allow us to determine $Q_1 u$, $P_1 u$, $Q_3 v$, $P_3 v$, $\overline{v}_2(x)$, $\overline{u}_2(x)$. The unknown value x_2 enters the expressions for Q_2u, P_2u linearly. Note that $\overline{v}_2(x)$ and $\overline{u}_2(x)$ are discontinuous at the point x_0.

Equations analogous to (2.4.28) for x_i, $i = 2, 3, \ldots$, are obtained in the next order approximations and the whole procedure described above is repeated. The boundary functions Π_i and R_i are determined in the standard way. For example, for $\Pi_1 u(\tau_0)$ we obtain the equation

$$\frac{d^2\Pi_1 u}{d\tau_0^2} = \overline{f}_u(0)\Pi_1 u$$

subject to the boundary conditions

$$\Pi_1' u(0) = -\overline{u}_0'(0), \qquad \Pi_1 u(\infty) = 0.$$

Here $\overline{f}_u(0) = f_u(\overline{u}_0(0), \overline{v}_0(0))$. By virtue of (2.4.13), $\overline{u}_0'(0) = \varphi'(\overline{v}_0(0))\overline{v}_0'(0) = 0$. Therefore $\Pi_1 u = 0$.

For $\Pi_3 v(\tau_0)$ we obtain the equation

$$\frac{d^2\Pi_3 v}{d\tau_0^2} = \overline{g}_u(0)\Pi_1 u,$$

and hence, taking into account that $\Pi_1 u = 0$, we have $\Pi_3 v = 0$.

Continuing the process, we will find that all the functions $\Pi_i u$, $i \geq 1$, and $\Pi_i v$, $i \geq 3$, are identically zero. The functions R_i, which can be determined analogously to the Π-functions, are also identically zero.

2.4.3 The estimate with respect to discrepancy

Suppose that $x_0, x_1, \ldots, x_n$ are defined and the terms of the asymptotic series (2.4.7) and (2.4.8) are determined up to the order n inclusively. In addition, suppose that $Q_i u$ and $P_i u$ for $i = n+1$ and $n+2$ are also defined. Recall that $\overline{v}_i(x)$ and $\overline{u}_i(x)$ are discontinuous at the point x_0 for $i \geq 2$.

Let us introduce the notation $X_n = \sum_{k=0}^{n} \mu^k x_k$ and set $\tau' = (x - X_n)/\mu$. We will use the expansion (2.4.7) for $0 \leq x \leq X_n$ and the expansion (2.4.8) for $X_n \leq x \leq l$. To be able to do that we must change the argument of the Q- and P-functions to τ' and also "translate" the discontinuity of the functions $\overline{v}_i(x)$, $\overline{u}_i(x)$ from the point x_0 to the point X_n. This can be accomplished as follows.

Let, e.g., $X_n > x_0$. In this case we will change the function $\overline{v}_i$ in the interval $x_0 \leq x < X_n$ defining it there as the solution of the differential equation for $\overline{v}_i$ with the initial conditions $\overline{v}_i(x_0)$ and $\overline{v}_i'(x_0)$ equal to $\overline{v}_i(x_0 - 0)$ and $\overline{v}_i'(x_0 - 0)$, respectively. The new "changed" function $\overline{v}_i(x)$ (and also the function $\overline{u}_i(x)$ expressed through $\overline{v}_i(x)$ by the known formula) will now have its discontinuity, not at the point x_0, but at X_n.

Let us denote the nth partial sums of the series (2.4.7) and (2.4.8) by U_n and V_n:

$$U_n = \begin{cases} \displaystyle\sum_{k=0}^{n} \mu^k (\overline{u}_k(x) + Q_k u(\tau')), & 0 \leq x < X_n, \\ \displaystyle\sum_{k=0}^{n} \mu^k (\overline{u}_k(x) + P_k u(\tau')), & X_n \leq x \leq l, \end{cases}$$

and V_n has a similar representation. Let us set

$$\tilde{V}_n = \begin{cases} V_n + \mu^{n+1} Q_{n+1} v(\tau') + \mu^{n+2} Q_{n+2} v(\tau'), & 0 \le x < X_n, \\ V_n + \mu^{n+1} P_{n+1} v(\tau') + \mu^{n+2} P_{n+2} v(\tau'), & X_n \le x \le l. \end{cases}$$

From the method of construction of U_n and $\tilde{V}_n$, it follows that they satisfy (2.4.3) with an accuracy of the order μ^{n+1} at all the points of the interval $0 \le x \le l$, except at the point X_n:

$$\mu^2 U_n'' - f(U_n, \tilde{V}_n) = O(\mu^{n+1}), \qquad \tilde{V}_n'' - g(U_n, \tilde{V}_n) = O(\mu^{n+1}) \tag{2.4.29}$$

(if we take V_n instead of $\tilde{V}_n$ then the right-hand side in the second equality will be of the order $O(\mu^{n+1} + \mu^{n-1}\exp(-\kappa|\tau'|)))$.

At X_n, the functions U_n and $\tilde{V}_n$ have discontinuities such that

$$\begin{aligned} \sigma_0 &= U_n(X_n + 0) - U_n(X_n - 0) = O(\mu^{n+1}), \\ \sigma_1 &= U_n'(X_n + 0) - U_n'(X_n - 0) = O(\mu^n), \\ &\tilde{V}_n(X_n + 0) - \tilde{V}_n(X_n - 0) = O(\mu^{n+1}), \\ &\tilde{V}_n'(X_n + 0) - \tilde{V}_n'(X_n - 0) = O(\mu^{n+1}). \end{aligned} \tag{2.4.30}$$

Using the equalities (2.4.30) in (2.4.29), we obtain

$$\sigma_2 = U_n''(X_n + 0) - U_n''(X_n - 0) = \frac{1}{\mu^2}[f(U_n(X_n + 0), \tilde{V}_n(X_n + 0))$$

$$-f(U_n(X_n - 0), \tilde{V}_n(X_n - 0)) + O(\mu^{n+1})] = O(\mu^{n-1}),$$

and analogously, $\tilde{V}_n''(X_n + 0) - \tilde{V}_n''(X_n - 0) = O(\mu^{n+1})$.

Let us introduce the function

$$\tilde{u}(\tau') = \begin{cases} (A\tau'^2 + B\tau' + C)\exp(\tau'), & \tau' < 0, \\ 0\,, & \tau \ge 0. \end{cases}$$

Here we will choose the coefficients A, B, C in such a way that $\tilde{u}(-0) = \sigma_0$, $\tilde{u}'(-0) = \mu\sigma_1$, $\tilde{u}''(-0) = \mu^2\sigma_2$. It can be easily seen that we should take $C = \sigma_0$, $B = \mu\sigma_1 - C$, and $A = (\mu^2\sigma_2 - 2B - C)/2$. Clearly, A, B and C are $O(\mu^{n+1})$, so $\tilde{u} = O(\mu^{n+1})$. Further, $\tilde{u}' = O(\mu^{n+1})$ and $\tilde{u}'' = O(\mu^{n+1})$ for $\tau' \ne 0$. Let us set

$$\tilde{U}_n = \begin{cases} U_n + \tilde{u}, & 0 \le x \le X_n, \\ U_n, & X_n \le x \le l. \end{cases}$$

Then $\tilde{U}_n \in \mathbf{C}^2[0, l]$. "Improving" the function $\tilde{V}_n$ in a similar way, we obtain $\tilde{V}_n \in \mathbf{C}^2[0, l]$.

Now the main result can be formulated as follows.

Theorem 2.6 *If the conditions 1^0–5^0 are satisfied, the functions $\tilde{U}_n$ and $\tilde{V}_n$ satisfy system* (2.4.3) *in the entire interval $0 \le x \le l$ as well as the boundary conditions* (2.4.4) *with an accuracy of the order μ^{n+1}.*

2.4.4 The second type of "bursting" point

The second type of the points x_* is connected with equation $\overline{v}_0'(x_0) = 0$ (cf. (2.4.26)).

4_1^0. *Let the equation* $\overline{v}_0'(x_0) = 0$ *have a solution* $x_0 \in (0, l)$ *and suppose* $F(\overline{v}_0(x_0)) \neq 0$ *with* $\overline{v}_0''(x_0) \neq 0$.

In this case, it follows from the form of function $f_1(\tau)$ in (2.4.23) that the functions Q_1u and P_1u do not depend on x_1, and therefore after defining x_0, the functions Q_0u, P_0u, Q_1u, P_1u, Q_2v, P_2v, Q_3v, P_3v, $\overline{v}_1(x)$, $\overline{u}_1(x)$ will be completely defined.

In the next step we obtain the equation $Q_2u(0) = P_2u(0)$ for x_1, which can be transformed to

$$F(\overline{v}_0(x_0))\overline{v}_0''(x_0)x_1 = k_1, \tag{2.4.31}$$

where k_1 is known. By virtue of condition 4_1^0, x_1 is uniquely defined and, in turn, we determine Q_2u, P_2u, Q_4v, P_4v, $\overline{v}_2(x)$, $\overline{u}_2(x)$.

The procedure for defining later terms of the asymptotics is the same. For x_i $(i \geq 2)$, we obtain equations of the type (2.4.31). As earlier, functions $\overline{v}_i(x)$, $\overline{u}_i(x)$ have discontinuities at the point x_0 for $i \geq 2$. The functions $\tilde{U}_n$ and $\tilde{V}_n$, constructed in the same way as in § 2.4.3, satisfy the system (2.4.3) in the entire interval $0 \leq x \leq l$ and the boundary conditions (2.4.4) with an accuracy of the order μ^{n+1}.

2.4.5 The third type of "bursting" point

Consider now the case when condition $2'$ holds: problem (2.4.12), (2.4.13) has a solution $\overline{v}_0 = v_0 = \text{const}$ for $v_0 \in I$ (here I is the interval from condition 1^0). In this case, $\overline{u}_0 = \varphi(v_0) = \varphi_0 = \text{const}$, $f(\varphi_0, v_0) = 0$, and $g(\varphi_0, v_0) = 0$, i.e., the point (φ_0, v_0) is a stationary point of the system (2.4.3). We note that the solution $u = \varphi_0$, $v = v_0$ also satisfies conditions (2.4.4). However, we are interested in a different solution, which has the form (2.4.7), (2.4.8).

The functions Q_0u and P_0u are defined as in § 2.4.2, and since v_0 and φ_0 are now known quantities, the functions Q_0u and P_0u are completely determined, while the value of x_0 is still unknown. We note also that Q_0u and P_0u satisfy

$$P_0u(\tau) = Q_0u(-\tau), \qquad \tau \geq 0. \tag{2.4.32}$$

For $\overline{v}_1$ we have (2.4.20) with conditions (2.4.21), (2.4.22). Note that in this case $A = \overline{g}_v + \overline{g}_u\varphi'(v_0) = \text{const}$.

$3'$. *Let* $A = k^2 > 0$ (as we will see later in § 2.4.6, exactly this case holds for a well-known "Brusselator" model).

Solving the problem (2.4.20) – (2.4.22), we obtain

$$\overline{v}_1(x) = q \begin{cases} \cosh(k(x_0 - l))\cosh(kx), & 0 \leq x \leq x_0, \\ \cosh(kx_0)\cosh(k(x - l)), & x_0 \leq x \leq l, \end{cases} \tag{2.4.33}$$

where $q = -2Q_2'v(0)/k\sinh(kl)$ is known, and $\overline{u}_1$ is expressed (as before) by $\overline{u}_1 = \varphi'(v_0)\overline{v}_1$.

For $Q_1u(\tau)$ we have the problem (2.4.23), (2.4.24) with $f_1(\tau) = \overline{v}_1(x_0)h_1(\tau)$, for

$$h_1(\tau) = f_v(\varphi_0 + Q_0u(\tau), v_0) + f_u(\varphi_0 + Q_0u(\tau), v_0)\varphi'(v_0). \tag{2.4.34}$$

The existence of the solution $Q_1u(\tau)$ and its exponential decay as $\tau \to -\infty$ can be proved as in § 2.4.2, and for $Q_1u(0)$ we obtain (cf. (2.4.25))

$$Q_1u(0) = -\frac{\overline{v}_1(x_0)}{f(\psi_0, v_0)} \int_{-\infty}^{0} h_1(\tau)Q_0'u(\tau)d\tau. \tag{2.4.35}$$

An analogous expression can be obtained for $P_1u(0)$:

$$P_1u(0) = -\frac{\overline{v}_1(x_0)}{f(\psi_0, v_0)} \int_{\infty}^{0} H_1(\tau)P_0'u(\tau)d\tau,$$

where

$$H_1(\tau) = f_v(\varphi_0 + P_0u(\tau), v_0) + f_u(\varphi_0 + P_0u(\tau), v_0)\varphi'(v_0). \tag{2.4.36}$$

By virtue of (2.4.32) $H_1(\tau) = h_1(-\tau)$, $P_0'u(\tau) = -Q_0'u(-\tau)$ for $\tau \geq 0$. Hence, for $Q_1u(\tau)$ and $P_1u(\tau)$ a relation of the type (2.4.32) also holds; in particular, $Q_1u(0) = P_1u(0)$. Since $\overline{u}_1(x)$ is a continuous function, the equality $Q_1u(0) = P_1u(0)$ provides the continuity of u at the point x_* in the first-order approximation: $\overline{u}_1(x_0 - 0) + Q_1u(0) = \overline{u}_1(x_0 + 0) + P_1u(0)$. We note that the value of x_0 is still undefined. This distinguishes the case of condition $2'$ from that of condition 2^0, considered in §§ 2.4.2–2.4.4, where $Q_1u(0) = P_1u(0)$ provided the equation for x_0. Now the equation for x_0 appears only in the next step, during the construction of the terms of the second order.

For $\overline{u}_2(x)$ and $\overline{v}_2(x)$ we obtain the equations

$$\overline{u}_2 = \varphi'(v_0)\overline{v}_2 + B\overline{v}_1^2(x), \qquad \overline{v}_2'' = k^2\overline{v}_2 + D\overline{v}_1^2(x). \tag{2.4.37}$$

Here B and D are known, e.g., $B = -\frac{1}{2}\overline{f}_u^{-1}(\overline{f}_{uu}\varphi'^2(v_0) + 2\overline{f}_{uv}\varphi'(v_0) + \overline{f}_{vv})$. The boundary conditions for $\overline{v}_2$ have the form

$$\overline{v}_2'(0) = \overline{v}_2'(l) = 0. \tag{2.4.38}$$

From the condition that v and v' are continuous at the point x_*, we obtain the equalities

$$\overline{v}_2(x_0 + 0) - \overline{v}_2(x_0 - 0) = -[\overline{v}_1'(x_0 + 0) - \overline{v}_1'(x_0 - 0)]x_1, \tag{2.4.39}$$

$$\overline{v}_2'(x_0 + 0) - \overline{v}_2'(x_0 - 0) = Q_3'v(0) - P_3'v(0). \tag{2.4.40}$$

The functions Q_3v and P_3v entering (2.4.40) are defined analogously to Q_2v and P_2v, and they satisfy a relation of the type (2.4.32).

We will return to the problem for $\overline{v}_2(x)$ later (after determining x_0). Now let us consider the problem for $Q_2u(\tau)$:

$$\frac{d^2Q_2u}{d\tau^2} = f_u\, Q_2u + f_2(\tau), \qquad -\infty < \tau \leq 0,$$

$$Q_2'u(0) = -\overline{u}_1'(x_0 - 0), \qquad Q_2u(-\infty) = 0.$$

The function f_2 can be written as $f_2(\tau) = h_1(\tau)\tau\overline{v}_1'(x_0 - 0) + \tilde{f}_2(\tau)$, where $h_1(\tau)$ is given by (2.4.34), and

$$\tilde{f}_2(\tau) = h_1(\tau)[\overline{v}_2(x_0 - 0) + \overline{v}_1'(x_0 - 0)x_1] + f_v\, Q_2 v(\tau) + \frac{1}{2} f_{uu}[\overline{u}_1(x_0) + Q_1 u(\tau)]^2$$

$$+ f_{uv}[\overline{u}_1(x_0) + Q_1 u(\tau)]\overline{v}_1(x_0) + \frac{1}{2} f_{vv}\,\overline{v}_1^2(x_0)$$

(here all the partial derivatives of the function f are evaluated at the point $(\varphi_0 + Q_0 u(\tau), v_0)$).

As in (2.4.35), we obtain

$$Q_2 u(0) = -\frac{1}{f(\psi_0, v_0)} \int_{-\infty}^{0} f_2(\tau) Q_0' u(\tau)\, d\tau.$$

For $P_2 u(0)$ we obtain an expression of the same type:

$$P_2 u(0) = -\frac{1}{f(\psi_0, v_0)} \int_{\infty}^{0} F_2(\tau) P_0' u(\tau)\, d\tau,$$

where $F_2(\tau) = H_1(\tau)\tau\overline{v}_1'(x_0 + 0) + \tilde{F}_2(\tau)$; $H_1(\tau)$ is defined by (2.4.36), and $\tilde{F}_2(\tau)$ can be obtained from the expression for $\tilde{f}_2(\tau)$ when we change Q-functions to P-functions, and $x_0 - 0$ to $x_0 + 0$. Hence (taking into account (2.4.39)), we have $\tilde{F}_2(\tau) = \tilde{f}_2(-\tau)$ for $\tau \geq 0$.

Splitting $Q_2 u(0)$ into two terms corresponding to the two summands in the expression for $f_2(\tau)$, and making the change of variable of integration $\tau = \int_{\psi_0 - \varphi_0}^{Q_0 u} ds/\Phi(s)$ in the first term, we obtain

$$Q_2 u(0) = -\frac{1}{f(\psi_0, v_0)} [F(v_0)\overline{v}_1'(x_0 - 0) + \alpha].$$

Here $F(v_0)$ is defined by (2.4.27) and $\alpha = \int_{-\infty}^{0} \tilde{f}_2(\tau) Q_0' u(\tau)\, d\tau$.

In a similar way, we obtain

$$P_2 u(0) = -\frac{1}{f(\psi_0, v_0)} [-F(v_0)\overline{v}_1'(x_0 + 0) + \alpha].$$

The condition that function u be continuous at the point x_* leads in the second-order approximation to the equality

$$\overline{u}_2(x_0 - 0) + \overline{u}_1'(x_0 - 0)x_1 + Q_2 u(0) = \overline{u}_2(x_0 + 0) + \overline{u}_1'(x_0 + 0)x_1 + P_2 u(0).$$

From here, taking into account the relations $\overline{u}_1'(x) = \varphi'(v_0)\overline{v}_1'(x)$, $\overline{u}_2(x) = \varphi'(v_0)\overline{v}_2(x) + B\overline{v}_1^2(x)$, and continuity of the function $\overline{v}_1(x)$ at x_0, we obtain

$$\varphi'(v_0)[\overline{v}_2(x_0 + 0) - \overline{v}_2(x_0 - 0) + (\overline{v}_1'(x_0 + 0) - \overline{v}_1'(x_0 - 0))x_1] + P_2 u(0) - Q_2 u(0) = 0.$$

However, the expression in the square brackets is zero by virtue of (2.4.39). Thus, we arrive at the equation

$$P_2 u(0) - Q_2 u(0) = 0.$$

Substituting the expressions for $P_2u(0)$ and $Q_2u(0)$, we obtain

$$\frac{1}{f(\psi_0, v_0)} F(v_0)[\overline{v}_1'(x_0+0) + \overline{v}_1'(x_0-0)] = 0.$$

Recall that the first type of "bursting" point was defined in § 2.4.2 through the equation $F(v_0) = 0$. Now we will assume that the following condition is satisfied:

4′. $F(v_0) \neq 0$.

Then we obtain the equation

$$\overline{v}_1'(x_0+0) + \overline{v}_1'(x_0-0) = 0$$

for x_0. Using the expression (2.4.33) for $\overline{v}_1(x)$, we arrive at

$$\sinh(k(2x_0 - l)) = 0$$

and, therefore, $x_0 = l/2$.

We note that $\overline{v}_1(x)$ is the even function with respect to the point $x_0 = l/2$ (nonsmooth at this point), while P_2u and Q_2u satisfy a relation of the type (2.4.32).

Coefficient x_1 of the series (2.4.8) is defined in the next step. The condition that u be continuous at the point x_* leads in the third-order approximation to the equation $P_3u(0) - Q_3u(0) = 0$. As can be shown, this equation might be transformed to

$$\overline{v}_2'\left(\frac{l}{2}+0\right) + \overline{v}_2'\left(\frac{l}{2}-0\right) = \delta x_1, \tag{2.4.41}$$

where $\delta = -2k^2\overline{v}_1(x_0)$.

Let us now return to (2.4.37) for $\overline{v}_2(x)$ with the additional conditions (2.4.38)–(2.4.40), in which we set $x_0 = l/2$. Let us add to these conditions (2.4.41). This "extra" condition can be satisfied by appropriate choice of x_1. Let us perform the corresponding calculations. We set $\overline{v}_2 = w + \tilde{v}_2$, where w is the solution of the problem

$$w'' = k^2 w + D\overline{v}_1^2(x), \qquad w'(0) = w'(l) = 0.$$

Evidently, $w(x)$ is an even function with respect to the point $l/2$, and since $w(x)$ has a derivative at the point $l/2$, $w'(l/2) = 0$. Hence, for $\tilde{v}_2(x)$ we have the homogeneous equation $\tilde{v}_2'' = k^2\tilde{v}_2$ with the additional conditions

$$\tilde{v}_2'(0) = \tilde{v}_2'(l) = 0, \qquad \tilde{v}_2\left(\frac{l}{2}+0\right) - \tilde{v}_2\left(\frac{l}{2}-0\right) = \beta x_1,$$

$$\tilde{v}_2'\left(\frac{l}{2}+0\right) - \tilde{v}_2'\left(\frac{l}{2}-0\right) = \gamma, \qquad \tilde{v}_2'\left(\frac{l}{2}+0\right) + \tilde{v}_2'\left(\frac{l}{2}-0\right) = \delta x_1.$$

Here

$$\beta = \overline{v}_1'\left(\frac{l}{2}-0\right) - \overline{v}_1'\left(\frac{l}{2}+0\right),$$

and $\gamma = Q_3'v(0) - P_3'v(0)$. Some simple calculations lead to the equation

$$\left(\beta + \frac{\delta}{k}\coth\frac{kl}{2}\right) x_1 = 0$$

for x_1.

$5'$. *Suppose* $\beta + (\delta/k)\coth(kl/2) \neq 0$.

Then $x_1 = 0$. In this case, then, the function $\overline{v}_2(x)$ is even with respect to the point $l/2$.

Analogous situation takes place in all the higher-order approximations: $x_i = 0$, $i = 2, 3, \ldots$. Therefore

$$x_* = x_0 = l/2.$$

Functions $\overline{v}_i(x)$ and hence $\overline{u}_i(x)$ are even with respect to the point $x_* = l/2$, and the functions $Q_i u$, $P_i u$, and $Q_i v$, $P_i v$ satisfy relations of the type (2.4.32) and possess exponential estimates of the types (2.4.17), (2.4.19).

Let us introduce the notation

$$\tilde{U}_n = \begin{cases} \sum_{k=0}^{n} \mu^k(\overline{u}_k(x) + Q_k u(\tau)) + \mu^{n+1} Q_{n+1} u(\tau), & 0 \le x \le \dfrac{l}{2}, \\[2ex] \sum_{k=0}^{n} \mu^k(\overline{u}_k(x) + P_k u(\tau)) + \mu^{n+1} P_{n+1} u(\tau), & \dfrac{l}{2} \le x \le l; \end{cases}$$

$$\tilde{V}_{n+1} = \begin{cases} \sum_{k=0}^{n+1} \mu^k(\overline{v}_k(x) + Q_k v(\tau)) + \mu^{n+2} Q_{n+2} v(\tau), & 0 \le x \le \dfrac{l}{2}, \\[2ex] \sum_{k=0}^{n+1} \mu^k(\overline{v}_k(x) + P_k v(\tau)) + \mu^{n+2} P_{n+2} v(\tau), & \dfrac{l}{2} \le x \le l. \end{cases}$$

Theorem 2.7 *If the conditions* 1^0, $2'$–$5'$ *are satisfied, then the functions* $\tilde{U}_n$ *and* $\tilde{V}_{n+1}$ *are continuous in the interval* $[0, l]$ *together with their derivatives (up to second order inclusively) and satisfy* (2.4.3) *with an accuracy of the order* μ^{n+1}, *i.e.*

$$\mu^2 \tilde{U}_n'' - f(\tilde{U}_n, \tilde{V}_{n+1}) = O(\mu^{n+1}), \qquad \tilde{V}_{n+1}'' - g(\tilde{U}_n, \tilde{V}_{n+1}) = O(\mu^{n+1});$$

they also satisfy the boundary conditions (2.4.4) *with an accuracy of the order* μ^N, *where* N *is any natural number.*

The statement of the theorem follows directly from the method of construction of $\tilde{U}_n$ and $\tilde{V}_{n+1}$. We should also note that inclusion of the terms $\mu^{n+1} Q_{n+1} u(\tau)$ and $\mu^{n+1} P_{n+1} u(\tau)$ into expression for $\tilde{U}_n$ provides the continuity of $\tilde{U}_n'$ at the point $l/2$ (for $n = 0$ these terms may not be included since $Q_1' u(0) = P_1' u(0)$). The terms $\mu^i Q_i v(\tau)$ and $\mu^i P_i v(\tau)$ ($i = n+1$, $n+2$) are included in $\tilde{V}_{n+1}$ in order that the second equation of the system (2.4.3) be satisfied with accuracy of order μ^{n+1}. Without them the accuracy would be of the order $\mu^{n+1} + \mu^{n-1}\exp(-\kappa|\tau|)$. Along with these terms, $\mu^{n+1}\overline{v}_{n+1}(x)$ is also added to provide the continuity of $\tilde{V}_{n+1}'$ at the point $l/2$. The boundary conditions (2.4.4) are satisfied with an arbitrary accuracy (of the order μ^N) since $\overline{v}_i'(0) = \overline{v}_i'(l) = \overline{u}_i'(0) = \overline{u}_i'(l) = 0$, and Q- and P-functions satisfy the exponential decay estimates. If we multiply the Q- and P-functions by smoothing functions which are equal to zero outside of some vicinity of the point $x_* = l/2$, then the resulting $\tilde{U}_n$ and $\tilde{V}_{n+1}$ will satisfy the boundary conditions exactly.

We may note now that since $\overline{u}_i$ and $\overline{v}_i$ are even functions with respect to $x_* = l/2$, and the functions Q_i and P_i satisfy the relations of the type (2.4.32), the asymptotic expansions (2.4.7), (2.4.8) are also even with respect to $x_* = l/2$. Hence, we are coming to an important conclusion. Let us divide the interval $[0, l]$ into an arbitrary number m of subintervals of equal length. In each of these subintervals we can construct (2.4.7), (2.4.8) with a spike in the middle of subinterval. Since these asymptotics are even with respect to the middle point of each such subinterval and their derivatives at the ends of subinterval are equal to zero, the asymptotic expansions constructed in different subintervals are smoothly matched at the boundary points of neighboring subintervals. Thus, one may construct the asymptotic solution of (2.4.3), (2.4.4) with an arbitrary number of spikes that are distributed uniformly in the interval $[0, l]$. In practice, however, we need to take into account the following fact. The larger the number m of points where spikes take place, the smaller μ should be taken for the asymptotics to be effective, since for the asymptotic algorithm to work successfully in the subinterval of the length l/m, μ should be much smaller than this length: $\mu \ll l/m$.

2.4.6 "Brusselator" model

Consider the "Brusselator" model, which describes a chemical reaction with two intermediate substances (cf. Nicolis and Prigogine [110]). The nondimensional concentrations u and v of these substances satisfy the system

$$\frac{\partial u}{\partial t} = D_1 \triangle u + A - (B+1)u + u^2 v, \qquad \frac{\partial v}{\partial t} = D_2 \triangle v + Bu - u^2 v.$$

Here D_1, D_2, A and B are positive constants, and $\triangle$ is the Laplace operator. If $D_1 \ll D_2$, then setting $D_1 = \mu^2$, $D_2 = 1$, we obtain the following system of equations for the stationary case (in the one-dimensional problem, where $\triangle = d^2/dx^2$):

$$\mu^2 \frac{d^2 u}{dx^2} = -A + (B+1)u - u^2 v, \qquad \frac{d^2 v}{dx^2} = -Bu + u^2 v. \tag{2.4.42}$$

This system is of the type (2.4.3), where $f(u,v) = -A + (B+1)u - u^2 v$ and $g(u,v) = -Bu + u^2 v$. Let us investigate this system in the interval $[0, l]$ with the boundary conditions (2.4.4).

The equation $f(u,v) = 0$, i.e., $-A + (B+1)u - u^2 v = 0$, has two positive solutions with respect to u for $v \in I = (0 < v < (B+1)^2/4A)$ (the conditions $u > 0$, $v > 0$ follow from the physical meaning of the variables u and v):

$$u = \frac{B+1-\sqrt{(B+1)^2 - 4Av}}{2v} \equiv \varphi(v), \qquad \text{and}$$

$$u = \frac{B+1+\sqrt{(B+1)^2 - 4Av}}{2v} \equiv \chi(v).$$

Evidently, $\varphi(v) < \chi(v)$, and

$$f_u(\varphi(v), v) = \sqrt{(B+1)^2 - 4Av} > 0, \qquad f_u(\chi(v), v) = -\sqrt{(B+1)^2 - 4Av} < 0.$$

The expression for the function $\psi(v)$ defined by the formula (2.4.11) has the form

$$\psi(v) = \frac{B+1+2\sqrt{(B+1)^2 - 4Av}}{2v}.$$

Thus, the condition 1^0 is satisfied.

The boundary value problem (2.4.12), (2.4.13) for $\overline{v}_0$ can be written in the form

$$\overline{v}_0'' = \frac{-2A\overline{v}_0 + B + 1 - \sqrt{(B+1)^2 - 4A\overline{v}_0}}{2\overline{v}_0} \tag{2.4.43}$$

$$\overline{v}_0'(0) = \overline{v}_0'(l) = 0.$$

It can be easily seen that if $B > 1$ the right-hand side of the equation above is strictly negative for all $\overline{v}_0 < (B+1)^2/4A$, and therefore the boundary value problem (2.4.43) cannot have a solution. If $0 < B \leq 1$, then this problem has the unique solution

$$\overline{v}_0 = B/A = \text{const}.$$

Thus, the problem (2.4.43) does not have solutions other than the constant one, and therefore the system (2.4.42) cannot have "bursting" points of the first or second type.

The solution $\overline{v}_0 = B/A$ for $0 < B < 1$ satisfies the condition $\overline{v}_0 \in I$. Thus, for $0 < B < 1$ we obtain

$$\overline{v}_0 = B/A = v_0, \qquad \overline{u}_0 = \varphi(v_0) = A = \varphi_0,$$

i.e. condition $2'$ is satisfied. It can be easily verified that condition $4'$ is also satisfied. Indeed, the function $F(\overline{v}_0)$ is defined by the formula (2.4.27). In our case

$$f_v(u, v_0) = -u^2 \leq 0.$$

Hence, the function under the integral in the expression (2.4.27) for $F(v_0)$ does not change sign, so $F(v_0) \neq 0$.

Equation (2.4.16) for Q_0u is

$$\begin{aligned} Q_0'u &= \sqrt{2\int_A^{A+Q_0u}\left[-A + (B+1)\xi - \frac{B}{A}\xi^2\right] d\xi} \\ &= Q_0u\sqrt{(1-B) - \frac{2}{3}\frac{B}{A}Q_0u}. \end{aligned}$$

Solving this equation with the initial condition

$$Q_0u(0) = \psi_0 - \varphi_0 = 3(1-B)A/(2B),$$

we obtain

$$Q_0u(\tau) = \frac{6\alpha^2 A}{B}\frac{\exp(\alpha\tau)}{(1+\exp(\alpha\tau))^2}, \qquad \tau \leq 0,$$

where $\alpha = \sqrt{1-B} > 0$.

The expression for $P_0u(\tau)$ is obtained from that for $Q_0u(\tau)$ by changing τ to $-\tau$. Thus, the terms of the asymptotic expansion to the zeroth order are found explicitly.

For the terms of the asymptotic approximation to the first order, we have

$$\overline{u}_1 = \frac{A^2}{\alpha^2}\overline{v}_1, \qquad \overline{v}_1'' = \frac{A^2}{\alpha^2}\overline{v}_1,$$

and therefore $k^2 = A^2/\alpha^2 > 0$, i.e., condition $3'$ is satisfied.

Computation of $Q_2'v(0)$ gives $Q_2'v(0) = 3\alpha A/B$. After that we obtain $\overline{v}_1(x)$ explicitly by formula (2.4.33) (with $k = A/\alpha$):

$$\overline{v}_1(x) = -\frac{3\alpha^2}{B\sinh(kl/2)}\begin{cases}\cosh(kx), & 0 \le x \le l/2,\\ \cosh(k(x-l)), & l/2 \le x \le l.\end{cases}$$

Let us check that condition $5'$ is satisfied. The computation of the quantities

$$\beta = \overline{v}_1'\left(\frac{l}{2}-0\right) - \overline{v}_1'\left(\frac{l}{2}+0\right) \quad \text{and} \quad \delta = -2k^2\overline{v}_1\left(\frac{l}{2}\right)$$

yields

$$\beta = \frac{-6\alpha A}{B} \quad \text{and} \quad \delta = \frac{6A^2\coth(kl/2)}{B}.$$

Consequently,

$$\beta + \frac{\delta}{k}\coth\frac{kl}{2} = \frac{6\alpha A}{B}\left(\coth^2\frac{kl}{2} - 1\right) \neq 0,$$

and $5'$ holds.

Since for the "Brusselator" model, conditions 1^0, $2'$–$5'$ all hold, one can construct the asymptotic solution with an arbitrary number of "bursting" points that are distributed uniformly throughout the interval $[0, l]$.

Exercises

1. For the problem

$$\mu\frac{d^2y}{dx^2} = -y(y+1), \qquad \frac{dy}{dx}(-1) = \frac{dy}{dx}(1) = 0,$$

construct the zeroth-order approximation for the solution that has a spike at $x = 0$.

2. Consider a problem

$$\mu^2\frac{d^2u}{dx^2} = -u(v-u), \qquad \frac{d^2v}{dx^2} = u + v - 1,$$

$$u'(0) = u'(1) = v'(0) = v'(1) = 0.$$

(a) Check that for this problem the conditions analogous to ones for the "Brusselator" model are satisfied, and thus, the solutions with arbitrary number of "bursting" points within the interval $[0, 1]$ are possible.

(b) Construct the zeroth-order approximation for the solution which has a spike at $x = 0.5$.

2.4.7 On stability of contrast structures

The contrast structures investigated in the previous subsections were *stationary* solutions of the corresponding system of parabolic equations. Similarly, the contrast structures obtained in Butuzov and Vasil'eva [38] are stationary solutions of the parabolic equation (2.4.2). The following question arises: Since the actual solution $u(x,t,\mu)$ of (2.4.2) can approach only *stable* stationary solutions as $t \to \infty$, which types of contrast structures are stable?

We are interested in stability of the contrast structure $z(x,\mu)$ in the sense of Lyapunov, i.e. whether for any $\varepsilon > 0$ there exists $\delta(\varepsilon)$ such that if

$$\sup_x |u(x,0,\mu) - z(x,\mu)| < \delta,$$

the inequality

$$\sup_{x,t} |u(x,t,\mu) - z(x,\mu)| < \varepsilon$$

holds for $t \gg 0$. Some general principles for studying stability properties of stationary solutions of parabolic equations are discussed, e.g., in Henry [61]. Using the ideas of the first method of Lyapunov, the investigation of stability in the case of one parabolic equation (2.4.2) can be reduced to studying the spectrum of the following Sturm–Liouville problem:

$$\mu^2\psi'' = [F_z(z(x,\mu),x) + \lambda]\psi, \qquad \psi(0) = \psi(1) = 0.$$

Here, boundary conditions of Dirichlet type are taken. Stationary solution $z(x,\mu)$ will be stable if for all the eigenvalues $\lambda_1 > \lambda_2 > \cdots > \lambda_k > \cdots$ of the Sturm–Liouville problem the inequality $\lambda_k < 0$, $k = 1, 2, \ldots$, holds. If at least $\lambda_1 > 0$, then the solution is unstable.

Along these lines, we can come to the following conclusions.

1. The spike-type contrast structures (Fig. 2.11) are *not stable* since for them $\lambda_1 = O(1) > 0$. The addition of the second "slow" equation, as in (2.4.3), cannot change the sign of λ_1.

2. Contrast structures of threshold-type (see Fig. 2.7) have $\lambda_1 = O(\mu)$, i.e., $\lambda_1 = \mu\lambda_{11} + \mu^2\lambda_{12} + \cdots$, and for λ_{11} we have

$$\lambda_{11} = \frac{\displaystyle\int_{\varphi_1(x_{00})}^{\varphi_3(x_{00})} F_x(z, x_{00})dz}{\displaystyle\int_{\varphi_1(x_{00})}^{\varphi_3(x_{00})} \left(2\int_{\varphi_1(x_{00})}^{z} F(\xi, x_{00})d\xi\right)^{\frac{1}{2}} dz}, \tag{2.4.44}$$

where x_{00} is the leading term in the expansion for the transition point coordinate $x_0 = x_{00} + \mu x_{01} + \ldots$ (x_0 is the abscissa of the point where $z(x,\mu)$ intersects the curve $\varphi_2(x)$). In Fig. 2.7 the point x_0 is denoted by T. Under condition 5^0 (see p. 58), the value of λ_{11} differs from 0 (note that the numerator in (2.4.44) is equal, up to notations, to the expression in 5^0, which is nonzero). The sign of λ_{11} is defined by the sign of the numerator and may be either positive or negative. The presence of the second "slow" equation might change the sign of λ_1.

These and some other results on stability of contrast structures are discussed in Vasil'eva [144]–[146]. Various questions related to stability of contrast structures were also addressed by a number of authors (see, e.g., Hale and Sakamoto [58] and the references therein).

Chapter 3

Singularly Perturbed Partial Differential Equations

3.1 The method of Vishik–Lyusternik

3.1.1 The statement of the problem and an algorithm for constructing the asymptotic solution

A quite general approach for solving linear partial differential equations with small parameters multiplying the highest derivatives was proposed in the fundamental work by Vishik and Lyusternik [163]. We will illustrate the idea of this approach (known in the literature as the method of Vishik–Lyusternik) on a simple example of an elliptic equation in a bounded domain with smooth boundary.

Consider the equation

$$L_\varepsilon u \equiv \varepsilon^2 \bigtriangleup u - k^2(x,y)u = f(x,y,\varepsilon), \qquad (x,y) \in \Omega \tag{3.1.1}$$

subject to the Dirichlet boundary condition

$$u|_{\partial\Omega} = 0. \tag{3.1.2}$$

Here $\varepsilon > 0$ is a small parameter, $\bigtriangleup = \partial^2/\partial x^2 + \partial^2/\partial y^2$ is the Laplace operator, Ω is a bounded planar domain. The boundary of the domain is assumed to be a sufficiently smooth curve $\partial\Omega$. Functions $k(x,y)$ and $f(x,y,\varepsilon)$ are also sufficiently smooth and $k(x,y) > 0$ in $\overline{\Omega} = \Omega + \partial\Omega$. The smoothness of k, f, $\partial\Omega$ and the negative sign of the coefficient $-k^2(x,y)$ guarantee the existence of unique classical solution $u(x,y,\varepsilon)$ of the problem (3.1.1), (3.1.2). Our goal is to construct asymptotic expansion (uniform in $\overline{\Omega}$) of this solution with respect to a parameter ε.

The asymptotic expansion of the solution of (3.1.1), (3.1.2) consists of regular and boundary layer parts: $u = \overline{u} + \Pi$. (Recall the same structure of the asymptotic solution of the problems discussed in Chapter 2.) To construct the boundary layer functions we need to use the new (local) coordinates in the vicinity of the boundary $\partial\Omega$. Let the equations describing $\partial\Omega$ have the form

$$x = \varphi(l), \quad y = \psi(l), \quad 0 \le l \le l_0\ ,$$

where l is a parameter. When parameter l increases from 0 to l_0 the point $(\varphi(l),\ \psi(l))$ moves along $\partial\Omega$ in such a way that the domain Ω remains to the left. In a δ-vicinity

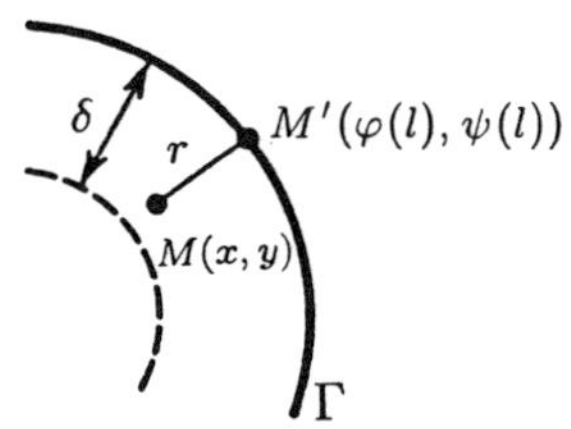

Figure 3.1: *New variables* (r, l) *in the vicinity of the boundary.*

of the curve $\partial\Omega$ we introduce the new local variables as follows. Let the point $M(x, y)$ belonging to $\overline{\Omega}$ lie on the normal to $\partial\Omega$ passing through the point $M'(\varphi(l),\ \psi(l))$ (see Fig. 3.1). We denote the distance MM' by r. The pair of numbers (r, l) will be the new coordinates of the point M. If the δ-vicinity $\{(0 \le r \le \delta) \times (0 \le l \le l_0)\}$ is sufficiently narrow, i.e. δ is small enough, the normals to $\partial\Omega$ drawn through various points of $\partial\Omega$ do not intersect within this neighborhood. Hence, in this region there exists a one-to- one correspondence between the two coordinates (x, y) and (r, l), expressed by relations

$$x = \varphi(l) - r\frac{\psi'(l)}{\sqrt{\varphi'^2(l) + \psi'^2(l)}}, \qquad y = \psi(l) + r\frac{\varphi'(l)}{\sqrt{\varphi'^2(l) + \psi'^2(l)}}.$$

In a δ-neighborhood of $\partial\Omega$, we have the following expression for the operator L_ε in terms of the new variables (r, l):

$$L_\varepsilon = \varepsilon^2\left(\frac{\partial^2}{\partial r^2} + \alpha(r,l)\frac{\partial^2}{\partial l^2} + \beta(r,l)\frac{\partial}{\partial r} + \gamma(r,l)\frac{\partial}{\partial l}\right) - k^2(x(r,l), y(r,l)),$$

where

$$\alpha(r,l) = \left(\frac{\partial l}{\partial x}\right)^2 + \left(\frac{\partial l}{\partial y}\right)^2, \quad \beta(r,l) = \frac{\partial^2 r}{\partial x^2} + \frac{\partial^2 r}{\partial y^2}, \quad \text{and} \quad \gamma(r,l) = \frac{\partial^2 l}{\partial x^2} + \frac{\partial^2 l}{\partial y^2}.$$

Let us now stretch the variable r: $r = \varepsilon\rho$. Expanding coefficients in the expression for L_ε in a power series in ε, we obtain

$$L_\varepsilon = \left(\frac{\partial^2}{\partial \rho^2} - k^2(l)\right) + \sum_{j=1}^{\infty} \varepsilon^j L_j\ ,$$

where $k^2(l) = k^2(\varphi(l), \psi(l))$, and L_j are linear differential operators containing the differentiations $\partial/\partial\rho$, $\partial/\partial l$, $\partial^2/\partial l^2$.

We seek the asymptotic solution of (3.1.1),(3.1.2) in the form

$$u(x, y, \varepsilon) = \overline{u} + \Pi = \sum_{i=0}^{\infty} \varepsilon^i(\overline{u}_i(x, y) + \Pi_i(\rho, l)). \tag{3.1.3}$$

Substituting (3.1.3) into (3.1.1), (3.1.2), and expanding $f(x, y, \varepsilon)$ into a power series in ε, we arrive at equalities

$$\{\varepsilon^2 \triangle - k^2(x,y)\} \sum_{i=0}^{\infty} \varepsilon^i \overline{u}_i(x,y)$$

$$+ \left\{ \left(\frac{\partial^2}{\partial \rho^2} - k^2(l) \right) + \sum_{j=1}^{\infty} \varepsilon^j L_j \right\} \sum_{i=0}^{\infty} \varepsilon^i \Pi_i(\rho, l) = \sum_{i=0}^{\infty} \varepsilon^i f_i(x,y), \tag{3.1.4}$$

$$\sum_{i=0}^{\infty} \varepsilon^i (\overline{u}_i(l) + \Pi_i(0,l)) = 0 \,. \tag{3.1.5}$$

Here $\overline{u}_i(l) = \overline{u}_i(\varphi(l), \psi(l))$. Equating coefficients of like powers of ε in (3.1.4) separately for terms depending on (x, y) and on (ρ, l), we obtain equations for the coefficients of the asymptotic series (3.1.3).

For the regular terms of the asymptotic expansion, we have

$$-k^2(x,y)\overline{u}_0 = f_0(x,y), \qquad -k^2(x,y)\overline{u}_1 = f_1(x,y),$$

$$\Delta \overline{u}_{i-2} - k^2(x,y)\overline{u}_i = f_i(x,y), \qquad i = 2, 3, \dots \,.$$

Hence,

$$\overline{u}_0 = -f_0(x,y)/k^2(x,y), \qquad \overline{u}_1 = -f_1(x,y)/k^2(x,y),$$

$$\overline{u}_i = -(f_i(x,y) - \Delta \overline{u}_{i-2})/k^2(x,y), \qquad i = 2, 3, \dots \,.$$

The regular part of the asymptotic expansion does not in general satisfy the boundary condition (3.1.2). In the terminology of the paper of Vishik and Lyusternik [163], the regular terms introduce a *discrepancy* in the boundary conditions. The purpose of the boundary functions $\Pi_k(\rho, l)$ is to compensate for this discrepancy. The equality (3.1.5) shows that the boundary functions together with the regular terms must satisfy the boundary condition (3.1.2).

For the boundary functions we have the following ordinary differential equations (l enters as a parameter)

$$\frac{\partial^2 \Pi_i}{\partial \rho^2} - k^2(l)\Pi_i = \pi_i(\rho, l), \qquad \rho \geq 0, \tag{3.1.6}$$

where $\pi_0 = 0$ and $\pi_i(\rho, l) = -\sum_{j=1}^{i} L_j \Pi_{i-j}(\rho, l)$, $i \geq 1$. Therefore the right-hand sides of the equations for Π_i are expressed recursively through the Π_j with $j < i$. The boundary conditions at $\rho = 0$ follow from (3.1.5):

$$\Pi_i(0, l) = -\overline{u}_i(l). \tag{3.1.7}$$

In addition, we require that the Π-functions decay as $\rho \to \infty$:

$$\Pi_i(\infty, l) = 0. \tag{3.1.8}$$

The explicit solutions to the problems (3.1.6) – (3.1.8) can be found successively. For example,

$$\Pi_0(\rho, l) = -\overline{u}_0(l)\exp(-k(l)\rho),$$

and the other functions $\Pi_i(\rho, l)$ are products of polynomials in ρ and $\exp(-k(l)\rho)$. Consequently, all the Π-functions possess the exponential estimate

$$|\Pi_i(\rho, l)| \leq c\exp(-\kappa\rho),$$

where, as before, we denote by c and κ appropriate positive constants that, in general, are different for different Π-functions.

Note that although formally Π-functions are defined for $\rho \geq 0$, they in fact make sense only for $0 \leq \rho \leq \delta/\varepsilon$, i.e., in a δ-vicinity of $\partial\Omega$ where we introduced the local coordinates (r, l). The standard method of smooth continuation of these functions onto the entire domain $\overline{\Omega}$ can be described as follows. We multiply the constructed Π_k's by a smoothing (infinitely differentiable) function $\Psi(r)$ such that $\Psi(r) = 1$ when $0 \leq r \leq \delta/3$, $0 \leq \Psi(r) \leq 1$ when $\delta/3 \leq r \leq 2\delta/3$, and $\Psi(r) = 0$ when $r \geq 2\delta/3$. Now we consider $\Pi_k(\rho, l)\Psi(r)$ as our new Π-functions. These new boundary functions are uniquely defined in the entire domain $\overline{\Omega}$. They coincide with the original Π_k-functions for $0 \leq r \leq \delta/3$ and they are asymptotically smaller than any power of ε for $r \geq \delta/3$:

$$|\Pi_i(\rho, l)| \leq c\exp\left(-\frac{\kappa\delta}{3\varepsilon}\right) = o(\varepsilon^N)$$

for any $N > 0$ when $r \geq \delta/3$. Thus, the procedure introduced above changes each of the Π-functions by an amount that is less than any power of ε. In the problems considered below, without mentioning this fact specially, we will presume (where necessary) that the boundary functions are defined in the entire domain with the help of the continuation procedure, similar to that which has been just introduced.

Definition 3.1 *We will say that the Π-functions are defined by the operator $L_\Pi = \partial^2/\partial\rho^2 - k^2(l)$ and the boundary conditions* (3.1.7),(3.1.8). *We will call L_Π a boundary layer operator in the vicinity of $\partial\Omega$.*

3.1.2 Estimation of the remainder term

Let $U_n(x, y, \varepsilon)$ denote the nth partial sum of the series (3.1.3):

$$U_n(x, y, \varepsilon) = \sum_{i=0}^{n} \varepsilon^i(\overline{u}_i(x, y) + \Pi_i(\rho, l)).$$

Theorem 3.1 *The series* (3.1.3)*is the asymptotic power series for the solution $u(x, y, \varepsilon)$ of the problem* (3.1.1), (3.1.2)*in the closed domain $\overline{\Omega}$ as $\varepsilon \to 0$, i.e., the following estimate holds:*

$$\max_{\overline{\Omega}} |u(x, y, \varepsilon) - U_n(x, y, \varepsilon)| = O(\varepsilon^{n+1}).$$

Proof. We set $w(x, y, \varepsilon) = u(x, y, \varepsilon) - U_n(x, y, \varepsilon)$. Substituting $u = U_n + w$ into (3.1.1), (3.1.2), we obtain the following boundary value problem for the remainder term w:

$$\varepsilon^2 \Delta w - k^2(x, y)w = h(x, y, \varepsilon), \qquad (x, y) \in \Omega,$$

$$w|_{\partial\Omega} = 0,$$

where $h(x,y,\varepsilon) = f(x,y,\varepsilon) - \varepsilon^2 \triangle U_n + k^2(x,y)U_n$. It follows from the method of construction of U_n that $h(x,y,\varepsilon) = O(\varepsilon^{n+1})$ in Ω.

Suppose that (x_0, y_0) is the point of Ω where $w(x,y,\varepsilon)$ attains its maximal positive value. Then $\Delta w \leq 0$ at this point, and the equation for w implies that

$$w(x_0,y_0,\varepsilon) \leq \frac{|h(x_0,y_0,\varepsilon)|}{k^2(x_0,y_0)} = O(\varepsilon^{n+1}).$$

In a similar way, if (x_1,y_1) is the point of Ω where $w(x,y,\varepsilon)$ attains its minimal negative value, then

$$|w(x_1,y_1,\varepsilon)| \leq \frac{|h(x_1,y_1,\varepsilon)|}{k^2(x_1,y_1)} = O(\varepsilon^{n+1}).$$

It follows from the above inequalities that

$$\max_{\overline{\Omega}} |w(x,y,\varepsilon)| = \max_{\overline{\Omega}} |u - U_n| = O(\varepsilon^{n+1}).$$

This completes the proof.

Example. Consider a problem

$$\varepsilon^2 \triangle u - (1+x^2)u = -x, \qquad \Omega = \{(x,y): \ x^2+y^2 < 1\}, \qquad u|_{\partial\Omega} = 0.$$

It can be easily verified that

$$\overline{u}_0(x,y) = \frac{x}{1+x^2};$$

equations of $\partial\Omega$ have the form:

$$x = \cos l, \qquad y = \sin l \quad 0 \leq l \leq 2\pi;$$

and

$$\Pi_0(\rho, l) = -\frac{\cos l}{1+\cos^2 l}\exp(-\sqrt{1+\cos^2 l}\ \rho), \qquad \text{where } \rho = r/\varepsilon.$$

The relation between (x,y) and (r,l) is given by

$$x = (1-r)\cos l, \qquad y = (1-r)\sin l, \qquad \text{and hence}$$

$$r = 1 - \sqrt{x^2+y^2}, \qquad \cos l = \frac{x}{\sqrt{x^2+y^2}}.$$

Exercise

Find the terms of the zeroth-order approximation for the solutions of the following problems:

(a) $\varepsilon^2 \triangle u - (1+y^2)u = x, \qquad \Omega = \{(x,y): \ x^2+y^2<1\}, \qquad u|_{\partial\Omega} = 0.$

(b) $\varepsilon^2 \triangle u - (1+x^2+y^2)u = -y, \qquad \Omega = \{(x,y): \ x^2+y^2<1\}, \qquad u|_{\partial\Omega} = 0.$

3.1.3 Concluding Remarks

We illustrated the method of Vishik–Lyusternik on the example of a linear elliptic partial differential equation. Note that even for this simple problem it is impossible to find the solution explicitly. At the same time the asymptotic approximation of the solution can be easily found with any required accuracy.

The method of Vishik–Lyusternik is well developed not only for elliptic but also for parabolic and hyperbolic partial differential equations; some nonlinear problems can be solved using this method as well. The main advantage of the method is its simplicity. Usually the boundary functions are defined as solutions of ordinary differential equations, where the stretched distance along the normal to the boundary (the variable ρ in our case) plays the role of the independent variable. In some problems, though, the boundary layer is described by more complicated equations, e.g. this is the case for parabolic equations. A survey of results obtained by the Vishik-Lyusternik method with many references can be found in Trenogin [134].

3.2 Corner boundary functions

3.2.1 Elliptic partial differential equations in a rectangle

The asymptotic solution of the problem considered in the previous section was constructed under an important assumption: the boundary $\partial\Omega$ of the domain Ω is supposed to be a smooth curve. The normal to the boundary exists at any point of $\partial\Omega$ and the boundary functions were constructed as the solutions of some ordinary differential equations with derivatives taken along these normals.

In the case when the boundary of the domain is no longer smooth, but contains corner points, the structure of the asymptotic solutions becomes more complicated in the vicinity of these points. The boundary functions constructed in the previous section are not sufficient to describe the asymptotic behavior of the solution near the corners. We need to introduce a new type of boundary function, *corner boundary functions*, in the vicinity of the corner points. We begin our discussion with an example for the same elliptic equation as in § 3.1, but now domain Ω will be a rectangle, whose boundary has four corner points.

Consider the problem

$$\varepsilon^2 \bigtriangleup u - k^2(x,y)u = f(x,y,\varepsilon), \tag{3.2.1}$$

$$(x,y) \in \Omega = (0 < x < a) \times (0 < y < b),$$

$$u|_{\partial\Omega} = 0. \tag{3.2.2}$$

We keep the same assumptions on functions $k(x,y)$ and $f(x,y,\varepsilon)$ as in § 3.1.

We will seek the asymptotic expansion of the solution of (3.2.1), (3.2.2) in the form

$$u = \overline{u} + \Pi + P. \tag{3.2.3}$$

Here $\overline{u} = \sum_{i=0}^{\infty} \varepsilon^i \overline{u}_i(x,y)$ is the regular part of the asymptotic solution that can be determined in absolutely the same manner as in § 3.1. The notation Π is used for the boundary functions playing an important role near the sides of the rectangle, and the notation P is used for the corner boundary functions important in the vicinities of the vertices of the rectangle. Corresponding to the sides of the rectangle, Π can be represented as a sum of four terms:

$$\Pi = \overset{(1)}{\Pi} + \overset{(2)}{\Pi} + \overset{(3)}{\Pi} + \overset{(4)}{\Pi} = \sum_{i=0}^{\infty} \varepsilon^i \left(\overset{(1)}{\Pi}_i (x,\eta) + \overset{(2)}{\Pi}_i (\xi, y) + \overset{(3)}{\Pi}_i (x, \eta_*) + \overset{(4)}{\Pi}_i (\xi_*, y) \right),$$

where $\eta = y/\varepsilon$, $\xi = x/\varepsilon$, $\eta_* = (b-y)/\varepsilon$, $\xi_* = (a-x)/\varepsilon$ are the boundary layer variables. In the vicinity of the corresponding side of the rectangle each of them plays the same role as the variable $\rho = r/\varepsilon$ near $\partial\Omega$ in § 3.1. To avoid describing the standard procedure of defining Π-functions for each side (this procedure is quite similar to the one described in § 3.1), we will name only the corresponding boundary layer operators and boundary conditions.

The functions $\overset{(1)}{\Pi}_i (x,\eta)$, representing the boundary layer near the side $y = 0$, are defined by the boundary layer operator

$$\frac{\partial^2}{\partial \eta^2} - k^2(x,0), \qquad \eta > 0,$$

and the boundary conditions

$$\overset{(1)}{\Pi}_i (x,0) = -\bar{u}_i(x,0), \qquad \overset{(1)}{\Pi}_i (x,\infty) = 0.$$

Thus, the $\overset{(1)}{\Pi}$-functions compensate for the discrepancy introduced by the regular part of the asymptotic expansion to the boundary condition on the side $y = 0$.

For $\overset{(1)}{\Pi}_0 (x,\eta)$, we obtain

(3.2.4)
$$\overset{(1)}{\Pi}_0 (x,\eta) = -\bar{u}_0(x,0)\exp(-k(x,0)\eta),$$

and the explicit expressions for $\overset{(1)}{\Pi}_i (x,\eta)$ can be found successively for $i = 1, 2, \dots$. All these functions satisfy the exponential estimate

(3.2.5)
$$|\overset{(1)}{\Pi}_i (x,\eta)| \le c\exp(-\kappa\eta).$$

In a similar way, the functions $\overset{(2)}{\Pi}_i (\xi, y)$ are defined by the boundary layer operator

$$\frac{\partial^2}{\partial \xi^2} - k^2(0,y), \qquad \xi > 0,$$

and the boundary conditions

$$\overset{(2)}{\Pi}_i (0,y) = -\bar{u}_i(0,y), \qquad \overset{(2)}{\Pi}_i (\infty,y) = 0.$$

For $\overset{(2)}{\Pi}_0 (\xi, y)$, we obtain

(3.2.6)
$$\overset{(2)}{\Pi}_0 (\xi, y) = -\bar{u}_0(0,y)\exp(-k(0,y)\xi).$$

The functions $\overset{(2)}{\Pi}_i(\xi, y)$ $(i = 1, 2, \ldots)$ can also be found explicitly and possess estimates of the type (3.2.5):

$$|\overset{(2)}{\Pi}_i(\xi, y)| \leq c\exp(-\kappa\xi).$$

The boundary functions $\overset{(3)}{\Pi}_i(x, \eta_*)$ and $\overset{(4)}{\Pi}_i(\xi_*, y)$ are defined in a similar way.

Notice that the boundary functions $\overset{(1)}{\Pi}_i(x, \eta)$, compensating for the discrepancy in the boundary condition on the side $y = 0$, introduce additional discrepancies in the boundary conditions on the sides $x = 0$ and $x = a$. These discrepancies are essential in the vicinities of the corner points $(0, 0)$ and $(a, 0)$ and, according to (3.2.5), they decay exponentially as η increases. Similar discrepancies are introduced by the functions $\overset{(2)}{\Pi}_i(\xi, y)$ in the conditions on the sides $y = 0$ and $y = b$, by the functions $\overset{(3)}{\Pi}_i(x, \eta_*)$ in the conditions on $x = 0$ and $x = a$, and by functions $\overset{(4)}{\Pi}_i(\xi_*, y)$ in the conditions on $y = 0$ and $y = b$.

To compensate for these discrepancies we introduce corner boundary functions. One corner boundary function must correspond to each of the vertices of the rectangle, and therefore P contains terms of four types:

$$\begin{aligned} P &= \overset{(1)}{P} + \overset{(2)}{P} + \overset{(3)}{P} + \overset{(4)}{P} \\ &= \sum_{i=0}^{\infty} \varepsilon^i \left(\overset{(1)}{P}_i(\xi, \eta) + \overset{(2)}{P}_i(\xi, \eta_*) + \overset{(3)}{P}_i(\xi_*, \eta_*) + \overset{(4)}{P}_i(\xi_*, \eta) \right). \end{aligned}$$

The functions $\overset{(1)}{P}_i$, e.g., are needed to compensate for the errors introduced by $\overset{(1)}{\Pi}$-functions in the boundary condition along $x = 0$ and by $\overset{(2)}{\Pi}$-functions in the boundary condition on $y = 0$. The equations for the functions $\overset{(1)}{P}_i(\xi, \eta)$ can be obtained from the original equation (3.2.1) (more precisely, from the homogeneous equation corresponding to (3.2.1)) in a standard way. We must introduce the new variables $\eta = y/\varepsilon$, $\xi = x/\varepsilon$, expand the coefficient $k^2(\varepsilon\xi, \varepsilon\eta)$ into a power series in ε, and equate coefficients of like powers of ε on both sides of the homogeneous equation corresponding to (3.2.1). This leads to the following equations $(i = 0, 1, 2, \ldots)$:

$$\text{(3.2.7)} \qquad \frac{\partial^2 \overset{(1)}{P}_i}{\partial\xi^2} + \frac{\partial^2 \overset{(1)}{P}_i}{\partial\eta^2} - k^2(0, 0)\overset{(1)}{P}_i(\xi, \eta) = p_i(\xi, \eta), \qquad \xi > 0, \qquad \eta > 0.$$

Here the functions $p_i(\xi, \eta)$ are expressed recursively through $\overset{(1)}{P}_j(\xi, \eta)$ with subscripts $j < i$, in particular $p_0(\xi, \eta) = 0$. Since the purpose of $\overset{(1)}{P}_i(\xi, \eta)$ is to compensate for the discrepancies in the boundary conditions introduced by the ordinary boundary functions $\overset{(1)}{\Pi}_i$ and $\overset{(2)}{\Pi}_i$, we should prescribe additional conditions for (3.2.7) in the form

$$\text{(3.2.8)} \qquad \overset{(1)}{P}_i(0, \eta) = -\overset{(1)}{\Pi}_i(0, \eta), \qquad \overset{(1)}{P}_i(\xi, 0) = -\overset{(2)}{\Pi}_i(\xi, 0).$$

Further, we demand that the $\overset{(1)}{P}$-functions tend to zero when either of the boundary layer variables tends to infinity:

$$\overset{(1)}{P}_i(\xi,\eta) \to 0 \quad \text{as} \quad (\xi+\eta) \to \infty. \tag{3.2.9}$$

Solutions of problems (3.2.7)–(3.2.9) can be successively expressed in explicit form using a Green's function. In the next part of this subsection we will discuss how to construct $\overset{(1)}{P}_0$, and show that the $\overset{(1)}{P}$-functions possess the exponential estimate

$$|\overset{(1)}{P}_i(\xi,\eta)| \le \exp(-\kappa(\xi+\eta)). \tag{3.2.10}$$

The boundary functions $\overset{(l)}{P}_i$, $l = 2, 3, 4$, playing important role in the vicinities of the other vertices of the rectangle Ω, are defined in absolutely the same way and possess similar exponential estimates.

Consider the problem for $\overset{(1)}{P}_0(\xi,\eta)$, i.e., (3.2.7) and the boundary conditions (3.2.8), (3.2.9) with $i = 0$. By virtue of expressions (3.2.4) and (3.2.6), the boundary conditions (3.2.8) can be rewritten as

$$\overset{(1)}{P}_0(0,\eta) = q\exp(-k\eta), \qquad \overset{(1)}{P}_0(\xi,0) = q\exp(-k\xi),$$

where $q = u_0(0,0)$ and $k = k(0,0)$. Introducing the new dependent variable v by the formula

$$\overset{(1)}{P}_0(\xi,\eta) = q\exp(-k(\xi+\eta)) + v(\xi,\eta), \qquad \xi > 0,\ \eta > 0, \tag{3.2.11}$$

we obtain the following problem for $v(\xi,\eta)$:

$$\frac{\partial^2 v}{\partial \xi^2} + \frac{\partial^2 v}{\partial \eta^2} - k^2(0,0)v = -qk^2\exp(-k(\xi+\eta)), \qquad \xi > 0,\ \eta > 0,$$

$$v(0,\eta) = 0, \qquad v(\xi,0) = 0,$$

$$v(\xi,\eta) \to 0 \quad \text{as} \quad (\xi+\eta) \to \infty.$$

The solution of this problem has the form (cf., e.g., Tikhonov and Samarskii [132])

$$v(\xi,\eta) = \int_0^\infty \int_0^\infty G(\xi,\eta,\xi_0,\eta_0)qk^2\exp(-k(\xi_0+\eta_0))d\xi_0 d\eta_0.$$

Here

$$G(\xi,\eta,\xi_0,\eta_0) = \frac{1}{2\pi}(K_0(kR_1) + K_0(kR_2) - K_0(kR_3) - K_0(kR_4))$$

is the Green's function, with $K_0(z)$ being the Bessel function of imaginary argument (cf. Tikhonov and Samarskii [132]),

$$R_1 = \sqrt{(\xi-\xi_0)^2 + (\eta-\eta_0)^2}, \qquad R_2 = \sqrt{(\xi+\xi_0)^2 + (\eta+\eta_0)^2},$$

$$R_3 = \sqrt{(\xi-\xi_0)^2+(\eta+\eta_0)^2}, \quad \text{and} \quad R_4 = \sqrt{(\xi+\xi_0)^2+(\eta-\eta_0)^2}.$$

To estimate $v(\xi,\eta)$, we will use a rough but adequate estimate of the function $K_0(z)$ for $z>0$:

$$0 < K_0(z) \le c(|\ln z|+1)\exp(-z),$$

where $c>0$ is a constant. This estimate follows from the asymptotics of $K_0(z)$ for small and large z. We also use the inequality $K_0(z_1) \le K_0(z_2)$ for $z_1 \ge z_2$. Then we have

$$G(\xi,\eta,\xi_0,\eta_0) \le \frac{2}{\pi}K_0(kR_1) \le c(|\ln R_1|+1)\exp(-kR_1).$$

Taking into account the inequality

$$R_1 = \sqrt{(\xi-\xi_0)^2+(\eta-\eta_0)^2} \ge \frac{1}{2}(|\xi-\xi_0|+|\eta-\eta_0|),$$

we obtain

$$|v(\xi,\eta)| \le \int_0^\infty\int_0^\infty c(|\ln R_1|+1)\exp\left(-\frac{k}{2}R_1\right)$$

$$\times \exp\left(-\frac{k}{4}(|\xi-\xi_0|+|\eta-\eta_0|)-k(\xi_0+\eta_0)\right)d\xi_0 d\eta_0.$$

It can be easily seen that

$$-\frac{k}{4}(|\xi-\xi_0|+|\eta-\eta_0|)-k(\xi_0+\eta_0) \le -\kappa(\xi+\eta),$$

for any κ such that $0<\kappa\le k/4$. Finally, taking into account the inequality

$$\int_0^\infty\int_0^\infty (|\ln R_1|+1)\exp\left(-\frac{k}{2}R_1\right)d\xi_0 d\eta_0 \le c,$$

we obtain the estimate for $v(\xi,\eta)$:

$$|v(\xi,\eta)| \le c\exp\left(-\kappa(\xi+\eta)\right).$$

It follows from (3.2.11) that $\overset{(1)}{P}_0(\xi,\eta)$ satisfies the same estimate, i.e. (3.2.10).

For the other functions $\overset{(1)}{P}_i(\xi,\eta)$ $(i\ge 1)$ this estimate can be easily shown by induction. For $\overset{(1)}{P}_i(\xi,\eta)$ we have equation (3.2.7) where the right-hand side depends linearly on the $\overset{(1)}{P}_j(\xi,\eta)$ with $j<i$. Consequently, if $\overset{(1)}{P}_j(\xi,\eta)$ for $j<i$ possess estimates of the type (3.2.10), the same bound will hold for $p_i(\xi,\eta)$. But then, following the steps undertaken above, an estimate, similar to that shown for $\overset{(1)}{P}_0(\xi,\eta)$, can be obtained for $\overset{(1)}{P}_i(\xi,\eta)$.

The exponential estimates for the $\overset{(l)}{P}_i$, $l=2,3,4$ can be derived analogously.

Note that, due to the estimate (3.2.10), the functions $\overset{(1)}{P}_k(\xi,\eta)$ are essential only in some small vicinity of the corner point $(0,0)$ of the rectangle Ω. These functions vanish exponentially as their arguments move away from the point $(0,0)$. This explains the name of these functions — *corner boundary functions.*

In what follows it is convenient to introduce the following definition.

Definition 3.2 *We will say that the* $\overset{(1)}{P}$*-functions are defined by the operator* $L_P = \partial^2/\partial\xi^2 + \partial^2/\partial\eta^2 - k^2(0,0)$ *and the boundary conditions* (3.2.8), (3.2.9). *We will call* L_P *the boundary layer operator in the vicinity of the vertex* $(0,0)$.

The proof of following theorem is absolutely similar to the proof of the corresponding theorem from § 3.1.

Theorem 3.2 *The series* (3.2.3) *is asymptotic series for the solution* $u(x,y,\varepsilon)$ *of problem* (3.2.1), (3.2.2) *in the rectangle* $\overline{\Omega}$ *as* $\varepsilon \to 0$.

Exercise

Find the zeroth-order terms of the asymptotic solution for

$$\varepsilon^2\left(\frac{\partial^2 u}{\partial x^2}+\frac{\partial^2 u}{\partial y^2}\right)-(1+x^2)u=1,$$

$$(x,y)\in\Omega=(0<x<1)\times(0<y<1),$$

subject to boundary conditions

(a) $u|_{\partial\Omega}=0$; (b) $\left.\frac{\partial u}{\partial x}\right|_{x=0,x=1}=\left.\frac{\partial u}{\partial y}\right|_{y=0,y=1}=0$;

(c) $\left.\frac{\partial u}{\partial x}\right|_{x=0,x=1}=0,\quad u|_{y=0,y=1}=0.$

3.2.2 Other elliptic equations with a corner boundary layer

Equation (3.2.1) does not contain first-order derivatives. The boundary layer structure of the asymptotics might change substantially if terms with lower-order derivatives (even with small factors multiplying them) are added to (3.2.1). As an example, consider the equation

$$\varepsilon^2 \triangle u-\varepsilon^\alpha A(x,y)\frac{\partial u}{\partial y}-k^2(x,y)u=f(x,y,\varepsilon)$$

in the rectangle $\Omega=(0<x<a)\times(0<y<b)$, with the boundary conditions (3.2.2). Here, in comparison with (3.2.1), the term $-\varepsilon^\alpha A(x,y)\partial u/\partial y$ is added.

For $\alpha \geq 1$ the asymptotic structure of the solution is similar to that described in § 3.2.1. In the case $0<\alpha<1$, however, the boundary layer operators in the vicinities of the sides $y=0$ and $y=b$ as well as near the vertices of Ω will change substantially. Their form will now depend on the signs of the coefficient $A(x,y)$ on these sides. Let

$$A(x,0)>0,\qquad A(x,b)>0,$$

and $\alpha=1/2$. In this case the asymptotic expansion is a series in powers of $\sqrt{\varepsilon}$. The boundary functions $\overset{(1)}{\Pi}_i(x,\eta)$ $(\eta=y/\sqrt{\varepsilon})$ in the vicinity of the side $y=0$ are defined by the boundary layer operator $-A(x,0)\partial/\partial\eta-k^2(x,0)$ $(\eta>0)$ and the boundary condition $\overset{(1)}{\Pi}_i(x,0)=-\overline{u}_i(x,0)$, where $\overline{u}_i(x,y)$ are the regular terms of the asymptotic solution. The boundary functions $\overset{(3)}{\Pi}_i(x,\eta_*)$, with $\eta_*=(b-y)/\varepsilon^{3/2}$, near the side $y=b$ are defined by the boundary layer operator $\partial^2/\partial{\eta_*}^2+A(x,b)\partial/\partial\eta_*$ $(\eta_*>0)$ and the boundary conditions $\overset{(3)}{\Pi}_i(x,0)=-\overline{u}_i(x,b)$, $\overset{(3)}{\Pi}_i(x,\infty)=0$.

As we see, the boundary layer operators in the neighborhoods of the sides $y = 0$ and $y = b$ are different from those of § 3.2.1. The other distinguishing feature of this problem is the presence of the boundary layer variables of different scales. Near the side $y = 0$ the variable y is stretched by $\sqrt{\varepsilon}$ and near the side $y = b$ it is stretched by $\varepsilon^{3/2}$.

In the vicinities of the sides $x = 0$ and $x = a$ the boundary layer operators do not change. For example, the boundary functions $\overset{(2)}{\Pi}_i(\xi, y)$ $(\xi = x/\varepsilon)$ near the side $x = 0$ are defined, as before, by the boundary layer operator $\partial^2/\partial\xi^2 - k^2(0, y)$ $(\xi > 0)$ and the boundary conditions $\overset{(2)}{\Pi}_i(0, y) = -\bar{u}_i(0, y)$, $\overset{(2)}{\Pi}_i(\infty, y) = 0$.

Substantial changes take place in the corner boundary layer operators. Boundary functions $\overset{(1)}{P}_i(\xi, \eta)$ near the vertex $(0, 0)$ are now defined by the parabolic operator

$$\frac{\partial^2}{\partial\xi^2} - A(0,0)\frac{\partial}{\partial\eta} - k^2(0,0) \qquad (\xi > 0, \eta > 0)$$

and boundary conditions

$$\overset{(1)}{P}_i(\xi, 0) = -\overset{(2)}{\Pi}_i(\xi, 0), \qquad \overset{(1)}{P}_i(0, \eta) = -\overset{(1)}{\Pi}_i(0, \eta).$$

A similar parabolic boundary layer operator appears in the vicinity of the vertex $(a, 0)$.

Two types of corner boundary layer functions appear near each of the vertices $(0, b)$ and (a, b). Let us consider them, e.g., for the vertex $(0, b)$. Corner boundary functions of the first type (we will denote them by $\overset{(2)}{R}_i(\xi, \eta_*)$) are defined by the ordinary differential operator $\partial^2/\partial{\eta_*}^2 + A(0, b)\partial/\partial\eta_*$ $(\eta_* > 0)$ and the boundary conditions $\overset{(2)}{R}_i(\xi, 0) = -\overset{(2)}{\Pi}_i(\xi, b)$, $\overset{(2)}{R}_i(\xi, \infty) = 0$. They introduce a discrepancy in the boundary condition at $x = 0$. Finally, corner boundary functions $\overset{(2)}{P}_i(\zeta, \eta_*)$ $(\zeta = x/\varepsilon^{3/2})$ are defined with the help of the elliptic operator

$$\frac{\partial^2}{\partial\zeta^2} + \frac{\partial^2}{\partial{\eta_*}^2} + A(0,b)\frac{\partial}{\partial\eta_*} \qquad (\zeta > 0,\ \eta_* > 0)$$

and the boundary conditions

$$\overset{(2)}{P}_i(\zeta, 0) = 0, \qquad \overset{(2)}{P}_i(0, \eta_*) = -[\overset{(3)}{\Pi}_i(0, \eta_*) + \overset{(2)}{R}_i(0, \eta_*)],$$

$$\overset{(2)}{P}_i(\zeta, \eta_*) \to 0 \quad \text{as} \quad (\zeta + \eta_*) \to \infty.$$

All the boundary functions possess exponential decay estimates of the type (3.2.5), (3.2.10) with respect to the boundary layer variables.

Note that the corner boundary functions $\overset{(1)}{P}_0(\xi, \eta)$ and $\overset{(1)}{P}_1(\xi, \eta)$ have unbounded second derivatives with respect to η in the vicinity of the point$(0, 0)$. Likewise for the functions $\overset{(4)}{P}_i(\xi_*, \eta)$ $(i = 0, 1)$ in the vicinity of the vertex $(a, 0)$. These second derivatives enter the right-hand sides of the equations for $\overset{(1)}{P}_2$ and $\overset{(1)}{P}_3$, respectively. In turn, the second derivatives of the functions $\overset{(1)}{P}_2$ and $\overset{(1)}{P}_3$ with respect to η enter the right-hand sides of the equations for the higher-order terms. Thus, the order of the singularities present will grow with growth of the order of $\overset{(1)}{P}$-functions. This does not allow one to prolong the iterative process of defining functions $\overset{(1)}{P}_i$ (as well as $\overset{(4)}{P}_i$) beyond the first step. If we restrict the asymptotics only to terms of the zeroth order, then their sum will give a uniform asymptotic approximation of the solution in $\overline{\Omega}$ with accuracy of the order $\sqrt{\varepsilon}$. If we exclude some small (but fixed as $\varepsilon \to 0$) vicinities of the vertices $(0,0)$ and $(a,0)$, then in the remaining part of the rectangle $\overline{\Omega}$, the nth partial sum of the constructed expansion gives the approximation to the solution with an accuracy of order $\varepsilon^{(n+1)/2}$.

Asymptotic expansions with corner boundary functions also were constructed for solutions of some elliptic systems in Butuzov and Nikitin [32], [34] and Butuzov and Udodov [35].

Exercise

Find the zeroth-order approximation to the solution of the problem (see Shih and Kellogg [126])

$$\varepsilon^2\left(\frac{\partial^2 u}{\partial x^2}+\frac{\partial^2 u}{\partial y^2}\right)-\varepsilon^\alpha\frac{\partial u}{\partial x}-u=-1,$$

$$(x,y)\in\Omega=(0<x<1)\times(0<y<1),$$

$$u|_{\partial\Omega}=0,$$

where (a) $\alpha=0$, (b) $\alpha=1/2$.

3.2.3 Parabolic equations with the corner boundary layer

Singularly perturbed parabolic systems of equations of the type

$$\varepsilon\left(\frac{\partial u}{\partial t}-a\,\triangle\,u\right)=f(u,v,x,t,\varepsilon),\qquad \frac{\partial v}{\partial t}-b\,\triangle\,v=g(u,v,x,t,\varepsilon), \tag{3.2.12}$$

appear naturally when describing chemical reactions with diffusion (here $x=(x_1,x_2,x_3)$ is used as a notation for three spatial variables). The components of the vector functions u and v are the concentrations of the reacting substances. The small parameter ε is a quantity inversely proportional to the constants of the fast reactions. Systems of similar kind appear also in some other applied problems (see Chapter 4). We will consider the construction of an asymptotic expansion in ε for the solution of the boundary value problem for (3.2.12). To simplify the description of the algorithm, we will restrict our study to the case when the slow variables are absent (i.e., the second equation in (3.2.12) will be omitted), and x is a one-dimensional spatial variable. Further, for convenience, let us denote the small parameter before the derivatives by ε^2.

Consider a system of equations

$$\varepsilon^2\left(\frac{\partial u}{\partial t}-a(x,t)\frac{\partial^2 u}{\partial x^2}\right)=f(u,x,t,\varepsilon), \tag{3.2.13}$$

$$(x,t)\in\Omega=(0<x<1)\times(0<t\le T),$$

where u is an m-dimensional vector function, subject to the initial condition

$$u(x,0,\varepsilon) = \varphi(x), \tag{3.2.14}$$

and Neumann boundary conditions

$$\frac{\partial u}{\partial x}(0,t,\varepsilon) = \frac{\partial u}{\partial x}(1,t,\varepsilon) = 0. \tag{3.2.15}$$

(Such boundary conditions are common in problems of chemical kinetics.)

Let the following conditions be satisfied.

1^0. *Functions $a(x,t) > 0$, $f(u,x,t,\varepsilon)$ and $\varphi(x)$ are sufficiently smooth (they should be $n+2$ times continuously differentiable to enable us to construct the asymptotic expansion to the nth-order), and conditions* (3.2.14)*and* (3.2.15)*are matched at the corner points* $(0,0)$ *and* $(1,0)$ *of the rectangle* Ω: $\varphi'(0) = \varphi'(1) = 0$.

2^0. *The reduced equation $f(\overline{u},x,t,0) = 0$ (obtained from* (3.2.13)*for $\varepsilon = 0$) has a solution $\overline{u} = \overline{u}_0(x,t)$ in the rectangle $\Omega = (0 \le x \le 1) \times (0 \le t \le T)$.*

Note that (because of nonlinearity) this equation might also have some other solutions.

3^0. *All the eigenvalues $\lambda_i(x,t)$ $(i = 1,\dots,m)$ of the matrix*

$$\overline{f}_u(x,t) = f_u(\overline{u}_0(x,t),x,t,0)$$

have negative real parts: $\mathrm{Re}\lambda_i(x,t) < 0$, $(x,t) \in \overline{\Omega}$.

4^0. *The solution $\Pi_0(x,\tau)$ of the initial value problem*

$$\frac{\partial \Pi_0}{\partial \tau} = f(\overline{u}_0(x,0) + \Pi_0, x, 0, 0), \qquad \tau \ge 0, \tag{3.2.16}$$

$$\Pi_0(x,0) = \varphi(x) - \overline{u}_0(x,0)$$

exists for $\tau \ge 0$ and satisfies the condition $\Pi_0(x,\infty) = 0$ (x enters merely as a parameter, $0 \le x \le 1$).

Condition 4^0 implies that the initial value $\varphi(x) - \overline{u}_0(x,0)$ belongs to the domain of attraction of the rest point $\Pi_0 = 0$ for (3.2.16). This rest point is asymptotically stable by virtue of 3^0. The conditions presented above are absolutely similar to conditions 1^0 – 4^0 from § 2.1.2. In fact, (3.2.16) is the analogue of the associated system (2.1.7) from § 2.1.

Under conditions 1^0 – 4^0, we will construct the asymptotic solution of (3.2.13)–(3.2.15) in the form

$$u = \overline{u} + \Pi + Q + Q^* + P + P^*$$

$$= \sum_{i=0}^{\infty} \varepsilon^i (\overline{u}_i(x,t) + \Pi_i(x,\tau) + Q_i(\xi,t) + Q_i^*(\xi_*,t) + P_i(\xi,\tau) + P_i^*(\xi_*,\tau)). \tag{3.2.17}$$

Here $\tau = t/\varepsilon^2$, $\xi = x/\varepsilon$ and $\xi_* = (1-x)/\varepsilon$ are boundary layer variables; $\overline{u}_i$ are coefficients of the regular series; Π_i, Q_i and Q_i^* are the boundary functions describing the boundary layers in the vicinities of the sides $t = 0$, $x = 0$ and $x = 1$, respectively.

The corner boundary functions P_i and P_i^* are needed to describe the boundary layers near the vertices $(0,0)$ and $(1,0)$.

Let us substitute the series (3.2.17) into (3.2.13) and change the right-hand side $f(\overline{u}+\Pi+Q+Q^*+P+P^*,x,t,\varepsilon)$ into an expression of the type (3.2.17):

$$\overline{f}+\Pi f+Qf+Q^*f+Pf+P^*f, \tag{3.2.18}$$

where

$$\overline{f}=f(\overline{u},x,t,\varepsilon), \qquad \Pi f=[f(\overline{u}+\Pi,x,t,\varepsilon)-f(\overline{u},x,t,\varepsilon)]\,|_{t=\varepsilon^2\tau},$$

$$Qf=[f(\overline{u}+Q,x,t,\varepsilon)-f(\overline{u},x,t,\varepsilon)]\,|_{x=\varepsilon\xi},$$

$$Pf=\left[f(\overline{u}+\Pi+Q+P,x,t,\varepsilon)-\Pi f-Qf-\overline{f}\right]\Big|_{t=\varepsilon^2\tau,x=\varepsilon\xi};$$

Q^*f and P^*f are similar to Qf and Pf. Now, expanding the terms of (3.2.18) into power series in ε and equating coefficients of like powers of ε separately for regular functions and for each type of boundary function, we obtain the equations for the terms of the asymptotic expansion. Additional conditions are obtained by substituting (3.2.17) into (3.2.14) and (3.2.15). We have

$$\overline{u}(x,0,\varepsilon)+\Pi(x,0,\varepsilon)=\varphi(x), \tag{3.2.19}$$

$$Q(\xi,0,\varepsilon)+P(\xi,0,\varepsilon)=0, \qquad Q^*(\xi_*,0,\varepsilon)+P^*(\xi_*,0,\varepsilon)=0, \tag{3.2.20}$$

$$\frac{\partial\overline{u}}{\partial x}(0,t,\varepsilon)+\frac{1}{\varepsilon}\frac{\partial Q}{\partial\xi}(0,t,\varepsilon)=0, \qquad \frac{\partial\overline{u}}{\partial x}(1,t,\varepsilon)-\frac{1}{\varepsilon}\frac{\partial Q^*}{\partial\xi_*}(0,t,\varepsilon)=0, \tag{3.2.21}$$

$$\frac{\partial\Pi}{\partial x}(0,\tau,\varepsilon)+\frac{1}{\varepsilon}\frac{\partial P}{\partial\xi}(0,\tau,\varepsilon)=0, \qquad \frac{\partial\Pi}{\partial x}(1,\tau,\varepsilon)-\frac{1}{\varepsilon}\frac{\partial P^*}{\partial\xi_*}(0,\tau,\varepsilon)=0. \tag{3.2.22}$$

For $\overline{u}_0(x,t)$ we obtain the equation

$$\overline{f}_0\equiv f(\overline{u}_0,x,t,0)=0,$$

which coincides with the reduced equation. Hence, $\overline{u}_0=\overline{u}_0(x,t)$. For $\overline{u}_i$ $(i\geq 1)$ we have linear equations of the type

$$\overline{f}_u(x,t)\overline{u}_i=f_i(x,t).$$

Here the $f_i(x,t)$ are expressed recursively through the $\overline{u}_j(x,t)$ with $j<i$. Hence, $\overline{u}_i=\overline{f}_u^{-1}(x,t)f_i(x,t)$. Note that $\overline{f}_u^{-1}(x,t)$ exists by virtue of condition 3^0.

For $\Pi_0(x,\tau)$ we obtain the equation

$$\frac{\partial\Pi_0}{\partial\tau}=\Pi_0 f\equiv f(\overline{u}_0(x,0)+\Pi_0,x,0,0), \qquad \tau\geq 0,$$

with the initial condition

$$\Pi_0(x,0) = \varphi(x) - \overline{u}_0(x,0).$$

This condition follows from (3.2.19). The problem for $\Pi_0(x,\tau)$ evidently coincides with problem (3.2.16) from condition 4^0. The following exponential estimate holds for $\Pi_0(x,\tau)$ by virtue of 3^0 and 4^0:

$$\| \Pi_0(x,\tau) \| \leq c\exp(-\kappa\tau). \tag{3.2.23}$$

Subsequent $\Pi_i(x,\tau)$ $(i \geq 1)$ are defined as solutions of the linear initial value problems

$$\frac{\partial \Pi_i}{\partial \tau} = f_u(x,\tau)\Pi_i + \pi_i(x,\tau),$$

$$\Pi_i(x,0) = -\overline{u}_i(x,0), \tag{3.2.24}$$

where

$$f_u(x,\tau) = f_u(\overline{u}_0(x,0) + \Pi_0(x,\tau), x, 0, 0). \tag{3.2.25}$$

Here the $\pi_i(x,\tau)$ are expressed recursively through $\Pi_j(x,\tau)$, $j < i$. We will use the notation $\Phi(x,\tau)$ for the fundamental matrix of the corresponding homogeneous equation ($\Phi(x,0) = I$, the identity $(m \times m)$-matrix). By virtue of condition 3^0, the following estimate holds:

$$\| \Phi(x,\tau)\Phi^{-1}(x,\tau_0) \| \leq c\exp(-\kappa(\tau-\tau_0)), \qquad 0 \leq x \leq 1,\ 0 \leq \tau_0 \leq \tau. \tag{3.2.26}$$

Using this estimate and the explicit expression for $\Pi_i(x,\tau)$ which has the form

$$\Pi_i(x,\tau) = -\Phi(x,\tau)\overline{u}_i(x,0) + \int_0^\tau \Phi(x,\tau)\Phi^{-1}(x,\tau_0)\pi_i(x,\tau_0)d\tau_0,$$

it is easy to prove by induction that all the Π-functions possess an exponential estimate of the type (3.2.23).

For $Q_0(\xi,t)$ we obtain the equation

$$-a(0,t)\frac{\partial^2 Q_0}{\partial \xi^2} = Q_0 f \equiv f(\overline{u}_0(0,t) + Q_0, 0, t, 0), \qquad \xi \geq 0,$$

where t enters as a parameter $(0 \leq t \leq T)$, with the boundary condition

$$\frac{\partial Q_0}{\partial \xi}(0,t) = 0.$$

This condition follows from the first relation in (3.2.21). We will impose as a second boundary condition

$$Q_0(\infty,t) = 0.$$

Then $Q_0(\xi,t)=0$, and for $Q_i(\xi,t)$ $(i\geq 1)$ we obtain linear boundary value problems of the type

$$-a(0,t)\frac{\partial^2 Q_i}{\partial\xi^2}=\overline{f}_u(0,t)Q_i+q_i(\xi,t), \tag{3.2.27}$$

$$\frac{\partial Q_i}{\partial\xi}(0,t)=-\frac{\partial\overline{u}_{i-1}}{\partial x}(0,t),\qquad Q_i(\infty,t)=0, \tag{3.2.28}$$

where the $q_i(\xi,t)$ are expressed recursively through $Q_j(\xi,t)$, $j<i$. The roots of the characteristic equation for (3.2.27) are $\pm\sqrt{-\lambda_k(0,t)/a(0,t)}$, $k=1,2,\ldots,m$, where $\lambda_k(x,t)$ are the eigenvalues of $\overline{f}_u(x,t)$. By virtue of condition 3^0 m roots have negative and m roots have positive real parts. Therefore the homogeneous equation corresponding to (3.2.27) has m exponentially decreasing and m exponentially increasing solutions as $\xi\to\infty$. This provides the unique solvability of the problems (3.2.27), (3.2.28). The solutions can be found successively in explicit form up to an arbitrary order by transforming matrix $\overline{f}_u(0,t)$ to Jordan canonical form. All these functions have the exponential estimate

$$\| Q_i(\xi,t)\|\leq c\exp(-\kappa\xi).$$

Functions $Q_i^*(\xi_*,t)$ are defined analogously to the Q-functions and possess a similar exponential estimate.

As in the case of elliptic equations, the corner boundary functions $P_k(\xi,\tau)$ and $P_k{}^*(\xi_*,\tau)$ are needed to compensate for the discrepancies introduced by the Π-functions in the boundary conditions at $x=0$ and $x=1$ as well as by the Q- and Q^*-functions in the initial condition at $t=0$. Equalities (3.2.20) and (3.2.22) reflect this purpose of corner boundary functions. The initial and boundary conditions for P_i and P_i^* can be found from these equalities. Further, we demand that the P-functions should approach zero as each of the boundary layer variables tends to infinity.

For $P_0(\xi,\tau)$ we obtain the following problem:

$$\frac{\partial P_0}{\partial\tau}-a\frac{\partial^2 P_0}{\partial\xi^2}=P_0f\equiv f(\overline{u}_0(0,0)+\Pi_0(0,\tau)+P_0,0,0,0)$$

$$-f(\overline{u}_0(0,0)+\Pi_0(0,\tau),0,0,0),\qquad \xi>0,\ \tau>0,$$

$$P_0(\xi,0)=-Q_0(\xi,0)=0,\qquad \frac{\partial P_0}{\partial\xi}(0,\tau)=0,$$

where $a=a(0,0)$. It can be easily seen that the unique solution of this problem is trivial. For $P_i(\xi,\tau)$ $(i\geq 1)$ we have the linear problems

$$\frac{\partial P_i}{\partial\tau}-a\frac{\partial^2 P_i}{\partial\xi^2}-f_u(0,\tau)P_i=p_i(\xi,\tau),$$

$$P_i(\xi,0)=-Q_i(\xi,0),\qquad \frac{\partial P_i}{\partial\xi}(0,\tau)=-\frac{\partial\Pi_{i-1}}{\partial x}(0,\tau), \tag{3.2.29}$$

where $f_u(0,\tau)$ is defined by (3.2.25), and the $p_i(\xi,\tau)$ are expressed recursively through Π_j, Q_j $(j \le i)$ and P_j $(j < i)$. Notice that the initial and boundary conditions for $P_i(\xi,\tau)$ are matched at the corner point $(0,0)$, i.e., $\partial Q_i/\partial\xi(0,0) = \partial\Pi_{i-1}/\partial x(0,0)$. This follows from the equalities $\partial Q_i/\partial\xi(0,0) = -\partial\overline{u}_{i-1}/\partial x(0,0)$ (see (3.2.28)) and $\Pi_{i-1}(x,0) = -\overline{u}_{i-1}(x,0)$ (see (3.2.24)); for $i=1$ the condition $\varphi'(0)=0$ should also be taken into account. Functions $P_i(\xi,\tau)$ can be found in explicit form

$$P_i(\xi,\tau) = g_i(\xi,\tau) + \int_0^\tau \int_0^\infty G(\xi,\tau,\xi_0,\tau_0)h_i(\xi_0,\tau_0)d\xi_0 d\tau_0,$$

where $g_i(\xi,\tau)$ is an arbitrary sufficiently smooth function satisfying conditions (3.2.29), e.g.

$$\begin{aligned} g_i(\xi,\tau) = &\; -Q_i(\xi,0)\exp(-\kappa\tau) \\ &+ \left[\frac{\partial\Pi_{i-1}}{\partial x}(0,\tau) - \frac{\partial\Pi_{i-1}}{\partial x}(0,0)\exp(-\kappa\tau)\right]\frac{1}{\kappa}\exp(-\kappa\xi); \end{aligned}$$

$$h_i(\xi,\tau) = p_i(\xi,\tau) - \frac{\partial g_i}{\partial\tau} + a\frac{\partial^2 g_i}{\partial\xi^2} + f_u(0,\tau)g_i.$$

The Green's matrix $G(\xi,\tau,\xi_0,\tau_0)$ has the form

$$G(\xi,\tau,\xi_0,\tau_0) = \Phi(0,\tau)\Phi^{-1}(0,\tau_0)\frac{1}{2\sqrt{\pi a(\tau-\tau_0)}}$$

$$\times\left\{\exp\left[-\frac{(\xi-\xi_0)^2}{4a(\tau-\tau_0)}\right] + \exp\left[-\frac{(\xi+\xi_0)^2}{4a(\tau-\tau_0)}\right]\right\}.$$

Here $\Phi(0,\tau)$ is the same fundamental matrix as in (3.2.26). By virtue of inequality (3.2.26) the following estimate for the Green's matrix holds:

$$\| G(\xi,\tau,\xi_0,\tau_0) \| \le \frac{c}{\sqrt{\tau-\tau_0}}\exp(-\kappa(|\xi-\xi_0| + \tau - \tau_0)).$$

Using this estimate it is easy to prove that all the P-functions possess the exponential estimate

$$\| P_i(\xi,\tau) \| \le c\exp\left(-\kappa(\xi+\tau)\right).$$

Functions $P_i^*(\xi_*,\tau)$ are defined analogously to P-functions and possess the similar estimate.

We have completed the construction of the asymptotic solution (3.2.17).

Let us formulate the following theorem.

Theorem 3.3 *Under conditions* 1^0–4^0 *and for sufficiently small* ε *the problem* (3.2.13)–(3.2.15) *has a unique solution* $u(x,t,\varepsilon)$ *and the series* (3.2.17) *is the asymptotic expansion of this solution in the rectangle* $\overline{\Omega}$ *as* $\varepsilon \to 0$.

Omitting details, let us present the outline of the proof. The equation for the remainder term $w = u - U$ (where $U = U_n + \varepsilon^{n+1}(Q_{n+1} + Q^*_{n+1} + P_{n+1} + P^*_{n+1})$ and U_n is the nth partial sum of the series (3.2.17)) can be written in the form

$$\varepsilon^2 \left(\frac{\partial w}{\partial t} - a(x,t) \frac{\partial^2 w}{\partial x^2} \right) - f_u(x,t,\varepsilon) w = h(w,x,t,\varepsilon), \tag{3.2.30}$$

where

$$f_u(x,t,\varepsilon) = f_u(\overline{u}_0(x,t) + \Pi_0(x,t/\varepsilon), x, t, 0),$$

and

$$h(w,x,t,\varepsilon) = f(U+w,x,t,\varepsilon) - \varepsilon^2 \left(\frac{\partial U}{\partial t} - a(x,t) \frac{\partial^2 U}{\partial x^2} \right) - f_u(x,t,\varepsilon) w.$$

Function h, nonlinear with respect to w, possesses the following two properties similar to the properties of the function g in § 1.3.1.

1*.

$$h(0,x,t,\varepsilon) = f(U,x,t,\varepsilon) - \varepsilon^2 \left(\frac{\partial U}{\partial t} - a(x,t) \frac{\partial^2 U}{\partial x^2} \right) = O(\varepsilon^{n+1})$$

in Ω, as follows from the method of construction of U.

2*. If $\| w_k(x,t,\varepsilon) \| \leq c_1 \varepsilon, \ k = 1,2$, then there exist numbers $c_2 > 0$ and $\varepsilon_0 > 0$ such that for $0 < \varepsilon \leq \varepsilon_0$ the following inequality holds:

$$\sup_{\overline{\Omega}} \| h(w_2,x,t,\varepsilon) - h(w_1,x,t,\varepsilon) \| \leq c_2 \varepsilon \sup_{\overline{\Omega}} \| w_2 - w_1 \| .$$

Using the Green's matrix, let us pass on from (3.2.30) with homogeneous conditions of types (3.2.14) and (3.2.15) to the equivalent integral equation

$$\begin{aligned} w(x,t,\varepsilon) &= \textstyle\int_0^t \int_0^1 G(x,t,x_0,t_0,\varepsilon) h(w(x_0,t_0,\varepsilon),x_0,t_0,\varepsilon) dx_0 dt_0 \\ &\equiv L(w,x,t,\varepsilon). \end{aligned} \tag{3.2.31}$$

By virtue of condition 3^0 we have the the Green's matrix estimate

$$\| G(x,t,x_0,t_0,\varepsilon) \| \leq \frac{c}{\varepsilon^2} \frac{1}{\sqrt{t-t_0}} \exp\left(-\kappa \frac{t-t_0}{\varepsilon^2} \right) \exp\left[-\kappa \frac{(x-x_0)^2}{t-t_0} \right].$$

It follows from this, together with the properties 1* and 2* of $h(w,x,t,\varepsilon)$, that the integral operator $L(w,x,t,\varepsilon)$ possesses similar properties.

Applying the method of successive approximations to (3.2.31), we obtain

$$w_0 = 0, \ w_{i+1} = L(w_i,x,t,\varepsilon), \ i = 1,2,\ldots.$$

We can prove (as in § 1.3.1), using the two properties of operator L, that the unique solution $w(x,t,\varepsilon)$ of (3.2.31) exists for sufficiently small ε, and has the estimate

$$\max_{\overline{\Omega}} \| w(x,t,\varepsilon) \| = O(\varepsilon^{n+1}),$$

from which it follows that $\max_{\overline{\Omega}}|u - U_n| = O(\varepsilon^{n+1})$.

In a similar way the asymptotic solution can be constructed for the case of the system (3.2.12) containing both fast variables (function u) and slow variables (function v). However, the addition of the slow variables leads to the situation when the asymptotics can be constructed only in the zeroth- and first-order approximations. This is related to the fact that the initial and boundary conditions for the function $\overline{v}_2(x,t)$ (the coefficient by ε^2 in the regular series for $v(x,t,\varepsilon)$) will not be matched at the corner points $(0,0)$ and $(1,0)$. As a result, the function $\overline{v}_2(x,t)$ will be nonsmooth in $\overline{\Omega}$. The derivatives $\partial\overline{v}_2/\partial t$ and $\partial^2\overline{v}_2/\partial x^2$ will be unbounded in the vicinities of the corner points (they will have the singularity of the order $1/\sqrt{t}$). The order of singularity will grow with each step of the iteration process of defining the terms of the asymptotic solution, and this does not allow us to use the algorithm described above further than the first step.

Exercise
Consider the equation

$$\varepsilon^2 \left(\frac{\partial u}{\partial t} - \frac{\partial^2 u}{\partial x^2} \right) = -(1+x^2)u + 1,$$

$$(x,t) \in \Omega = (0 < x < 1) \times (0 < t < T),$$

subject to conditions

(a) $u(x,0) = \dfrac{\partial u}{\partial x}(0,t) = \dfrac{\partial u}{\partial x}(1,t) = 0$;

(b) $u(x,0) = u(0,t) = u(1,t) = 0$.

Define the zeroth- and the first-order approximations to the solution of the problem with conditions (a); construct the the zeroth-order approximation to the solution of the problem with conditions (b).

The method of boundary functions is well developed also for equations of hyperbolic type (cf. Butuzov [16]), for the systems of parabolic equations (cf. Butuzov and Kalachev [18]), for partial differential equations in the multidimensional case (cf. Butuzov [17]), as well as for difference equations (cf. Butuzov [15]). For differentiability properties of solutions of (3.2.1), (3.2.2) in a rectangle see, e.g., Han and Kellogg [59]. The method works successfully for a variety of applied problems (see Chapter 4 where some applications are discussed).

3.3 The smoothing procedure

3.3.1 Statement of the problem and an algorithm for construction of the asymptotic solution

The construction of asymptotic solutions for many problems involving partial differential equations often encounters the difficulty connected with nonsmoothness of some of the terms of the asymptotic expansion. We already discussed one such case in § 3.2.3, where some terms of the asymptotic solution were defined as the solutions of parabolic equations and had unbounded derivatives. As a result, we could not apply the standard procedure for defining the terms of the second and higher orders.

Here we will consider the case where the nonsmoothness of the terms in the asymptotic series has a different nature. It appears when some of the terms are defined as solutions of the first-order partial differential equations. Thus, they might be nonsmooth, and even discontinuous on a characteristic emerging from the corner point of the boundary. Usually, in such situations one has to use matching to construct the asymptotic expansions (see, e.g., Il'in [67]). On the other hand, the zeroth- and the first-order approximations to the solution, which are usually sufficient for practical purposes, can be obtained in a somewhat simpler way, using a *smoothing procedure* for the nonsmooth terms.

Consider

$$L_\varepsilon u \equiv \varepsilon \frac{\partial u}{\partial t} + b(x)\frac{\partial u}{\partial x} - \varepsilon^2 a(x)\frac{\partial^2 u}{\partial x^2} = f(u,x,t,\varepsilon), \tag{3.3.1}$$

$$(x,t) \in \Omega = (0 < x < 1) \times (0 < t < T),$$

$$u(x,0,\varepsilon) = \varphi(x), \tag{3.3.2}$$

$$\frac{\partial u}{\partial x}(0,t,\varepsilon) = 0, \tag{3.3.3}$$

$$\frac{\partial u}{\partial x}(1,t,\varepsilon) = 0. \tag{3.3.4}$$

Here u is a scalar function, and $\varepsilon > 0$ is a small parameter.

We suppose that the following condition is satisfied.

1^0. *Functions $a(x)$, $b(x)$, $\varphi(x)$ and $f(u,x,t,\varepsilon)$ are sufficiently smooth and $a(x) > 0$, $b(x) > 0$ for $0 \le x \le 1$.*

The initial and boundary conditions are not assumed to be matched, in particular, we allow $\varphi'(0) \neq 0$. Other conditions will be imposed during the construction of the asymptotics.

We will construct the asymptotic solution of (3.3.1) – (3.3.4) with an accuracy of order ε^2 in the form

$$U(x,t,\varepsilon) = \sum_{i=0}^{1} \varepsilon^i (\overline{u}_i(x,t) + \tilde{\Pi}_i(x,\tau,\xi) + S_i(\xi,\tau)). \tag{3.3.5}$$

The purpose and method of defining each of the terms in the expression for U, as well as the form of the boundary layer variables τ and ξ, are described below.

The main term $\overline{u}_0(x,t)$ of the regular part of the asymptotics is defined as the solution of the reduced equation

$$b(x)\frac{\partial \overline{u}_0}{\partial x} = f(\overline{u}_0,x,t,0), \qquad 0 \le x \le 1, \tag{3.3.6}$$

with the initial condition $\partial\overline{u}_0/\partial x(0,t) = 0$ (as follows from (3.3.3)). Setting $x = 0$ in (3.3.6) and taking into account the initial condition, we obtain the equation

$$f(\overline{u}_0(0,t), 0, t, 0) = 0 \tag{3.3.7}$$

for $\overline{u}_0(0,t)$.

2^0. *Let the equation* (3.3.7)*have a solution (root)* $\overline{u}_0(0,t) = \alpha(t)$ *such that*

$$f_u(\alpha(t), 0, t, 0) < 0, \qquad 0 \leq t \leq T. \tag{3.3.8}$$

Note that due to nonlinearity, equation (3.3.7) might have several such roots. The unique root $\alpha(t)$ can be chosen as follows.

Consider an auxiliary problem

$$\frac{\partial z}{\partial \tau} = f(z, 0, 0, 0), \qquad \tau \geq 0; \quad z(0) = \varphi(0), \tag{3.3.9}$$

where $\varphi(0)$ is the value of $\varphi(x)$ at $x = 0$ (cf. (3.3.2)). By virtue of (3.3.8) the point $z = \alpha(0)$ is an asymptotically stable rest point of (3.3.9).

3^0. *Let* $\varphi(0)$ *belong to the domain of attraction of the rest point* $z = \alpha(0)$, *i.e., suppose the solution* $z(\tau)$ *of the problem* (3.3.9)*exists and satisfies* $z(\tau) \to \alpha(0)$ *as* $\tau \to \infty$.

This condition allows us to uniquely choose the appropriate root of (3.3.7). Now to find the function $\overline{u}_0(x,t)$, we must solve (3.3.6) with the initial condition $\overline{u}_0(0,t) = \alpha(t)$.

4^0. *Let* (3.3.6)*with the initial condition* $\overline{u}_0(0,t) = \alpha(t)$ *have a solution in the interval* $0 \leq x \leq 1$.

Function $\overline{u}_1(x,t)$ can be found as the solution of the linear differential equation

$$b(x)\frac{\partial \overline{u}_1}{\partial x} = \overline{f}_u(x,t)\overline{u}_1 + f_1(x,t), \qquad 0 \leq x \leq 1, \tag{3.3.10}$$

with the initial condition $\partial\overline{u}_1/\partial x(0,t) = 0$. Here

$$\overline{f}_u(x,t) = f_u(\overline{u}_0(x,t), x, t, 0) \quad \text{and} \quad f_1(x,t) = \overline{f}_\varepsilon(x,t) - \frac{\partial \overline{u}_0}{\partial t}$$

are known functions. Substituting $x = 0$ into (3.3.10) and taking into account the initial condition, we obtain the initial value $\overline{u}_1(0,t) = -f_1(0,t)/\overline{f}_u(0,t)$. The solution of (3.3.10) with this initial condition is defined uniquely.

Functions $\overline{u}_0(x,t)$ and $\overline{u}_1(x,t)$ introduce a discrepancy in the initial condition (3.3.2). To compensate for it we must construct boundary functions $\Pi_0(x,\tau)$ and $\Pi_1(x,\tau)$, where $\tau = t/\varepsilon$. It should be mentioned that the functions $\tilde{\Pi}_i(x,\tau,\xi)$ entering the asymptotic series (3.3.5) are smoothed boundary functions. They will be introduced only after constructing of the ordinary boundary functions Π_0 and Π_1.

For Π_0 we obtain the first-order partial differential equation

$$\frac{\partial \Pi_0}{\partial \tau} + b(x)\frac{\partial \Pi_0}{\partial x} = \Pi_0 f \equiv f(\overline{u}_0(x,0) + \Pi_0, x, 0, 0) - f(\overline{u}_0(x,0), x, 0, 0), \tag{3.3.11}$$

$$(x,\tau) \in D = (0 \leq x \leq 1) \times (0 \leq \tau < \infty),$$

with additional conditions

$$\Pi_0(x,0) = \varphi(x) - \overline{u}_0(x,0), \tag{3.3.12}$$

$$\frac{\partial \Pi_0}{\partial x}(0,\tau) = 0. \tag{3.3.13}$$

The characteristic

$$\tau = B(x) = \int_0^x \frac{dx}{b(x)},$$

emerging from the corner point $(0,0)$, divides the domain D into two parts. For $\tau \le B(x)$ the solution is defined by condition (3.3.12).

5^0. *Let* (3.3.11) *with initial condition* (3.3.12) *have a solution* $\Pi_0 = \overset{(1)}{\Pi}_0(x,\tau)$ *in the region* $\{0 \le x \le 1,\quad 0 \le \tau \le B(x)\}$.

To construct the solution for $\tau \ge B(x)$, let us first find $\Pi_0(0,\tau)$. For this we substitute $x = 0$ into (3.3.11) and use condition (3.3.13). We obtain the following equation for $\Pi_0(0,\tau)$:

$$\frac{\partial \Pi_0(0,\tau)}{\partial \tau} = f(\overline{u}_0(0,0) + \Pi_0(0,\tau), 0, 0, 0), \qquad \tau \ge 0.$$

From (3.3.12) we have the initial condition $\Pi_0(0,0) = \varphi(0) - \overline{u}_0(0,0)$. The change of variables $z = \Pi_0(0,\tau) + \overline{u}_0(0,0) = \Pi_0(0,\tau) + \alpha(0)$ transforms the problem for $\Pi_0(0,\tau)$ into the problem (3.3.9) for z. By virtue of condition 3^0 the solution $\Pi_0(0,\tau)$ exists for $\tau \ge 0$ and $\Pi_0(0,\tau) \to 0$ as $\tau \to \infty$. Let us denote this solution by $\beta(\tau)$. Condition (3.3.8) provides the exponential estimate $|\beta(\tau)| \le c\exp(-\kappa\tau)$ for $\beta(\tau)$. To find $\Pi_0(x,\tau)$ for $\tau \ge B(x)$, we now have to solve (3.3.11) with the initial condition $\Pi_0(0,\tau) = \beta(\tau)$. We denote $\Pi_0(x,\tau)$ for $\tau \ge B(x)$ by $\overset{(2)}{\Pi}_0(x,\tau)$. Because of the exponential estimate for $\beta(\tau)$, $\overset{(2)}{\Pi}_0(x,\tau)$ has a similar estimate:

$$|\overset{(2)}{\Pi}_0(x,\tau)| \le c\exp(-\kappa\tau), \qquad 0 \le x \le 1, \quad \tau \ge B(x). \tag{3.3.14}$$

It is obvious that the same estimate also holds for the whole function $\Pi_0(x,\tau)$.

The function $\Pi_1(x,\tau)$ is defined as the solution of the linear problem

$$\frac{\partial \Pi_1}{\partial \tau} + b(x)\frac{\partial \Pi_1}{\partial x} = \Pi_1 f \equiv f_u(x,\tau)\Pi_1 + \pi_1(x,\tau), \qquad (x,\tau) \in D, \tag{3.3.15}$$

$$\Pi_1(x,0) = -\overline{u}_1(x,0), \qquad \frac{\partial \Pi_1}{\partial x}(0,\tau) = 0. \tag{3.3.16}$$

Here $f_u(x,\tau) = f_u(\overline{u}_0(x,0) + \Pi_0(x,\tau), x, 0, 0)$, and $\pi_1(x,\tau)$ is a known function which can be written in explicit form and has an estimate of the type (3.3.14). Function $\Pi_1(x,\tau)$ (like $\Pi_0(x,\tau)$) is defined for $\tau \le B(x)$ with the help of the first condition in (3.3.16), and for $\tau \ge B(x)$ with the help of the second condition in (3.3.16). Let us introduce the notation: $\Pi_1 = \overset{(1)}{\Pi}_1(x,\tau)$ for $\tau \le B(x)$ and $\Pi_1 = \overset{(2)}{\Pi}_1(x,\tau)$ for $\tau \ge B(x)$.

For $\overset{(2)}{\Pi}_1(x,\tau)$ (and thus, for the entire function $\Pi_1(x,\tau)$), an estimate of the type (3.3.14) holds.

Note that Π_0 and Π_1 are continuous functions in the domain D but their derivatives are discontinuous on the characteristic $\tau = B(x)$. This fact does not allow us to prolong the iteration process of constructing the Π-functions any further. Moreover, the functions Π_0 and Π_1 that we constructed do not satisfy (3.3.11) and (3.3.15), respectively, on the characteristic $\tau = B(x)$, and therefore they are not sufficient for obtaining the asymptotic approximation to the solution with an accuracy $O(\varepsilon^2)$.

Let us apply the smoothing procedure to the Π-functions. First, let us consider a smooth continuation of $\overset{(1)}{\Pi}_i(x,\tau)$ to the region $\tau > B(x)$ and a smooth continuation of $\overset{(2)}{\Pi}_i(x,\tau)$ to the region $\tau < B(x)$. The continuation of $\overset{(1)}{\Pi}_0(x,\tau)$ can be constructed, e.g., as follows: we continue smoothly the functions $b(x)$, $f(u,x,0,0)$, $\varphi(x)$ and $\overline{u}_0(x,0)$ to the region $x < 0$ preserving the sign of $b(x)$, and then solve (3.3.11) in the domain $\{x \le 1,\ \tau \ge 0\}$ with the initial condition (3.3.12). The resulting smooth solution coincides with $\overset{(1)}{\Pi}_0(x,\tau)$ in the region $\{0 \le x \le 1,\ 0 \le \tau \le B(x)\}$. In a similar manner the smooth continuations for the other $\overset{(k)}{\Pi}_i$-functions can be constructed.

Let us introduce the new independent variable $\xi = (B(x) - \tau)/\varepsilon$ and the new (smoothed) Π- functions:

$$\tilde{\Pi}_i(x,\tau,\xi) = \overset{(1)}{\Pi}_i(x,\tau)g(\xi) + \overset{(2)}{\Pi}_i(x,\tau)g(-\xi),$$

where

$$g(\xi) = \frac{1}{\sqrt{\pi}}\int_{-\infty}^{\xi} \exp(-s^2)ds.$$

Functions $\tilde{\Pi}_i$, as well as the continuations of $\overset{(1)}{\Pi}_i$ and $\overset{(2)}{\Pi}_i$, are smooth in the entire domain D. They differ from $\Pi_i(x,\tau)$ only outside an arbitrarily small vicinity of the characteristic $\tau = B(x)$ since $g(-\infty) = 0$ and $g(\infty) = 1$. Furthermore, for $|\tau - B(x)| \ge A\varepsilon|\ln\varepsilon|$ the difference $\tilde{\Pi}_i - \Pi_i$ is of the order ε^N, where N is arbitrarily large for sufficiently large A. However, near the characteristic $\tau = B(x)$, the functions $\tilde{\Pi}_i$ differ from Π_i by order ε. Therefore the substitution of the smooth functions $\tilde{\Pi}_i$ for the nonsmooth functions Π_i introduces discrepancies of the order ε into (3.3.1) and conditions (3.3.2), (3.3.3).

Simple, but quite cumbersome, calculations show that the discrepancy introduced by the function $\tilde{\Pi} = \tilde{\Pi}_0 + \varepsilon\tilde{\Pi}_1$ in the equation has the form

$$L_\varepsilon\tilde{\Pi} - \tilde{\Pi}f = \varepsilon q(\tau)(\xi^2 - 1)\exp(-\xi^2) + O(\varepsilon^2).$$

Here

$$\tilde{\Pi}f = f(\overline{u} + \tilde{\Pi}, x, t, \varepsilon) - f(\overline{u}, x, t, \varepsilon),$$

$$q(\tau) = \frac{2b(0)}{\sqrt{\pi}}\varphi'(0)\alpha(\tau)\exp\left\{\int_0^{B^{-1}(\tau)} [f_u(s)/b(s)]ds\right\},$$

$$\alpha(\tau) = a(B^{-1}(\tau))/b^2(B^{-1}(\tau)), \qquad f_u(s) = f_u(\overline{u}_0(s,0) + \Pi_0(s, B(s)), s, 0, 0),$$

and $B^{-1}(\tau)$ is the function inverse to $\tau = B(x)$.

Calculation of errors introduced in conditions (3.3.2) and (3.3.3) leads to the following:

for $t = 0,\ x \geq 0$ (or, $\tau = 0,\ \xi \geq 0$), we obtain

$$\overline{u}(x, 0) + \tilde{\Pi}(x, 0, \xi) - \varphi(x) = -\varepsilon b(0)\varphi'(0)\xi g(-\xi) + O(\varepsilon^2);$$

for $x = 0,\ t \geq 0$ (or, $\tau = -\varepsilon\xi, \quad \xi \leq 0$), we obtain

$$\left.\frac{\partial\tilde{\Pi}}{\partial x} + \frac{\partial\tilde{\Pi}}{\partial\xi}\frac{\partial\xi}{\partial x}\right|_{x=0} = \varphi'(0)(\xi g(\xi))' + O(\varepsilon)h(\xi),$$

where $h(\xi) = O(\xi(\xi g(\xi))')$.

To compensate for the leading parts of these discrepancies, we construct the function $S = S_0(\xi, \tau) + \varepsilon S_1(\xi, \tau)$. Corresponding to its purpose, S must satisfy the following equation with an accuracy of order ε^2:

$$L_\varepsilon S - Sf = -\varepsilon q(\tau)(\xi^2 - 1)\exp(-\xi^2),$$

where

$$Sf = f(\overline{u} + \tilde{\Pi} + S, x, t, \varepsilon) - f(\overline{u} + \tilde{\Pi}, x, t, \varepsilon).$$

The additional conditions for S are

$$S|_{\tau=0,\ \xi\geq 0} = \varepsilon b(0)\varphi'(0)\xi g(-\xi), \qquad \left.\frac{\partial S}{\partial\xi}\right|_{\tau=-\varepsilon\xi,\ \xi\leq 0} = -\varepsilon b(0)\varphi'(0)(\xi g(\xi))'.$$

Since the right-hand sides in the equation and in the initial conditions are of the order ε, $S_0 = 0$, and for S_1 we obtain the problem

$$\frac{\partial S_1}{\partial\tau} - \alpha(\tau)\frac{\partial^2 S_1}{\partial\xi^2} - f_u(B^{-1}(\tau))S_1 = -q(\tau)(\xi^2 - 1)\exp(-\xi^2), \tag{3.3.17}$$

$$S_1(\xi, 0)|_{\xi\geq 0} = b(0)\varphi'(0)\xi g(-\xi), \tag{3.3.18}$$

$$\left.\frac{\partial S_1}{\partial\xi}\right|_{\tau=-\varepsilon\xi,\ \xi\leq 0} = -b(0)\varphi'(0)(\xi g(\xi))', \tag{3.3.19}$$

Note that the operator $L_0 = \partial/\partial\tau - \alpha(\tau)\partial^2/\partial\xi^2$ is the leading part of the operator L_ε written in the variables ξ, τ, i.e. $L_\varepsilon = L_0 + \varepsilon L_1 + \ldots$.

The inconvenience of the boundary condition (3.3.19) is that it is given on a half-line $\tau = -\varepsilon\xi,\ \xi \leq 0$. Let us substitute it by a condition on the half-axis $\tau = 0,\ \xi \leq 0$:

$$\left.\frac{\partial S_1}{\partial\xi}(\xi, 0)\right|_{\xi\leq 0} = -b(0)\varphi'(0)(\xi g(\xi))'.$$

Integrating this expression with respect to ξ and taking into account that $S_1(0,0) = 0$ (cf. (3.3.18)), we obtain

$$S_1(\xi,0)|_{\xi\le 0} = -b(0)\varphi'(0)\xi g(\xi).$$

Therefore $S_1(\xi,\tau)$ is defined as the solution of (3.3.17) with the initial condition

$$S_1(\xi,0) = \begin{cases} b(0)\varphi'(0)\xi g(-\xi), & \xi \ge 0, \\ -b(0)\varphi'(0)\xi g(\xi), & \xi \le 0. \end{cases}$$

The solution of this problem can be found explicitly:

$$S_1(\xi,\tau) = v(\xi,\tau')\exp\left\{\int_0^{B^{-1}(\tau)} [f_u(s)/b(s)]ds\right\}, \tag{3.3.20}$$

where

$$v(\xi,\tau') = -\frac{b(0)\varphi'(0)}{\sqrt{\pi}}\left(\xi\int_{\xi/2\sqrt{\tau'}}^{\xi} \exp(-s^2)ds - \sqrt{\tau'}\exp\left(-\frac{\xi^2}{4\tau'}\right)\right),$$

$$\text{for} \quad \tau' = \int_0^\tau \alpha(s)ds.$$

The following estimate for S_1 can be easily derived from the expressions above:

$$|S_1(\xi,\tau)| \le c\exp\left(-\frac{\kappa t}{\varepsilon}\right), \qquad 0 \le x \le 1,\ 0 \le t \le T.$$

Thus, all terms of the asymptotic expansion (3.3.5) are determined.

Construction of the asymptotic terms to higher orders comes across the principal difficulties connected, in particular, with the errors appearing due to translation of the condition (3.3.19) to the half-axis $\tau = 0$, $\xi \le 0$.

3.3.2 Estimation of the remainder term

Theorem 3.4 *Under conditions* 1^0–5^0 *and for sufficiently small* ε, *problem* (3.3.1)–(3.3.4)*has a unique solution* $u(x,t,\varepsilon)$ *that satisfies the estimate*

$$\max_{\overline{\Omega}} |u(x,t,\varepsilon) - U(x,t,\varepsilon)| = O(\varepsilon^2).$$

Let us discuss the outline of the proof, omitting details. First, note that the constructed function $U(x,t,\varepsilon)$ does not satisfy the boundary condition (3.3.4). To compensate, we must add boundary functions near the side $x = 1$. We denote these functions by $R_i(\zeta,t)$, $\zeta = (1-x)/\varepsilon^2$. They are defined in the standard way with the help of the boundary layer operator $a(1)\partial^2/\partial\zeta^2 + b(1)\partial/\partial\zeta$ and decay exponentially as $\zeta \to \infty$. The leading term of this boundary layer part of the solution is of the order ε^2, i.e., $R_0 = R_1 = 0$. Therefore we need not include these functions in the final asymptotics for a remainder term of the order ε^2, but to prove the theorem it is worthwhile to add them to $U(x,t,\varepsilon)$. For the discrepancy introduced by these functions to be of the order ε^2, we should define $R = \varepsilon^2 R_2 + \varepsilon^3 R_3$.

We let $V(x,t,\varepsilon) = U + R$ and $v(x,t,\varepsilon) = u - V$. Substituting $u = V + v$ into (3.3.1), we obtain the equation

$$L_\varepsilon v - f_u(x,t,\varepsilon)v = h(v,x,t,\varepsilon), \tag{3.3.21}$$

for v, where

$$f_u(x,t,\varepsilon) = f_u(V(x,t,\varepsilon),x,t,\varepsilon) \qquad \text{and}$$

$$h = f(V+v,x,t,\varepsilon) - f_u(x,t,\varepsilon)v - \varepsilon\frac{\partial V}{\partial t} + \varepsilon^2 a(x)\frac{\partial^2 V}{\partial x^2} - b(x)\frac{\partial V}{\partial x}.$$

Function h possesses two properties analogous to properties 1* and 2* for the function h of § 3.2.3. The initial and boundary conditions for v are

$$v(x,0,\varepsilon) = O(\varepsilon^2),$$

$$\frac{\partial v}{\partial x}(0,t,\varepsilon) = O\left(\exp\left(-\frac{\kappa t}{\varepsilon^3}\right) + \varepsilon\exp\left(-\frac{\kappa t}{\varepsilon^2}\right)\right), \qquad \frac{\partial v}{\partial x}(1,t,\varepsilon) = O(\varepsilon^2). \tag{3.3.22}$$

Next, we divide the domain $\overline{\Omega}$ into two subdomains: $\omega = (0 \le x \le 1) \times (0 \le t \le A\varepsilon)$ and $\overline{\Omega} - \omega$. Here A is a sufficiently large number so that for $t \ge A\varepsilon$ and a sufficiently small ε, the inequality $f_u(0,t,\varepsilon) \le -\kappa < 0$ holds. Such a choice of A is possible by virtue of condition (3.3.8) and the estimate (3.3.14) for $\Pi_0(x,\tau)$. In each of the subdomains the method of successive approximations can be used to construct the solution of (3.3.21), (3.3.22), and the maximum principle together with the related method of barrier functions is used to obtain estimates of the successive approximations. As a result, we prove the existence and uniqueness of the solution as well as the estimate $\max_{\overline{\Omega}} |v(x,t,\varepsilon)| = O(\varepsilon^2)$. A detailed proof is presented in Butuzov and Mamonov [25].

Exercise

Using the smoothing procedure (where necessary), determine the zeroth- and the first-order approximations to the solutions of

$$\varepsilon\frac{\partial u}{\partial t} + \frac{\partial u}{\partial x} - \varepsilon^2\frac{\partial^2 u}{\partial x^2} = -(1+x)u,$$

$$(x,t) \in \Omega = (0 < x < 1) \times (0 < t < T),$$

$$\frac{\partial u}{\partial x}(0,t) = \frac{\partial u}{\partial x}(1,t) = 0,$$

with (a) $u(x,0) = 1$, (b) $u(x,0) = x$. (For the method to solve the first-order partial differential equations see § 3.6 or, e.g., Tikhonov, Vasil'eva, and Sveshnikov [133].)

3.3.3 Application of the smoothing procedure to some other problems

Consider an initial boundary value problem for

$$\frac{\partial u}{\partial t} + b(x)\frac{\partial u}{\partial x} - \varepsilon^2 a(x)\frac{\partial^2 u}{\partial x^2} = f(u,x,t,\varepsilon), \tag{3.3.23}$$

$$(x,t) \in \Omega = (0 < x < 1) \times (0 < t \leq T).$$

Here the regular terms $\overline{u}_0$ and $\overline{u}_1$ of the asymptotic approximation are nonsmooth on the characteristic

$$t = B(x) = \int_0^x \frac{dx}{b(x)}$$

of the reduced equation. The smoothing procedure in this case is analogous to that described in § 3.3.1. The function $\varepsilon S_1(\xi, t)$ for $\xi = (B(x) - t)/\varepsilon$ appears in the asymptotic expansion. This function is defined as the solution of a parabolic equation (as for the S-function of § 3.3.1). Note that the solution $u(x,t,\varepsilon)$ of the original equation (3.3.23) is an everywhere smooth function, in contrast to the solution $\overline{u}_0(x,t)$ of the reduced equation. Nonsmoothness of the limiting solution $\overline{u}_0(x,t)$ on the characteristic means that in a vicinity of characteristic there exists an internal layer. This layer, in fact, is described by the function $\varepsilon S_1(\xi, t)$, therefore we can call it an *internal layer function*. The smoothing procedure gives one the opportunity to construct an asymptotic solution, uniform in $\overline{\Omega}$, with a remainder of order ε^2.

Remark. If the initial condition $u(x,0,\varepsilon) = \varphi(x)$ is not matched with the boundary condition $u(0,t,\varepsilon) = \psi(t)$, i.e., if $\varphi(0) \neq \psi(0)$, then the solution $\overline{u}_0(x,t)$ of the reduced problem is discontinuous on the characteristic $t = B(x)$. (At the same time, the solution of the original problem is smooth, as before.) In this case, the S-function describing the internal layer already appears in the zeroth approximation.

Problems of convective diffusion with filtration in a porous media, when the convective mass transfer is faster than the diffusion, are modeled by equations of the type (3.3.23). The asymptotic solutions for some typical two-dimensional problems of nonstationary convective diffusion, occuring in the modeling of underground water pollution, are constructed using the smoothing procedure in Bomba [13].

When the Dirichlet boundary conditions are prescribed for parabolic equation (cf. Butuzov and Mamonov [24])

$$\varepsilon \frac{\partial u}{\partial t} + \varepsilon b(x) \frac{\partial u}{\partial x} - \varepsilon^4 a(x) \frac{\partial^2 u}{\partial x^2} = f(u,x,t,\varepsilon),$$

$$(x,t) \in \Omega = (0 < x < 1) \times (0 < t \leq T),$$

the corner boundary functions, needed to describe the boundary layer in the vicinity of the vertex $(0,0)$ of rectangle Ω, are nonsmooth. For example, the corner boundary function $P_0(\xi, \tau)$ when $b(0) > 0$ is defined as the solution of the problem

$$\frac{\partial P_0}{\partial \tau} + b(0) \frac{\partial P_0}{\partial \xi} = P_0 f \equiv f(\overline{u}_0(0,0) + \Pi_0(0,\tau) + Q_0(\xi,0) + P_0, 0,0,0)$$

$$-f(\overline{u}_0(0,0) + \Pi_0(0,\tau), 0,0,0) - f(\overline{u}_0(0,0) + Q_0(\xi,0), 0,0,0) + f(\overline{u}_0(0,0), 0,0,0),$$

$$(\xi > 0, \ \tau > 0),$$

$$P_0(\xi, 0) = -Q_0(\xi, 0), \qquad P_0(0,\tau) = -\Pi_0(0,\tau).$$

Here $\Pi_0(x,\tau)$ and $Q_0(\xi,t)$ are boundary functions (defined in the standard way) describing the boundary layers near the sides $t = 0$ and $x = 0$ of the rectangle Ω, respectively. The function $P_0(\xi,\tau)$ is not smooth on the characteristic $\xi = b(0)\tau$ (it has discontinuous derivatives), and the corner boundary function $P_1(\xi,\tau)$ in the next order approximation is discontinuous on this characteristic. The procedure described above gives the opportunity to smooth the functions $P_i(\xi,\tau)$ and to construct an asymptotic approximation for the solution, which is uniform in $\overline{\Omega}$, with a remainder term of order ε^2.

The smoothing procedure can also be applied to equations of elliptic and hyperbolic types. Different solutions for such equations obtained using the smoothing procedure are discussed in Butuzov and Nesterov [29]–[31]. It is interesting to mention that in all these problems the internal layer function (S-function) is defined as the solution of a parabolic equation, and it has the same form as (3.3.20).

3.4 Systems of equations in critical cases

3.4.1 Statement of a typical problem

As we mentioned in § 2.2, the case when the reduced problem has a family of solutions is called the *critical case*. The algorithm to construct the asymptotic solution for singularly perturbed system of partial differential equations in the critical case is in many ways similar to the algorithm for the system of such ordinary differential equations. Let us consider the example of the Dirichlet boundary value problem for a system of elliptic type (cf. Butuzov and Mamonov [23])

$$\varepsilon^2 \triangle u = A(x,y)u + \varepsilon^2 f(u,x,y,\varepsilon), \qquad (x,y) \in \Omega, \tag{3.4.1}$$

$$u|_{\partial\Omega} = \varphi(x,y). \tag{3.4.2}$$

Here $\triangle = \partial^2/\partial x^2 + \partial^2/\partial y^2$, u is an m- dimensional vector function, Ω is a bounded planar domain with boundary $\partial\Omega$.

Let the following conditions hold.

1^0. *Matrix A, functions f, g, and the boundary $\partial\Omega$ are sufficiently smooth.*

2^0. *Eigenvalues $\lambda_i(x,y)$ of the matrix $A(x,y)$ for $(x,y) \in \overline{\Omega}$ satisfy the following conditions:*

(a) $\lambda_i(x,y) = 0, \quad i = 1,\dots,k < m;$

(b) $\lambda_i(x,y) \neq -a^2 < 0, \quad i = k+1,\dots,m.$

It follows from condition 2^0(a) that, by virtue of $\det A(x,y) = 0$, the critical case holds and the reduced equation $A(x,y)\overline{u} = 0$ has a family of solutions. Condition 2^0(b) means that the eigenvalues $\lambda_i(x,y)$ $(i = k+1,\dots,m)$ do not take real negative values at points $(x,y) \in \overline{\Omega}$.

3^0. *k linearly independent eigenvectors $e_i(x,y)$ $(i = 1,\dots,k)$ correspond to the eigenvalue $\lambda = 0$ at every point $(x,y) \in \overline{\Omega}$.*

The multipicity of the other eigenvalues as well as the number of the eigenvectors corresponding to them are not important and might be different for different points in $\overline{\Omega}$.

To formulate the next condition, let us introduce a matrix $T(x,y)$ transforming $A(x,y)$ to a block-diagonal form:

$$T^{-1}(x,y)A(x,y)T(x,y) = \begin{pmatrix} \Theta & 0 \\ 0 & B(x,y) \end{pmatrix}.$$

Here $B(x,y)$ is an $(m-k)\times(m-k)$-matrix whose eigenvalues equal the eigenvalues $\lambda_i(x,y)$ $(i=k+1,\dots,m)$ of the matrix $A(x,y)$, and Θ is a $(k\times k)$-matrix of zeros by virtue of conditions 2^0a and 3^0.

4^0. *The boundary value problem*

$$\varepsilon^2 \triangle v = B(x,y)v, \qquad (x,y)\in\Omega; \qquad v|_{\partial\Omega}=0 \tag{3.4.3}$$

has only the trivial solution for small ε.

Other conditions will be formulated during the construction of the asymptotic expansion.

Remark 1. For conditions 2^0(b) and 4^0 to hold, it is sufficient that the symmetric matrix $(B(x,y)+B^*(x,y))$ is positive definite in every point $(x,y)\in\overline{\Omega}$ (here matrix B^* represents the transpose of B.)

Remark 2. The first k columns of the matrix $T(x,y)$ are the eigenvectors $e_i(x,y)$ $(i=1,\dots,k)$ of the matrix $A(x,y)$ corresponding to the zero eigenvalue. The first k rows of the matrix $T^{-1}(x,y)$ are the eigenvectors $g_j(x,y)$ $(j=1,\dots,k)$ of the matrix $A^*(x,y)$ corresponding to the zero eigenvalue. Let us denote the $(m\times k)$-matrix with columns $e_i(x,y)$ $(i=1,\dots,k)$ by $e(x,y)$ and the $(k\times m)$-matrix with rows $g_j(x,y)$ $(j=1,\dots,k)$ by $g(x,y)$. From the equality $T^{-1}T=I_m$ it follows that $g(x,y)e(x,y)=I_k$, where I_m and I_k are identity matrices of dimensions $m\times m$ and $k\times k$, respectively.

3.4.2 Construction of the asymptotic solution

The asymptotic expansion of the solution $u(x,y,\varepsilon)$ of problem (3.4.1), (3.4.2) will be constructed in the form

$$u(x,y,\varepsilon)=\overline{u}+\Pi=\sum_{i=0}^{\infty}\varepsilon^i(\overline{u}_i(x,y)+\Pi_i(\rho,l)), \tag{3.4.4}$$

where $\overline{u}$ is the regular part of the asymptotic solution, and Π is the boundary layer part; $\rho=r/\varepsilon$ is a stretched variable, (r,l) are the same local variables in the vicinity of $\partial\Omega$ as the ones introduced in § 3.1.1. Operator $L_\varepsilon=\varepsilon^2\triangle - A(x,y)$ in the (ρ,l) variables has the form

$$L_\varepsilon=\left(\frac{\partial^2}{\partial\rho^2}-A(l)\right)+\varepsilon L_1+\varepsilon^2 L_2+\cdots,$$

where L_1, L_2, ... are linear differential operators containing the derivatives $\partial/\partial\rho$, $\partial/\partial l$, $\partial^2/\partial l^2$; $A(l)=A(x,y)|_{\partial\Omega}$, and the notation $\varphi(l)$, $\overline{u}_0(l)$, $e(l)$, etc., will have analogous meaning.

In the standard way, substituting the series (3.4.4) into (3.4.1) and (3.4.2) and representing the function $f(\overline{u}+\Pi,x,y,\varepsilon)$ as $f=\overline{f}+\Pi f$, we obtain equations for the functions $\overline{u}_i$ and Π_i.

For $\overline{u}_0(x,y)$, we have

$$A(x,y)\overline{u}_0=0.$$

By virtue of condition 3^0, the general solution of this equation is

$$\overline{u}_0(x,y)=\sum_{i=1}^{k}\alpha_{i0}(x,y)e_i(x,y)=e(x,y)\alpha_0(x,y).$$

Here $\alpha_{i0}(x,y)$ are arbitrary scalar functions, $e_i(x,y)$ are the eigenvectors of the matrix $A(x,y)$ corresponding to $\lambda = 0$, and $\alpha_0(x,y)$ is the k-dimensional vector function with elements $\alpha_{i0}(x,y)$. In a similar way, we obtain

$$\overline{u}_1(x,y) = e(x,y)\alpha_1(x,y),$$

where $\alpha_1(x,y)$ is an arbitrary k-dimensional vector function.

For $\Pi_0(\rho, l)$, we obtain the equation

$$\frac{\partial^2 \Pi_0}{\partial \rho^2} = A(l)\Pi_0, \qquad \rho \geq 0,$$

where l enters as a parameter. From (3.4.2) we derive the boundary condition

$$\Pi_0(0,l) = \varphi(l) - \overline{u}_0(l) = \varphi(l) - e(l)\alpha_0(l)$$

at $\rho = 0$. Further, we demand, as usual, that the Π-functions approach zero when $\rho \to \infty$:

$$\Pi_i(\infty, l) = 0.$$

Let us make the change of variables

$$\Pi_0(\rho,l) = T(l)P(\rho,l) \equiv T(l)\begin{pmatrix} P_1 \\ P_2 \end{pmatrix},$$

where $T(l) = T(x,y)|_{\partial\Omega}$ is the matrix which transforms $A(l)$ to block-diagonal form, P_1 has k and P_2 has $m-k$ elements. For P_1, P_2 we obtain the equations

$$\frac{\partial^2 P_1}{\partial \rho^2} = 0, \qquad \frac{\partial^2 P_2}{\partial \rho^2} = B(l)P_2 \tag{3.4.5}$$

with boundary conditions

$$P(0,l) = T^{-1}(l)(\varphi(l) - e(l)\alpha_0(l)), \tag{3.4.6}$$

$$P(\infty, l) = 0.$$

From the first equation of (3.4.5) and the condition at infinity, we obtain $P_1(\rho,l) = 0$. Considering now the first k equalities in the vector equation (3.4.6), and taking into account the structure of the matrix $T^{-1}(l)$ (see Remark 2 in § 3.4.1), we obtain

$$\alpha_0(l) = g(l)\varphi(l).$$

Thus, for yet unknown function $\alpha_0(x,y)$ we have found the boundary condition

$$\alpha_0|_{\partial\Omega} = g(l)\varphi(l). \tag{3.4.7}$$

Next, we solve the second equation of (3.4.5) together with the boundary condition, consisting of the last $m-k$ components of relations (3.4.6), to find $P_2(\rho,l)$. By virtue of condition 2^0b, the function $P_2(\rho, l)$ is defined uniquely and has the exponential estimate $\| P_2(\rho, l) \| \le c\exp(-\kappa\rho)$. Evidently, a similar estimate also holds for $\Pi_0(\rho, l)$:

(3.4.8) $$\| \Pi_0(\rho, l) \| \le c\exp(-\kappa\rho).$$

The equation for $\alpha_0(x, y)$ is obtained from the solvability condition of the equation for $\overline{u}_2(x, y)$:

(3.4.9) $$A(x, y)\overline{u}_2 = \Delta\overline{u}_0 - f(\overline{u}_0, x, y, 0).$$

For solvability of this linear system of algebraic equations with $\det A(x, y) = 0$, it is necessary and sufficient that the right-hand side be orthogonal to the eigenvectors $g_j(x, y)$ $(j = 1, \ldots, k)$ of the matrix $A^*(x, y)$. This condition can be written in the form:

$$g(x, y)(\triangle(e(x, y)\alpha_0(x, y)) - f(e(x, y)\alpha_0(x, y), x, y, 0)) = 0.$$

Opening the brackets and taking into account the identity $ge = I_k$, we obtain the following semi-linear elliptic equation for $\alpha_0(x, y)$:

(3.4.10) $$L\alpha_0 + F(\alpha_0, x, y) = 0, \qquad (x, y) \in \Omega,$$

where

$$L = \triangle + 2\left(g\frac{\partial e}{\partial x}\right)\frac{\partial}{\partial x} + 2\left(g\frac{\partial e}{\partial y}\right)\frac{\partial}{\partial y} + g \triangle e, \qquad F = -gf(e\alpha_0, x, y, 0).$$

5^0. *Let* (3.4.10) *with the boundary condition* (3.4.7) *have a solution.*

So, the function $\overline{u}_0(x, y)$ is completely determined and the solution of (3.4.9) can be written in a form

$$\overline{u}_2(x, y) = e(x, y)\alpha_2(x, y) + \tilde{u}_2(x, y).$$

Here $\alpha_2(x, y)$ is an arbitrary k-dimensional vector function, and $\tilde{u}_2(x, y)$ is a known particular solution of (3.4.9).

The equation for $\overline{u}_3(x, y)$ has the form

$$A(x, y)\overline{u}_3 = \Delta\overline{u}_1 - \overline{f}_u(x, y)\overline{u}_1 - \overline{f}_\varepsilon(x, y),$$

where the matrix $\overline{f}_u$ and the vector function $\overline{f}_\varepsilon$ are computed at the point

$$(\overline{u}_0(x, y), x, y, 0).$$

The solvability condition for this equation provides the equation for $\alpha_1(x, y)$ (since α_1 enters the expression for $\overline{u}_1$):

(3.4.11) $$L_0\alpha_1 = g\overline{f}_\varepsilon(x, y),$$

where $L_0 = L - \overline{f}_u(x, y)$. Analogously to the case of $\alpha_0(l)$, the boundary condition for $\alpha_1(x, y)$ is defined along with the solution of the problem for $\Pi_1(\rho, l)$:

$$\frac{\partial^2 \Pi_1}{\partial \rho^2} = A(l)\Pi_1 + \pi_1(\rho, l),$$

$$\Pi_1(0, l) = -\overline{u}_1(l) = -e(l)\alpha_1(l), \qquad \Pi_1(\infty, l) = 0.$$

Here $\pi_1(\rho, l) = -L_1\Pi_0$ has an exponential estimate of the type (3.4.8).

The change of variables

$$\Pi_1(\rho, l) = T(l) \begin{pmatrix} Q_1 \\ Q_2 \end{pmatrix}$$

leads to the equations

$$\frac{\partial^2 Q_1}{\partial \rho^2} = q_1(\rho, l), \qquad \frac{\partial^2 Q_2}{\partial \rho^2} = B(l)Q_2 + q_2(\rho, l) \tag{3.4.12}$$

and boundary conditions

$$Q_1(0, l) = -\alpha_1(l), \tag{3.4.13}$$

$$Q_2(0, l) = -(T^{-1}(l))_2 e(l)\alpha_1(l) = 0, \tag{3.4.14}$$

$$Q_i(\infty, l) = 0, \qquad i = 1, 2. \tag{3.4.15}$$

Here $q_i(\rho, l) = (T^{-1}(l))_i \pi(\rho, l)$; the subindex $i = 1$ means that the first k rows of matrix $T^{-1}(l)$ are taken; when $i = 2$, the last $m - k$ rows of the matrix $T^{-1}(l)$ are taken. The appearance of zero in the right-hand side of (3.4.14) follows from the fact that $T^{-1}T = E_m$ and from the structure of the matrix T (see Remark 2 in § 3.4.1).

From the first equation of (3.4.12) and condition (3.4.15), we obtain

$$Q_1(\rho, l) = \int_\infty^\rho d\rho_0 \int_\infty^{\rho_0} q_1(\rho_1, l)d\rho_1.$$

This allows us to define from (3.4.13) the boundary value $\alpha_1(l)$ of the function $\alpha_1(x, y)$:

$$\alpha_1|_{\partial\Omega} = \int_0^\infty d\rho_0 \int_\infty^{\rho_0} q_1(\rho_1, l)d\rho_1. \tag{3.4.16}$$

Next, we solve the second equation of (3.4.12) with boundary conditions (3.4.14) and (3.4.15) to find $Q_2(\rho, l)$. Both functions Q_1 and Q_2 have estimates of the type (3.4.8). Thus, the same kind of estimate also holds for $\Pi_1(\rho, l)$. Now to find $\alpha_1(x, y)$ we need only to solve the linear elliptic equation (3.4.11) with boundary condition (3.4.16).

6^0. *Let the problem* (3.4.11), (3.4.16)*have a unique solution.*

In other words, we require that the operator L_0 with Dirichlet boundary conditions does not have a zero eigenvalue.

So, the terms of the first-order approximation have been determined. Higher-order terms $\overline{u}_i$ and Π_i $(i \geq 2)$ of the series (3.4.4) are defined analogously. All the Π-functions have the estimate (3.4.8).

3.4.3 Estimation of the remainder term

Let us introduce the manifold $S = S_1 \cup S_2$, where

$$S_1 = \{(u, x, y) : \ u = \overline{u}_0(x, y); \ (x, y) \in \Omega\},$$

$$S_2 = \{(u, x, y) : \ u = \overline{u}_0(l) + \Pi_0(\rho, l), \ \rho \geq 0; \ (x, y) \in \partial\Omega\}.$$

Theorem 3.5 *Under conditions 1^0—6^0, there exists a number $\delta > 0$ such that for sufficiently small ε problem (3.4.1), (3.4.2)has a unique solution $u(x, y, \varepsilon)$ in a δ- vicinity of the manifold S, and the series (3.4.4)is the asymptotic series for $u(x, y, \varepsilon)$ in the domain $\overline{\Omega}$ as $\varepsilon \to 0$, i.e., the following estimate holds:*

$$\max_{\overline{\Omega}} \| u(x, y, \varepsilon) - U_n(x, y, \varepsilon) \| = O(\varepsilon^{n+1}),$$

where $U_n(x, y, \varepsilon)$ is the nth partial sum of the series (3.4.4).

The outline of the proof of this theorem is similar to that of Theorem 3.3. The differential equations for the remainder term are transformed to the equivalent system of integral equations. Condition 4^0 and the estimate for the Green's function of the problem (3.4.3), proved in Butuzov and Udodov [35], are essential for performing such a transformation and using the method of successive approximations. The detailed proof of Theorem 3.5 is presented in Butuzov and Mamonov [23].

Exercise

Find the zeroth-order approximation to the solution of the system

$$\varepsilon^2 \left(\frac{\partial^2 u}{\partial x^2} + \frac{\partial^2 u}{\partial y^2} \right) = (1 + x^2)(u + v),$$

$$\varepsilon^2 \left(\frac{\partial^2 v}{\partial x^2} + \frac{\partial^2 v}{\partial y^2} \right) = (1 + x^2)(u + v),$$

$$\Omega = \{(x, y) : \ x^2 + y^2 < 1\},$$

subject to boundary conditions

(a) $u|_{\partial\Omega} = 0, \quad v|_{\partial\Omega} = 1;$

(b) $u|_{\partial\Omega} = x|_{\partial\Omega} = \cos\varphi,$ where φ is a polar angle, $v|_{\partial\Omega} = 0.$

Hint. The problem for the regular functions of the zeroth order in (b) can be easily solved in polar coordinates.

3.4.4 Some other systems in critical cases

If, in (3.4.1), we change the small factor ε^2 multiplying $f(u, x, t, \varepsilon)$ to ε and leave the term ε^2 in front of Δu on the left-hand side of the equation, the form of the asymptotic solution will change. Along with the regular terms $\overline{u}_i(x, y)$ and the boundary layer functions $\Pi_i(\rho, l)$, boundary layer functions of a different type will appear in the asymptotic expansion of the solution. These new functions will depend on the stretched variable $\tau = \rho/\sqrt{\varepsilon}$ and on l as a parameter. In this case the asymptotic series is constructed in powers of $\sqrt{\varepsilon}$. Detailed discussion of this problem is presented in Butuzov and Urazgil'dina [36], where the Dirichlet as well as the Neumann and the Robin boundary value problems are considered.

If the problem (3.4.1), (3.4.2) is solved in a domain whose boundary contains corner points, e.g., in a rectangle, then corner boundary functions appear in the asymptotic solution. Due to the nonsmoothness of the boundary, it is possible to construct asymptotic approximation, which is uniform in $\overline{\Omega}$, only to the zeroth order, with the remainder of order ε (for details see Butuzov and Nikitin [33]).

In Butuzov and Kalachev [19] the initial boundary value problem for a system of parabolic equations in the critical case was considered:

$$\varepsilon^2\left(\frac{\partial u}{\partial t} - a(x,t)\frac{\partial^2 u}{\partial x^2}\right) = A(x,t)u + \varepsilon^2 f(u,x,t,\varepsilon), \tag{3.4.17}$$

$$(x,t) \in \Omega = (0 < x < 1) \times (0 < t \leq T),$$

where u is an m-dimensional vector function. Eigenvalues of the matrix $A(x,t)$ are assumed to satisfy the following condition:

$$\lambda_i(x,t) = 0,\ i = 1,\ldots,k < m; \qquad \mathrm{Re}\lambda_i(x,t) < 0,\ i = k+1,\ldots,m.$$

The asymptotic solution to the zeroth order, uniform in $\overline{\Omega}$, was constructed for this problem, and the corresponding theorem on the estimation of the remainder was proved. The asymptotic approximation contains the regular term (one of the solutions of the reduced equation $A(x,t)\overline{u} = 0$), ordinary boundary functions describing the boundary layers in the vicinities of the sides $t = 0$, $x = 0$ and $x = l$ of the rectangle Ω, and corner boundary functions, describing the boundary layer near the corner points $(0,0)$ and $(l,0)$.

If the small factor ε^2 in the right-hand side of the equation (3.4.17) is substituted by ε, then ordinary and corner boundary functions of two types will appear in the asymptotic solution. Along with the boundary functions depending on the stretched variables $\tau_1 = t/\varepsilon^2$, $\xi_1 = x/\varepsilon$, and $\xi_{1*} = (l-x)/\varepsilon$ (as for (3.4.17)), there will be the boundary functions depending on the stretched variables $\tau_2 = t/\varepsilon$, $\xi_2 = x/\sqrt{\varepsilon}$ and $\xi_{2*} = (l-x)/\sqrt{\varepsilon}$. It is possible to construct the asymptotic approximation for the solution of such problem with Dirichlet boundary conditions with an accuracy of order ε.

3.4.5 Example: nonisothermal chemical reaction

So far, we discussed problems for partial differential equations in the critical case, where the reduced equation was an algebraic equation of the type $A\overline{u}_0 = 0$ and $\det A = 0$. The algorithm does not change substantially when the reduced equation is a nonlinear algebraic equation that has a family of solutions. (Examples of applied problems in the critical case, where the reduced equations are not algebraic but differential, are considered in Chapter 4, §§4.1, 4.2; see also § 3.5.2.)

Consider a nonisothermal chemical reaction (where one molecule of substance A transforms to one molecule of substance B: $A \to B$) that, in the case of one spatial dimension and in the presence of diffusion and thermo-conduction, is described by the system (cf. Butuzov and Kalachev [20])

$$\varepsilon\left(\frac{\partial u}{\partial t}-a\frac{\partial^2 u}{\partial x^2}\right)=-m\exp\left(\frac{\theta}{1+\beta\theta}\right)u,$$

(3.4.18)

$$\varepsilon\left(\frac{\partial \theta}{\partial t}-b\frac{\partial^2 \theta}{\partial x^2}\right)=n\exp\left(\frac{\theta}{1+\beta\theta}\right)u,$$

subject to the initial and boundary conditions

$$u|_{t=0}=u^0(x), \qquad \theta|_{t=0}=\theta^0(x)=0;$$

(3.4.19)

$$\left.\frac{\partial u}{\partial x}\right|_{x=0,x=1}=\left.\frac{\partial \theta}{\partial x}\right|_{x=0,x=1}=0.$$

Here u $(0\le u\le 1)$ is a relative concentration of substance A, θ is a nondimensionalized temperature, and a, b, m, n, and β are positive constants. The small parameter $0<\varepsilon\ll 1$ is inversely proportional to the time constant of the fast reaction. The boundary conditions correspond to the case when there are no fluxes of mass and heat through the boundaries.

The exact solution of the singularly perturbed problem (3.4.18), (3.4.19) cannot be found. Let us construct the leading terms of the asymptotic approximation of the solution in terms of the small parameter ε. The asymptotics will consist of a regular part and boundary functions describing the behavior of the solution in the vicinity of the initial moment of time and near the boundaries:

$$u(x,t,\varepsilon)=\overline{u}_0(x,t)+\varepsilon\overline{u}_1(x,t)+\Pi_0u(x,\tau)+\cdots,$$

with a similar expression for $\theta(x,t,\varepsilon)$; and for $\tau=t/\varepsilon$. Here we present only those asymptotic terms that will be important for our discussion.

For $\varepsilon=0$, system (3.4.18) reduces to one equation:

$$\exp\left(\frac{\overline{\theta}_0}{1+\beta\overline{\theta}_0}\right)\overline{u}_0=0.$$

Hence, $\overline{u}_0=0$, and $\overline{\theta}_0(x,t)$ is still unknown (therefore we have the critical case).

For the boundary functions $\Pi_0u(x,\tau)$ and $\Pi_0\theta(x,\tau)$, describing the solution near the initial moment of time, we obtain the system

$$\frac{\partial\Pi_0u}{\partial\tau}=-m\exp\left(\frac{\overline{\theta}_0(x,0)+\Pi_0\theta}{1+\beta(\overline{\theta}_0(x,0)+\Pi_0\theta)}\right)\Pi_0u,$$

$$\frac{\partial\Pi_0\theta}{\partial\tau}=n\exp\left(\frac{\overline{\theta}_0(x,0)+\Pi_0\theta}{1+\beta(\overline{\theta}_0(x,0)+\Pi_0\theta)}\right)\Pi_0u.$$

Comparing the right-hand sides of these equations, and using the notation $n/m=r$, we arrive at the equality

$$\frac{\partial\Pi_0\theta}{\partial\tau}=-r\frac{\partial\Pi_0u}{\partial\tau}.$$

Hence,

$$\Pi_0\theta(x,\tau) = -r\Pi_0 u(x,\tau) + c(x),$$

where $c(x)$ is arbitrary. Let us impose the usual condition for boundary functions at infinity: $\Pi_0\theta(x,\infty) = \Pi_0 u(x,\infty) = 0$. Then $c(x) = 0$, and

$$\Pi_0\theta(x,\tau) = -r\Pi_0 u(x,\tau). \tag{3.4.20}$$

From the initial conditions at the zeroth order, we have the following relations:

$$\overline{u}_0(x,0) + \Pi_0 u(x,0) = u^0(x),$$

$$\overline{\theta}_0(x,0) + \Pi_0\theta(x,0) = 0.$$

Taking into account $\overline{u}_0 = 0$ and (3.4.20), we obtain initial conditions for $\Pi_0 u$ and $\overline{\theta}_0$:

$$\Pi_0 u(x,0) = u^0(x), \tag{3.4.21}$$

$$\overline{\theta}_0(x,0) = ru^0(x). \tag{3.4.22}$$

By virtue of (3.4.20) and (3.4.22), the equation for $\Pi_0 u(x,\tau)$ is reduced to

$$\frac{\partial \Pi_0 u}{\partial \tau} = -m\exp\left(\frac{r(u^0(x) - \Pi_0 u)}{1 + \beta r(u^0(x) - \Pi_0 u)}\right)\Pi_0 u = -k(x, \Pi_0 u)\Pi_0 u.$$

Since $k(x, \Pi_0 u) > 0$ for any x and $\Pi_0 u$, this equation with the initial condition (3.4.21) has a unique solution for which the exponential estimate

$$|\Pi_0 u(x,\tau)| \le c\exp(-\kappa\tau) \qquad (0 \le x \le 1, \quad \tau \ge 0)$$

holds. The equation for $\Pi_0 u(x,\tau)$ can be solved implicitly:

$$\int_{u^0(x)}^{\Pi_0 u(x,\tau)} \exp\left(\frac{r(u^0(x) - s)}{1 + \beta r(u^0(x) - s)}\right) ds = -m\tau,$$

Function $\Pi_0\theta(x,\tau)$ is expressed through $\Pi_0 u(x,\tau)$ by (3.4.20) and also has an exponential decay estimate. Thus, the Π-functions at the zeroth order are defined, and for the yet unknown $\overline{\theta}_0(x,t)$, we have the initial condition (3.4.22). The equation for $\overline{\theta}_0(x,t)$ is obtained at the next step of the asymptotic algorithm (a typical situation for problems in the critical case) during investigation of the problem for $\overline{u}_1$ and $\overline{\theta}_1$:

$$0 = \overline{u}_1, \qquad \frac{\partial \overline{\theta}_0}{\partial t} - b\frac{\partial^2 \overline{\theta}_0}{\partial x^2} = n\exp\left(\frac{\overline{\theta}_0}{1 + \beta\overline{\theta}_0}\right)\overline{u}_1.$$

These equations follow from (3.4.18) and are independent of $\overline{\theta}_1$. Hence, $\overline{u}_1 = 0$, $\overline{\theta}_1$ is arbitrary, and $\overline{\theta}_0$ must satisfy the equation

$$\frac{\partial \overline{\theta}_0}{\partial t} - b\frac{\partial^2 \overline{\theta}_0}{\partial x^2} = 0. \tag{3.4.23}$$

Along with the initial condition (3.4.22) for $\overline{\theta}_0$, we should also determine the boundary conditions at $x = 0$ and $x = 1$. This can be done together with constructing the boundary functions describing the behavior of the asymptotic solution in the vicinities of $x = 0$ and $x = 1$ for $0 \le t \le T$. These boundary functions are determined similarly to Π-functions, and it can be shown that at the zeroth-order approximation they are identically zero. Thus, the boundary conditions for $\overline{\theta}_0(x,t)$ have the form

$$\left.\frac{\partial \overline{\theta}_0}{\partial x}\right|_{x=0,x=1} = 0. \tag{3.4.24}$$

The solution $\overline{\theta}_0(x,t)$ of (3.4.23) with initial condition (3.4.22) and boundary conditions (3.4.24) can be written as

$$\overline{\theta}_0(x,t) = \int_0^1 G(x,t,x') r u^0(x') dx',$$

where

$$G(x,t,x') = 1 + 2\sum_{k=1}^{\infty} \exp(-\pi^2 k^2 bt) \cos \pi k x \cdot \cos \pi k x'$$

is a corresponding Green's function (cf., e.g., Tikhonov and Samarskii [132]).

This completes the construction of the zeroth-order terms of the asymptotic solution. It can be shown that they approximate the exact solution with an accuracy of order $\sqrt{\varepsilon}$:

$$u(x,t,\varepsilon) = \Pi_0 u(x,\tau) + O(\sqrt{\varepsilon}),$$

$$\theta(x,t,\varepsilon) = \overline{\theta}_0(x,t) + \Pi_0\theta(x,\tau) + O(\sqrt{\varepsilon}).$$

Terms of the higher-order approximations can be constructed as well.

For singular perturbation analysis of some other chemical kinetics problems in the critical case see, e.g., Bobisud and Christenson [10].

Exercise

Consider a problem

$$\varepsilon^2\left(\frac{\partial p}{\partial t} - D_1\frac{\partial^2 p}{\partial x^2} + p\right) = -M_1(p-n),$$

$$\varepsilon^2\left(\frac{\partial n}{\partial t} - D_2\frac{\partial^2 n}{\partial x^2} + n\right) = M_2(p-n),$$

$$(x,t) \in \Omega = (0 < x < 1) \times (0 < t \le T),$$

$$p(x,0) = n(x,0) = 0,$$

$$-D_1\frac{\partial p}{\partial x}(0,t) + p(0,t) = A, \qquad -D_2\frac{\partial n}{\partial x}(0,t) + n(0,t) = A,$$

$$D_1\frac{\partial p}{\partial x}(1,t) + p(1,t) = 0, \qquad D_2\frac{\partial n}{\partial x}(1,t) + n(1,t) = 0.$$

(a) Derive the equation for the regular function of the zeroth-order $\overline{p}_0$.

(b) Determine the initial condition for $\overline{p}_0$. Show that all the boundary functions of the zeroth order are identically zero.

(c) Define the asymptotically correct *boundary* conditions for $\overline{p}_0$ (show that after the right choice of the boundary conditions for $\overline{p}_0$, the remaining discrepancy in the conditions, which is of the order $O(1)$, can be compensated for by the boundary functions of the first order). Solve the problem for $\overline{p}_0$ and, thus, for $\overline{n}_0$. Note that $p(x,t) = \overline{p}_0(x,t) + O(\varepsilon)$ and $n(x,t) = \overline{n}_0(x,t) + O(\varepsilon)$.

A similar approach can be used to asymptotically derive the *ambipolar diffusion equation* in physics of semiconductors (see Butuzov and Kalachev [22]).

3.5 Periodic solutions

3.5.1 Periodic solutions of parabolic equations

In the study of kinetic systems with distributed parameters described by equations of parabolic type, the problem of finding time-periodic solutions is often of interest. In such systems, each spatial point is a generator of oscillations, and the coupling between them is put into effect through diffusion. We will be interested in the case of small diffusion.

Consider the problem of finding a 2π-time-periodic solution for the scalar parabolic equation with Dirichlet boundary conditions:

$$\varepsilon^2 \frac{\partial^2 u}{\partial x^2} = \frac{\partial u}{\partial t} + s(x,t)u + \varepsilon F(u,x,t,\varepsilon) + f(x,t), \tag{3.5.1}$$

$$u(0,t) = u(l,t) = 0, \tag{3.5.2}$$

$$u(x,t) = u(x,t+2\pi). \tag{3.5.3}$$

We suppose that the functions $s(x,t)$, $f(x,t)$ and $F(u,x,t,\varepsilon)$ are 2π-periodic in t and sufficiently smooth for $0 \le x \le l$, $0 \le t \le 2\pi$, $|u| \le H = \text{const}$, $0 < \varepsilon \le \varepsilon_0$.

We will construct the asymptotic solution of the problem (3.5.1)–(3.5.3) in the form

$$u = \overline{u}(x,t,\varepsilon) + Q(\xi,t,\varepsilon) + Q^*(\xi_*,t,\varepsilon). \tag{3.5.4}$$

Here

$$\overline{u}(x,t,\varepsilon) = \overline{u}_0(x,t) + \varepsilon\overline{u}_1(x,t) + \cdots + \varepsilon^k\overline{u}_k(x,t) + \cdots$$

is the regular series;

$$Q(\xi,t,\varepsilon) = Q_0(\xi,t) + \varepsilon Q_1(\xi,t) + \cdots + \varepsilon^k Q_k(\xi,t) + \cdots$$

is the boundary layer series in the vicinity of $x = 0$ $(\xi = x/\varepsilon)$; and

$$Q^*(\xi_*,t,\varepsilon) = Q_0^*(\xi_*,t) + \varepsilon Q_1^*(\xi_*,t) + \cdots + \varepsilon^k Q_k^*(\xi_*,t) + \cdots$$

is the boundary layer series in the vicinity of $x = l$ $(\xi_* = (x-l)/\varepsilon)$.

In a standard way, substituting (3.5.4) into (3.5.1)–(3.5.3), we obtain the terms of (3.5.4).

For the terms of the regular series we have the equations

$$\frac{\partial \overline{u}_0}{\partial t} + s(x,t)\overline{u}_0 + f(x,t) = 0, \tag{3.5.5}$$

$$\frac{\partial \overline{u}_k}{\partial t} + s(x,t)\overline{u}_k + F_k(x,t) = 0, \qquad k = 1, 2, \ldots, \tag{3.5.6}$$

where $F_1 = F(\overline{u}_0, x, t, 0)$ and the F_k are expressed recursively through $\overline{u}_i$, $i < k$. Equations (3.5.6) are ordinary differential equations, where x enters as a parameter. To find the solutions of these equations we need only use the periodicity condition (3.5.3).

Let

$$a(x) = \frac{1}{2\pi}\int_0^{2\pi} s(x,t)dt > 0 \quad \textit{for} \quad 0 \leq x \leq l. \tag{3.5.7}$$

Consider (3.5.5). Its general solution has the form

$$\overline{u}_0 = c\exp\left(-\int_0^t s(x,t')dt'\right) - \int_0^t \exp\left(-\int_\tau^t s(x,t')dt'\right) f(x,\tau)d\tau.$$

Hence, using (3.5.7) and the periodicity condition $\overline{u}_0(x,0) = \overline{u}_0(x,2\pi)$, we obtain

$$\overline{u}_0 = \frac{\exp\left(-\int_0^t s(x,t')dt'\right)}{\exp(-2\pi a(x)) - 1}\int_0^{2\pi} \exp\left(-\int_\tau^{2\pi} s(x,t')dt'\right) f(x,\tau)d\tau$$

$$- \int_0^t \exp\left(-\int_\tau^t s(x,t')dt'\right) f(x,\tau)d\tau.$$

By the periodicity of s and f, the function $\overline{u}_0(x,t)$ is a 2π-periodic solution of (3.5.5).

The periodic solutions of (3.5.6) are constructed analogously.

Remark. If $s = s(x)$, the periodic solution of (3.5.5) is conveniently represented in the form of the Fourier series

$$\overline{u}_0(x,t) = \sum_{k=-\infty}^{\infty} \alpha_{0k}(x)\exp(ikt),$$

where $\alpha_{0k}(x) = -f_k(x)/(s(x) + ik)$, and $f_k(x)$ are the corresponding Fourier coefficients of the function $f(x,t)$. In the general case ($s = s(x,t)$), the change of variables

$$\overline{u}_0 = v\exp\left(\int_0^t b(x,t')dt'\right),$$

where $b(x,t) = a(x) - s(x,t)$, leads to the equation

$$\frac{\partial v}{\partial t} + a(x)v + f(x,t)\exp\left(-\int_0^t b(x,t')dt'\right) = 0,$$

and the solution of this equation is conveniently sought in the form of a Fourier series.

The boundary functions Q_k are defined by the following equations and auxiliary conditions:

$$\frac{\partial^2 Q_k}{\partial \xi^2} = \frac{\partial Q_k}{\partial t} + s(0,t)Q_k + q_k(\xi,t), \tag{3.5.8}$$

$$Q_k(0,t) = -\overline{u}_k(0,t), \qquad Q_k(\infty,t) = 0, \qquad Q_k(\xi,t) = Q_k(\xi,t+2\pi). \tag{3.5.9}$$

Here the q_k are expressed recursively through Q_i, $i < k$; in particular, $q_0 = 0$.

For $Q_0(\xi,t)$, we have the equation

$$\frac{\partial^2 Q_0}{\partial \xi^2} = \frac{\partial Q_0}{\partial t} + s(0,t)Q_0$$

subject to the additional conditions (3.5.9) with $k = 0$. Let us make the change of variables analogous to that presented in the remark above:

$$Q_0 = v_0 \exp\left(\int_0^t b(0,t')dt'\right),$$

where $b(0,t) = a(0) - s(0,t)$. Then for v_0 we obtain the problem

$$\frac{\partial^2 v_0}{\partial \xi^2} = \frac{\partial v_0}{\partial t} + a(0)v_0,$$

$$v_0(0,t) = -\overline{u}_0(0,t)\exp\left(-\int_0^t b(0,t')dt'\right) \equiv -\overline{v}_0(t), \tag{3.5.10}$$

$$v_0(\xi,t) = v_0(\xi,t+2\pi), \qquad v_0(\infty,t) = 0.$$

We seek the solution of this problem as the Fourier series

$$v_0 = \sum_{k=-\infty}^{\infty} \beta_{0k}(\xi)\exp(ikt). \tag{3.5.11}$$

For $\beta_{0k}(\xi)$ we obtain the boundary value problem

$$\beta''_{0k} = (a(0)+ik)\beta_{0k}, \quad \beta_{0k}(0) = -\overline{v}_{0k}, \quad \beta_{0k}(\infty) = 0. \tag{3.5.12}$$

Here $\overline{v}_{0k}$ is a Fourier coefficient of the function $\overline{v}_0(t)$. Equation (3.5.12) is a second-order ordinary differential equation with constant coefficients. The corresponding characteristic equation has two roots $\lambda_{1,2} = \pm\sqrt{a(0)+ik}$. By virtue of (3.5.7), one of the roots (let it be λ_2) has a negative real part and the other root has a positive real part. Therefore the solution of (3.5.12) is defined uniquely and has the form

$$\beta_{0k}(\xi) = -\overline{v}_{0k}\exp(\lambda_2(k)\xi).$$

Sufficient smoothness of s and f provides the smoothness of $\overline{u}_0(0,t)$ as well as $\overline{v}_0(t)$, which, in turn, guarantees any required order of decay of the Fourier coefficients

$\overline{v}_{0k}$. Let $|\overline{v}_{0k}| \leq c/k^3$. Then the series (3.5.11) converges uniformly and can be differentiated once term-wise with respect to t and twice with respect to ξ. Thus, the series (3.5.11) will in fact be the solution of the problem (3.5.10). Since $\mathrm{Re}\lambda_2(k)$ has the estimate $\mathrm{Re}\lambda_2(k) \leq -\sqrt{a(0)} = -\kappa$ uniformly in k, the following estimate for v_0 and, hence, for Q_0 holds:

$$|Q_0(\xi,t)| \leq c\exp(-\kappa\xi). \tag{3.5.13}$$

For what follows, it is worthwhile to mention that for sufficiently smooth functions s and f, an estimate similar to (3.5.13) will also hold for the derivatives of Q_0 with respect to t.

In contrast with (3.5.8) for $k = 0$, the (3.5.8) for $k = 1$ is nonhomogeneous with

$$q_1 = F(\overline{u}_0(0,t) + Q_0(\xi,t), 0, t, 0) - F(\overline{u}_0(0,t), 0, t, 0) - \frac{\partial s}{\partial x}(0,t)\,\xi\,Q_0(\xi,t)$$

$$= Q_0F - \frac{\partial s}{\partial x}(0,t)\,\xi\,Q_0(\xi,t).$$

After the change of variables

$$Q_1 = v_1 \exp\left(\int_0^t b(0,t')dt'\right),$$

we obtain the problem

$$\frac{\partial^2 v_1}{\partial \xi^2} = \frac{\partial v_1}{\partial t} + a(0)v_1 + \tilde{q}_1(\xi,t),$$

$$v_1(0,t) = -\overline{u}_1(0,t)\exp\left(-\int_0^t b(0,t')dt'\right) = -\overline{v}_1(t),$$

$$v_1(\xi,t) = v_1(\xi,t+2\pi), \qquad v_1(\infty,t) = 0,$$

where

$$\tilde{q}_1 = q_1 \exp\left(-\int_0^t b(0,t')dt'\right).$$

The solution of this problem is also sought in the form of a Fourier series (3.5.11) (where the subindex 0 is changed to a subindex 1). The equation for β_{1k} is also nonhomogeneous:

$$\beta_{1k}'' = (a(0) + ik)\beta_{1k} + \tilde{q}_{1k}(\xi), \quad \beta_{1k}(0) = -\overline{v}_{1k}, \quad \beta_{1k}(\infty) = 0,$$

so

$$\beta_{1k}(\xi) = -\overline{v}_{1k}\exp(\lambda_2(k)\xi) + \int_0^\infty G_k(\xi,\xi_0)\tilde{q}_{1k}(\xi_0)d\xi_0,$$

where G_k is a corresponding Green's function. A direct calculation shows that the Green's function satisfies the estimate

$$|G_k(\xi,\xi_0)| \leq c\exp(-\sqrt{a(0)}|\xi - \xi_0|).$$

For $\beta_{1k}(\xi)$ (as for $\beta_{0k}(\xi)$ above), it is important to obtain the estimate

$$|\beta_{1k}(\xi)| \leq \frac{c}{k^3}\exp(-\kappa\xi),$$

which will guarantee the required rate of convergence of the Fourier series for v_1 and, eventually, an estimate of the type (3.5.13) for Q_1. To show that estimate for β_{1k} holds, it will be sufficient to have

$$|\overline{v}_{1k}| \leq \frac{c}{k^3} \quad \text{and} \quad |\tilde{q}_{1k}(\xi)| \leq \frac{c}{k^3}\exp(-\kappa\xi).$$

The first inequality follows with sufficient smoothness of s, f, and F. Let us prove the estimate for $\tilde{q}_{1k}(\xi)$. We write

$$\tilde{q}_{1k}(\xi) = \frac{1}{2\pi}\int_0^{2\pi} \tilde{q}_1(\xi,t)\exp(-ikt)dt = -\frac{i}{2\pi k^3}\int_0^{2\pi}\frac{\partial^3\tilde{q}_1}{\partial t^3}\exp(-ikt)dt.$$

It follows that to prove the required estimate for $\tilde{q}_{1k}(\xi)$, it is sufficient to establish the inequality

$$\left|\frac{\partial^3\tilde{q}_1}{\partial t^3}\right| \leq c\exp(-\kappa\xi). \tag{3.5.14}$$

First, let us obtain such inequality for $\partial\tilde{q}_1/\partial t$. Since

$$\tilde{q}_1 = \left(Q_0F - \frac{\partial s}{\partial x}(0,t)\xi Q_0(\xi,t)\right)\exp\left(-\int_0^t b(0,t')dt'\right),$$

$$\begin{aligned}\frac{\partial\tilde{q}_1}{\partial t} = &\left\{\left[\frac{\partial Q_0F}{\partial t} - \frac{\partial^2 s}{\partial x\partial t}(0,t)\xi Q_0(\xi,t) - \frac{\partial s}{\partial x}(0,t)\xi\frac{\partial Q_0(\xi,t)}{\partial t}\right]\right.\\ &\left. -b(0,t)\left[Q_0F - \frac{\partial s}{\partial x}(0,t)\xi Q_0(\xi,t)\right]\right\}\exp\left(-\int_0^t b(0,t')dt'\right).\end{aligned} \tag{3.5.15}$$

Let us represent Q_0F in the form

$$Q_0F = Q_0(\xi,t)\int_0^1 F_u(\overline{u}_0(0,t)+\theta Q_0,0,t,0)d\theta.$$

Then

$$\frac{\partial Q_0F}{\partial t} = \frac{\partial Q_0(\xi,t)}{\partial t}\int_0^1 F_u d\theta$$

$$+Q_0(\xi,t)\int_0^1\left(F_{uu}\left(\frac{\partial\overline{u}_0(0,t)}{\partial t} + \theta\frac{\partial Q_0(\xi,t)}{\partial t}\right) + F_{ut}\right)d\theta.$$

Using the estimate (3.5.13) and a similar estimate for $\partial Q_0/\partial t$, which holds when s and f are sufficiently smooth, we obtain

$$\left|\frac{\partial Q_0F}{\partial t}\right| \leq c\exp(-\kappa\xi).$$

Finally, from the inequality above and (3.5.15) it follows that

$$\left|\frac{\partial \tilde{q}_1}{\partial t}\right| \le c\exp(-\kappa\xi).$$

For sufficiently smooth s, f, and F, expressions of the type (3.5.15) can also be obtained for $\partial^2\tilde{q}_1/\partial t^2$ and $\partial^3\tilde{q}_1/\partial t^3$, and hence, for the derivatives (in particular, for $\partial^3\tilde{q}_1/\partial t^3$) inequality (3.5.14) will follow.

Therefore the Fourier series for v_1 will converge uniformly and it can be differentiated term-wise once with respect to t and twice with respect to ξ. Further, for the function v_1, and thus for Q_1, an exponential estimate will hold: $|Q_1(\xi,t)| \le c\exp(-\kappa\xi)$.

The problems (3.5.8), (3.5.9) for $k \ge 2$ are studied analogously. The exponential estimate for the function q_k needed to obtain the exponential estimate for Q_k is easily proved by induction under assumption of sufficient smoothness of s, f and F. The boundary functions Q_k^* are constructed in a similar way.

The following theorem holds.

Theorem 3.6 *For sufficiently small ε, there exists a unique solution $u(x,t,\varepsilon)$ of problem* (3.5.1)–(3.5.3), *and* (3.5.4) *is the asymptotic series for $u(x,t,\varepsilon)$ as $\varepsilon \to 0$, i.e., the estimate*

$$\max_{0\le x\le l,\ 0\le t\le 2\pi} |u(x,t,\varepsilon) - U_n(x,t,\varepsilon)| = O(\varepsilon^{n+1}) \tag{3.5.16}$$

holds, where

$$U_n = \sum_{k=0}^{n} \varepsilon^k(\overline{u}_k + Q_k + Q_k^*)$$

is the nth partial sum of the series (3.5.4).

The theorem can be proved by the usual scheme (see, e.g., Theorem 3.3) with some differences in details. Let us discuss the main ideas of the proof (which is detailed in Vasil'eva [142], Vasil'eva and Butuzov [150], and Vasil'eva and Volkov [160], [161]).

For the remainder term $w = u - U_n$, we obtain the problem

$$\varepsilon^2\frac{\partial^2 w}{\partial x^2} = \frac{\partial w}{\partial t} + s(x,t)w + g(w,x,t,\varepsilon),$$

$$w(0,t) = w(l,t) = 0, \qquad w(x,t) = w(x,t+2\pi).$$

Here function g possesses the same two properties as function h in the proof of Theorem 3.3. Let us make the change of variables

$$w = v\exp\left(\int_0^t b(x,t')dt'\right).$$

Then for v we obtain the problem

$$\varepsilon^2\frac{\partial^2 v}{\partial x^2} = \frac{\partial v}{\partial t} + a(x)v + \varepsilon^2 d(x,t)\frac{\partial v}{\partial x} - h(v,x,t,\varepsilon), \tag{3.5.17}$$

$$v(0,t) = v(l,t) = 0, \qquad v(x,t) = v(x,t+2\pi), \tag{3.5.18}$$

where h possesses the same two properties as g, and d is expressed through s.

This problem is equivalent to the system of integral equations

$$v(x,t) = \int_0^l G(x,x_0,t,0)\varphi(x_0)dx_0 + \int_0^t dt_0 \int_0^l G(x,x_0,t,t_0)h(v,x_0,t_0,\varepsilon)dx_0, \tag{3.5.19}$$

$$\varphi(x) = \int_0^l G(x,x_0,2\pi,0)\varphi(x_0)dx_0 + \int_0^{2\pi} dt_0 \int_0^l G(x,x_0,2\pi,t_0)h(v,x_0,t_0,\varepsilon)dx_0. \tag{3.5.20}$$

Here $\varphi(x) = v(x,0)$, and G is a Green's function for (3.5.17) with boundary conditions $v(0,t) = v(l,t) = 0$ and initial condition $v(x,0)$. Equation (3.5.20) expresses the periodicity of $v(x,t)$. The unknown functions in the system (3.5.19), (3.5.20) are $v(x,t)$ and $\varphi(x)$.

Introducing the notation K_t and N_t for the corresponding linear integral operators, we can rewrite (3.5.19), (3.5.20) in an abbreviated form:

$$v(x,t) = K_t[\varphi] + N_t[h(v)], \tag{3.5.21}$$

$$\varphi(x) = K_{2\pi}[\varphi] + N_{2\pi}[h(v)]. \tag{3.5.22}$$

For a given h, (3.5.22) is a linear nonhomogeneous Fredholm integral equation of the second kind for $\varphi(x)$ with a continuous kernel. The corresponding homogeneous equation has only the trivial solution. Indeed, this is equivalent to the statement about the absence of the nontrivial solutions of (3.5.17), (3.5.18) for $h = 0$, which, in turn, is a consequence of the maximum principle, which holds by virtue of the condition $a(x) > 0$. Therefore (3.5.22) is uniquely solvable for $\varphi(x)$:

$$\varphi(x) = RN_{2\pi}[h(v)],$$

where R is a bounded linear operator. Substituting this expression for $\varphi(x)$ into the equation (3.5.21), we obtain

$$v(x,t) = K_t\left[RN_{2\pi}[h(v)]\right] + N_t[h(v)] \equiv L(v,x,t,\varepsilon).$$

Applying the method of successive approximations to this equation (as in the proof of Theorem 3.3), we can prove the existence and uniqueness of the solution and obtain the estimate

$$\max_{0\le x\le l,\ 0\le t\le 2\pi} |v(x,t)| = O(\varepsilon^{n+1}),$$

as well as (3.5.16).

Exercise
Find the zeroth-order approximation for the solution of

$$\varepsilon^2 \frac{\partial^2 u}{\partial x^2} = \frac{\partial u}{\partial t} + u + \sin t,$$

$$u(0,t) = u(l,t) = 0, \qquad u(x,t) = u(x, t+2\pi).$$

Let us consider some modifications and generalizations of the problem discussed.

(1) The algorithm described above and the method of estimating the remainder term can also be applied to (3.5.1), subject to periodicity conditions in t and Neumann boundary conditions

$$\frac{\partial u}{\partial x}(0,t) = 0, \ \frac{\partial u}{\partial x}(l,t) = 0, \tag{3.5.23}$$

as are typical of problems in chemical and biological kinetics. The asymptotics in some sense is even simpler now since the functions Q_0 and Q_0^* become zero (as in § 3.2.3).

For conditions (3.5.23), the algorithm can be extended to more general equations, where the nonlinearity is not small:

$$\varepsilon^2 \frac{\partial^2 u}{\partial x^2} = \frac{\partial u}{\partial t} + F(u, x, t, \varepsilon).$$

Instead of (3.5.5) and (3.5.6) with $k = 1$, we will now have the equations

$$\frac{\partial \overline{u}_0}{\partial t} + F(\overline{u}_0, x, t, 0) = 0, \tag{3.5.24}$$

$$\frac{\partial \overline{u}_1}{\partial t} + F_u(\overline{u}_0, x, t, 0)\overline{u}_1 + F_\varepsilon(\overline{u}_0, x, t, 0) = 0. \tag{3.5.25}$$

We assume that (3.5.24) has a periodic solution $\overline{u}_0(x,t)$, and that the corresponding variational equation (i.e., the homogeneous equation that we obtain from (3.5.25) when $F_\varepsilon = 0$) does not have any nontrivial periodic solutions. Then in the neighborhood of $\overline{u}_0(x,t)$ there exist no other periodic solutions of (3.5.24), and (3.5.25) has a unique periodic solution $\overline{u}_1(x,t)$. Under the assumption mentioned above, the problems for $\overline{u}_k(x,t)$ $(k > 1)$ will also be uniquely solvable.

The boundary layer function $Q_0(\xi, t)$ is defined as the solution of the problem

$$\begin{aligned} &\frac{\partial^2 Q_0}{\partial \xi^2} = \frac{\partial Q_0}{\partial t} + F(\overline{u}_0(0,t) + Q_0, 0, t, 0) - F(\overline{u}_0(0,t), 0, t, 0), \\ &\frac{\partial Q_0}{\partial \xi}(0,t) = 0, \quad Q_0(\infty, t) = 0, \quad Q_0(\xi, t) = Q_0(\xi, t+2\pi). \end{aligned} \tag{3.5.26}$$

Evidently, $Q_0 = 0$ is a solution of this problem. We demand that the corresponding variational equation

$$\frac{\partial^2 v}{\partial \xi^2} = \frac{\partial v}{\partial t} + F_u(\overline{u}_0(0,t), 0, t, 0)v$$

does not have any nontrivial periodic solutions satisfying the conditions

$$\frac{\partial v}{\partial \xi}(0,t) = 0, \qquad v(\infty, t) = 0.$$

Then in the neighborhood of $Q_0 = 0$ there are no other solutions of (3.5.26), and the problems for Q_i for $i > 0$ are uniquely solvable.

From the results obtained previously in this subsection it follows that our assumptions on the variational equations will be satisfied if (compare with (3.5.7))

$$\int_0^{2\pi} F_u(\overline{u}_0(x,t), x, t, 0)dt > 0, \qquad 0 \leq x \leq l.$$

(2) All the previous results can be extended without major difficulties to the case of several spatial variables, when the Laplace operator Δu (instead of $\partial^2 u/\partial x^2$) is used in (3.5.1). If the spatial domain has a boundary with the corner points (e.g., a rectangle), the expansion will contain boundary functions near each of the sides of a rectangle as well as corner boundary functions.

(3) Let u be a vector and $s(x,t)$ a matrix. A Lyapunov transformation (see, e.g. Iakubovich and Starzhinskii [65]) can be used to transform the system of equations with periodic coefficients to a system with constant coefficients. If we apply the Lyapunov transformation to our system (in this transformation x plays the role of a parameter), it will lead to the system with a matrix $A(x)$ that is in Jordan canonical form. In the simplest case, the matrix $A(x)$ will be diagonal. Making a transition from the equation for w to (3.5.17) we, in fact, performed the Lyapunov transformation for the scalar case. If, from the beginning, $s = s(x)$, then the diagonal elements of $A(x)$ are the eigenvalues of matrix $s(x)$. Therefore the transformed system will be split into equations of the type (3.5.17) coupled through small nonlinear terms. It follows that the results obtained above for the scalar equation can be extended also to systems of equations.

3.5.2 Critical cases

Earlier we defined the *critical case* as that for which the reduced equation has a family of solutions.

Consider the equation obtained from (3.5.1) when $s = 0$ and the factor multiplying F is ε^2:

(3.5.27)
$$\varepsilon^2 \frac{\partial^2 u}{\partial x^2} = \frac{\partial u}{\partial t} + \varepsilon^2 F(u, x, t, \varepsilon) + f(x,t).$$

Let us keep conditions (3.5.2) and (3.5.3). The reduced equation

(3.5.28)
$$\frac{\partial \overline{u}}{\partial t} + f(x,t) = 0$$

has the family of solutions

$$\overline{u} = -\int_0^t f(x,t')dt' + A(x),$$

where $A(x)$ is arbitrary. For this solution to be 2π-periodic in time, it is necessary and sufficient that the following condition holds:

$$\int_0^{2\pi} f(x,t')dt' = 0. \tag{3.5.29}$$

Let the condition (3.5.29)*be satisfied.* We will construct an asymptotic expansion of the solution to (3.5.27), (3.5.2), (3.5.3) in the form (3.5.4). For the leading term $\overline{u}_0(x,t)$ of the regular series, which satisfies (3.5.28) as well as the periodicity condition, we obtain

$$\overline{u}_0(x,t) = A_0(x) - \int_0^t f(x,t')dt', \tag{3.5.30}$$

where $A_0(x)$ is still unknown. For $\overline{u}_1(x,t)$, we have $\partial\overline{u}_1/\partial t = 0$, so $\overline{u}_1 = A_1(x)$, where $A_1(x)$ is also arbitrary. For $\overline{u}_2(x,t)$ we obtain

$$\frac{\partial \overline{u}_2}{\partial t} = \frac{\partial^2 \overline{u}_0}{\partial x^2} - F(\overline{u}_0, x, t, 0) = f_2(x,t),$$

where

$$f_2(x,t) = A_0''(x) - \int_0^t \frac{\partial^2 f}{\partial x^2}(x,t')dt' - F(A_0(x) - \int_0^t f(x,t')dt', x, t, 0). \tag{3.5.31}$$

Hence,

$$\overline{u}_2(x,t) = A_2(x) + \int_0^t f_2(x,t')dt',$$

where $A_2(x)$ is arbitrary. By virtue of the periodicity condition, $f_2(x,t)$ should satisfy a relation

$$\int_0^{2\pi} f_2(x,t')dt' = 0.$$

Substitution of (3.5.31) into this expression will provide the equation for $A_0(x)$, i.e.

$$A_0'' + \Phi(A_0, x) = 0, \tag{3.5.32}$$

with

$$\Phi(A_0, x) = -\frac{1}{2\pi}\int_0^{2\pi}\left[F(A_0(x) - \int_0^t f(x,t')dt', x, t, 0) + \int_0^t \frac{\partial^2 f}{\partial x^2}(x,t')dt'\right] dt.$$

To define $A_0(x)$ from (3.5.32), we need some additional conditions. They appear during the construction of the boundary functions Q_0 and Q_0^*. For $Q_0(\xi,t)$, we need

$$\frac{\partial^2 Q_0}{\partial \xi^2} = \frac{\partial Q_0}{\partial t}, \tag{3.5.33}$$

$$Q_0(0,t) = -\overline{u}_0(0,t), \quad Q_0(\infty,t) = 0, \quad Q_0(\xi,t) = Q_0(\xi,t+2\pi).$$

We represent Q_0 in the form of a Fourier series

$$Q_0 = \sum_{k=-\infty}^{\infty} \beta_{0k}(\xi)\exp(ikt),$$

where $\beta_{0k}(\xi) = -\overline{u}_{0k}\exp(\lambda_2(k)\xi)$, $\lambda_2(k) = \sqrt{ik}$, and $\mathrm{Re}\lambda_2(k) < 0$. Evidently, $\beta_{0k}(\infty) = 0$ for $k > 0$, but $\beta_{00}(\xi) = -\overline{u}_{00}\exp(\lambda_2(0)\xi) = -\overline{u}_{00}$. Thus, for condition $Q_0(\infty, t) = 0$ to hold, it is necessary to demand $\overline{u}_{00} = 0$, i.e.

$$\int_0^{2\pi} \overline{u}_0(0,t)dt = 0.$$

Substituting the expression (3.5.30) for $\overline{u}_0$ into this relation, we obtain the boundary condition for $A_0(x)$ at $x = 0$:

$$A_0(0) = \frac{1}{2\pi}\int_0^{2\pi} dt \int_0^t f(0,t')dt' \equiv A_0^0. \tag{3.5.34}$$

In a similar manner, during construction of the function $Q_0^*(\xi_*, t)$, we obtain a boundary condition for $A_0(x)$ at $x = l$:

$$A_0(l) = A_0^l. \tag{3.5.35}$$

Suppose that the boundary value problem (3.5.32), (3.5.34), (3.5.35)*is solvable.* Then, the leading terms of the series (3.5.4) are completely determined.

To construct terms of the first order, we must consider the equation for $\overline{u}_3$:

$$\frac{\partial \overline{u}_3}{\partial t} = \frac{\partial^2 \overline{u}_1}{\partial x^2} - F_u(\overline{u}_0, x, t, 0)\overline{u}_1 - F_\varepsilon(\overline{u}_0, x, t, 0) = f_3(x,t),$$

where

$$f_3(x,t) = A_1''(x) - F_u(\overline{u}_0, x, t, 0)A_1(x) - F_\varepsilon(\overline{u}_0, x, t, 0).$$

Hence,

$$\overline{u}_3(x,t) = A_3(x) - \int_0^t f_3(x,t')dt'.$$

The periodicity condition leads to the equality

$$\int_0^{2\pi} f_3(x,t')dt' = 0,$$

which can be rewritten as an equation for $A_1(x)$:

$$A_1'' - \left[\frac{1}{2\pi}\int_0^{2\pi} F_u\Big(A_0(x) - \int_0^t f(x,t')dt', x, t, 0\Big)dt\right] A_1 = h_1(x), \tag{3.5.36}$$

where

$$h_1(x) = \frac{1}{2\pi}\int_0^{2\pi} F_\varepsilon(\overline{u}_0(x,t), x, t, 0)dt.$$

Let us write the variational equation corresponding to (3.5.32) as

$$A'' + \frac{\partial \Phi}{\partial A_0}(A_0, x)A = 0. \tag{3.5.37}$$

Taking into account the expression for Φ, it can be easily seen that the homogeneous equation corresponding to (3.5.36) is just the variational equation (3.5.37).

Assume that the variational equation (3.5.37) *with zero boundary conditions* $A(0) = A(l) = 0$ *has only the trivial solution.*

The boundary conditions for A_1 appear during the construction of the functions Q_1 and Q_1^*. The equation for Q_1 is the same as for Q_0 (see (3.5.33)) and the boundary condition at $\xi = 0$ has the form $Q_1(0,t) = -\overline{u}_1(0,t) = -A_1(0)$. Analogously to (3.5.34), we obtain $A_1(0) = 0$. In absolutely the same way, during the construction of Q_1^* we get $A_1(l) = 0$. As a result, $Q_1 = Q_1^* = 0$, and for $A_1(x)$ we have (3.5.36) with zero boundary conditions. By virtue of our assumption on the variational equation (3.5.37), the boundary value problem for $A_1(x)$ has a unique solution.

Higher-order terms of expansion (3.5.4) are defined analogously. For any $k \geq 2$ we have

$$\overline{u}_k = A_k(x) + \int_0^t f_k(x, t')dt',$$

where the $f_k(x,t)$ are expressed through $A_{k-2}(x)$. The periodicity condition

$$\int_0^{2\pi} f_k(x, t')dt' = 0$$

provides a differential equation for $A_{k-2}(x)$. Boundary conditions $A_{k-2}(0)$ and $A_{k-2}(l)$ are defined during the construction of the boundary functions Q_{k-2} and Q_{k-2}^*. For each $A_k(x)$ $(k \geq 1)$, we obtain

$$\begin{gathered} A_k'' + \frac{\partial \Phi}{\partial A_0}(A_0, x)A_k = h_k(x), \\ A_k(0) = A_k^0, \qquad A_k(l) = A_k^l, \end{gathered} \tag{3.5.38}$$

where $h_k(x)$, A_k^0 and A_k^l are known. By virtue of the assumption on (3.5.37), this problem is uniquely solvable.

For Q_k and Q_k^* $(k \geq 2)$, we obtain equations of the type (3.5.33) with the nonhomogeneous terms. Their solutions are constructed in the form of Fourier series and satisfy the estimate $|Q_k| \leq c\exp(-\kappa\xi)$.

Equation (3.5.32) is quite complicated by itself. Therefore it is natural to introduce into (3.5.27) one more small parameter μ that would make it easier to study problem (3.5.32), (3.5.34), (3.5.35), as well as problems (3.5.38).

If we introduce the additional factor μ^2 before $\varepsilon^2 F(u, x, t, \varepsilon)$, the problem (3.5.32), (3.5.34), (3.5.35) will become regularly perturbed:

$$A_0'' + \mu^2 \Phi(A_0, x) = 0, \quad A_0(0) = A_0^0, \quad A_0(l) = A_0^l. \tag{3.5.39}$$

The $\mu = 0$ problem $\overline{A}_0'' = 0$, $\overline{A}_0(0) = A_0^0$, $\overline{A}_0(l) = A_0^l$ is uniquely solvable. For $\mu \neq 0$ sufficiently small, there exists a unique nearby solution of (3.5.39), which could be obtained as a power series in μ^2. With μ^2 present, all the problems (3.5.38) will also be solvable for sufficiently small μ.

If μ^2 is introduced in (3.5.27) as a factor before $\varepsilon^2(\partial^2 u/\partial x^2)$, then the problem (3.5.32), (3.5.34), (3.5.35) will become singularly perturbed:

$$\mu^2 A_0'' + \Phi(A_0, x) = 0, \quad A_0(0) = A_0^0, \quad A_0(l) = A_0^l.$$

We discussed such problems in § 2.3. Here solutions with boundary layers near $x = 0$ and $x = l$, as well as the solutions with internal layers, are possible. The variational problem (3.5.38) with the factor μ^2 before A_k'' is also solvable in these cases.

Therefore for the terms of the series (3.5.4), the asymptotic expansion in terms of the parameter μ can be constructed using known methods. Note, however, that it is possible to prove the existence of solution of (3.5.27), (3.5.2), (3.5.3) only under some additional assumptions.

Exercise

Construct the zeroth-order terms of the asymptotic solution for

$$\varepsilon^2 \frac{\partial^2 u}{\partial x^2} = \frac{\partial u}{\partial t} + \varepsilon^2 u + x \sin t,$$

$$u(0,t) = u(l,t) = 0, \qquad u(x,t) = u(x, t + 2\pi).$$

Now consider (3.5.1) in the case when u is a two-dimensional vector, i.e., (3.5.1) is a system of two scalar equations, and matrix s has the special form:

$$s = \begin{pmatrix} 0 & -1 \\ 1 & 0 \end{pmatrix}.$$

We will denote the components of the vector u by u and v, the components of vector F by F and G, and the components of vector f by f and g. Then the problem (3.5.1)–(3.5.3) can be rewritten as

$$\varepsilon^2 \frac{\partial^2 u}{\partial x^2} = \frac{\partial u}{\partial t} - v + \varepsilon^2 F(u, v, x, t, \varepsilon) + f(x,t),$$

(3.5.40)
$$\varepsilon^2 \frac{\partial^2 v}{\partial x^2} = \frac{\partial v}{\partial t} + u + \varepsilon^2 G(u, v, x, t, \varepsilon) + g(x,t),$$

$$u(0,t) = u(l,t) = v(0,t) = v(l,t) = 0,$$

$$u(x,t) = u(x, t + 2\pi), \qquad v(x,t) = v(x, t + 2\pi).$$

The reduced system

(3.5.41)
$$\frac{\partial \overline{u}}{\partial t} - \overline{v} = -f(x,t), \qquad \frac{\partial \overline{v}}{\partial t} + \overline{u} = -g(x,t)$$

with time periodic conditions for $\overline{u}$ and $\overline{v}$ has the family of solutions

$$\left(\begin{array}{c}\overline{u}\\ \overline{v}\end{array}\right)=A(x)\left(\begin{array}{c}\cos t\\ -\sin t\end{array}\right)+B(x)\left(\begin{array}{c}\sin t\\ \cos t\end{array}\right)+\left(\begin{array}{c}p(x,t)\\ q(x,t)\end{array}\right). \tag{3.5.42}$$

Here $A(x)$ and $B(x)$ are arbitrary functions and $\left(\begin{array}{c}p\\ q\end{array}\right)$ is a particular periodic solution of (3.5.41) which exists if the following conditions are satisfied:

$$\begin{aligned}&\int_0^{2\pi}(f(x,t)\cos t-g(x,t)\sin t)dt=0,\\ &\int_0^{2\pi}(f(x,t)\sin t+g(x,t)\cos t)dt=0.\end{aligned} \tag{3.5.43}$$

Relations (3.5.43) express the orthogonality of the right-hand side $\left(\begin{array}{c}-f\\ -g\end{array}\right)$ to the two linearly independent periodic solutions

$$\left(\begin{array}{c}\cos t\\ -\sin t\end{array}\right) \quad \text{and} \quad \left(\begin{array}{c}\sin t\\ \cos t\end{array}\right)$$

of the corresponding homogeneous system.

Suppose that *conditions* (3.5.43)*hold.* Thus, (3.5.40) is the problem in the critical case: the corresponding reduced problem has a family of solutions depending on two arbitrary functions $A(x)$ and $B(x)$.

It is convenient to represent p and q in the form of a Fourier series:

$$p=\sum_{k=-\infty}^{\infty}p_k(x)\exp(ikt), \qquad q=\sum_{k=-\infty}^{\infty}q_k(x)\exp(ikt).$$

Substituting these series into (3.5.41) and expanding f and g in the similar Fourier series, we obtain equations for the p_k and q_k:

$$ikp_k-q_k=-f_k, \qquad ikq_k+p_k=-g_k.$$

For $k\neq\pm1$, we then have

$$p_k=\frac{g_k+ikf_k}{1-k^2}, \qquad q_k=\frac{ikg_k-f_k}{1-k^2}.$$

If $k=1$ or $k=-1$, the two equations are equivalent and instead of two, we will have only one equation. Indeed, consider, e.g., the case $k=1$. Multiplying the second equation by i we can write the two equations as

$$ip_1-q_1=-f_1, \qquad ip_1-q_1=-ig_1. \tag{3.5.44}$$

The relations

$$f_1=\frac{1}{2\pi}\int_0^{2\pi}f(x,t)\exp(-it)dt=\frac{1}{2\pi}\int_0^{2\pi}(f\cos t-if\sin t)dt,$$

$$g_1 = \frac{1}{2\pi}\int_0^{2\pi}(g\cos t - ig\sin t)dt$$

and (3.5.43) imply the equality $f_1 = ig_1$. Thus, equations (3.5.44) coincide. Since we are interested in the particular solution, we can set $q_1 = f_1$ and $p_1 = 0$. In a similar manner, for $k = -1$ we can take $q_{-1} = f_{-1}$, $p_{-1} = 0$. Any other solution of the system (3.5.44) will differ from the one that we have just found by quantities that can be added to the first two terms in (3.5.42).

The algorithm for constructing the asymptotic solution of (3.5.40) is the same as for the scalar equation. We seek the asymptotic expansion in the form (3.5.4). Then $\overline{u}_0$ and $\overline{v}_0$ are expressed by the formula (3.5.42):

$$\begin{pmatrix} \overline{u}_0 \\ \overline{v}_0 \end{pmatrix} = A_0(x)\begin{pmatrix} \cos t \\ -\sin t \end{pmatrix} + B_0(x)\begin{pmatrix} \sin t \\ \cos t \end{pmatrix} + \begin{pmatrix} p(x,t) \\ q(x,t) \end{pmatrix}.$$

For $\overline{u}_2$ and $\overline{v}_2$, we obtain the system of equations

$$\text{(3.5.45)} \qquad \begin{aligned} \frac{\partial \overline{u}_2}{\partial t} - \overline{v}_2 &= \frac{\partial^2 \overline{u}_0}{\partial x^2} - F(\overline{u}_0, \overline{v}_0, x, t, 0), \\ \frac{\partial \overline{v}_2}{\partial t} + \overline{u}_2 &= \frac{\partial^2 \overline{v}_0}{\partial x^2} - G(\overline{u}_0, \overline{v}_0, x, t, 0). \end{aligned}$$

The solvability condition on the right-hand side of (3.5.45), similar to (3.5.43), provides a system for $A_0(x)$ and $B_0(x)$:

$$A_0'' + \Phi(A_0, B_0, x) = 0, \qquad B_0'' + \Psi(A_0, B_0, x) = 0.$$

Boundary conditions for A_0 and B_0 are obtained during the solution for the boundary functions. The construction of the series (3.5.4) can be continued and it does not differ much from the one described in § 3.5.2. For the next order approximations, we obtain linear nonhomogeneous equations for $A_k(x)$ and $B_k(x)$ (recall that $\overline{u}_k$ and $\overline{v}_k$ depend on A_k and B_k):

$$A_k'' + \frac{\partial \Phi}{\partial A_0}A_k + \frac{\partial \Phi}{\partial B_0}B_k = \varphi_k(x), \qquad B_k'' + \frac{\partial \Psi}{\partial A_0}A_k + \frac{\partial \Psi}{\partial B_0}B_k = \psi_k(x).$$

The construction algorithm for the asymptotic solution changes substantially in the so-called autonomous case, when the right-hand sides in (3.5.40) do not depend on t. Let us write the system in the form

$$\text{(3.5.46)} \qquad \begin{aligned} \varepsilon^2\frac{\partial^2 u}{\partial x^2} &= \frac{\partial u}{\partial t} - v + \varepsilon^2 F(u, v, x, \varepsilon), \\ \varepsilon^2\frac{\partial^2 v}{\partial x^2} &= \frac{\partial v}{\partial t} + u + \varepsilon^2 G(u, v, x, \varepsilon). \end{aligned}$$

The boundary conditions at $x = 0$ and $x = l$ will be the same as before:

$$\text{(3.5.47)} \qquad u(0,t) = u(l,t) = v(0,t) = v(l,t) = 0.$$

As for autonomous ordinary differential equations, the period of the solution of such problem is initially unknown. The reduced system (3.5.41) for $f = g = 0$ has a family of 2π-periodic solutions (3.5.42), where $p = q = 0$. We will seek the period of the original system (3.5.46) in the form of expansion in the powers of ε^2.

Using the Poincaré method, we introduce the new independent variable τ: $t = (1 + \varepsilon^2 g_1 + \varepsilon^4 g_2 + \ldots)\tau$. In variables x and τ (in contrast to system (3.5.40)) we construct the asymptotic solution to (3.5.42) in the form of a *regular series* in ε^2:

$$u = u_0(x, \tau) + \varepsilon^2 u_1(x, \tau) + \varepsilon^4 u_2(x, \tau) + \cdots,$$

$$v = v_0(x, \tau) + \varepsilon^2 v_1(x, \tau) + \varepsilon^4 v_2(x, \tau) + \cdots.$$

For u_0 and v_0, we obtain

$$\frac{\partial u_0}{\partial \tau} - v_0 = 0, \qquad \frac{\partial v_0}{\partial \tau} + u_0 = 0,$$

and hence,

$$\begin{pmatrix} u_0 \\ v_0 \end{pmatrix} = A_0(x) \begin{pmatrix} \cos\tau \\ -\sin\tau \end{pmatrix} + B_0(x) \begin{pmatrix} \sin\tau \\ \cos\tau \end{pmatrix},$$

where $A_0(x)$ and $B_0(x)$ are arbitrary. Equations for u_1 and v_1 have the form

$$\text{(3.5.48)} \qquad \begin{aligned} \frac{\partial u_1}{\partial \tau} - v_1 &= \frac{\partial^2 u_0}{\partial x^2} + v_0 g_1 - F(u_0, v_0, x, 0), \\ \frac{\partial v_1}{\partial \tau} + u_1 &= \frac{\partial^2 v_0}{\partial x^2} - u_0 g_1 - G(u_0, v_0, x, 0). \end{aligned}$$

Orthogonality conditions similar to (3.5.43) lead to the following equations for A_0 and B_0:

$$\text{(3.5.49)} \qquad A_0'' + g_1 B_0 = \Phi(A_0, B_0, x), \qquad B_0'' - g_1 A_0 = \Psi(A_0, B_0, x).$$

From (3.5.47) we have the boundary conditions

$$\text{(3.5.50)} \qquad A_0(0) = A_0(l) = B_0(0) = B_0(l) = 0.$$

Since the unknown g_1 enters (3.5.49), we must impose one more additional condition, e.g.,

$$\text{(3.5.51)} \qquad v(x_0, 0) = 0,$$

where x_0 is some point of the interval $(0, l)$. This yields the condition

$$\text{(3.5.52)} \qquad B_0(x_0) = 0.$$

Assume that we can define $A_0(x)$, $B_0(x)$ and g_1 from the system (3.5.49) with conditions (3.5.50), (3.5.52). Then by virtue of (3.5.48), we obtain

$$\begin{pmatrix} u_1 \\ v_1 \end{pmatrix} = A_1(x) \begin{pmatrix} \cos\tau \\ -\sin\tau \end{pmatrix} + B_1(x) \begin{pmatrix} \sin\tau \\ \cos\tau \end{pmatrix} + \begin{pmatrix} \tilde{u}_1 \\ \tilde{v}_1 \end{pmatrix}.$$

Here $A_1(x)$ and $B_1(x)$ are arbitrary functions and $\tilde{u}_1$, $\tilde{v}_1$ are known functions (the particular solution of (3.5.48)).

The equations for $A_1(x)$, $B_1(x)$ and g_2 are obtained in the next order approximation from the orthogonality condition for the system for u_2 and v_2. They will be the variational equations for (3.5.49):

$$
(3.5.53)\qquad
\begin{aligned}
A_1'' + g_1 B_1 &= \frac{\partial \Phi}{\partial A_0} A_1 + \frac{\partial \Phi}{\partial B_0} B_1 - g_2 B_0 + \varphi_1(x),\\
B_1'' - g_1 A_1 &= \frac{\partial \Psi}{\partial A_0} A_1 + \frac{\partial \Psi}{\partial B_0} B_1 + g_2 A_0 + \psi_1(x).
\end{aligned}
$$

We will not discuss this general statement of the problem in more detail. Instead, let us consider some interesting particular case of system (3.5.46) important for applications.

3.5.3 Example: reaction-diffusion system related to the Van der Pol equation

In the monograph of Romanovskii, Stepanov, and Chernavskii [122] devoted to mathematical models in biophysics and, in particular, to the mathematical description of biochemical reactions, the following system of equations is presented:

$$
(3.5.54)\qquad \varepsilon^2 \frac{\partial^2 u}{\partial x^2} = \frac{\partial u}{\partial t} - v, \qquad \varepsilon^2 \frac{\partial^2 v}{\partial x^2} = \frac{\partial v}{\partial t} + u - \varepsilon^2 (1 - u^2) v.
$$

It can be easily seen that this system is a particular case of (3.5.46). In the absence of diffusion (the terms with the second derivatives in x), system (3.5.54) can be transformed to the ordinary Van der Pol equation (see, e.g., Mitropol'skii [95] and Grasman [56])

$$
(3.5.55)\qquad \frac{\partial^2 u}{\partial t^2} = -u + \varepsilon^2 (1 - u^2) \frac{\partial u}{\partial t},
$$

well-known in the theory of nonlinear oscillations.

Let us impose the additional conditions (3.5.47), (3.5.51) and construct the asymptotic solution according to the scheme described in the previous subsection. The system (3.5.49) in this case has the form

$$
(3.5.56)\quad A_0'' + g_1 B_0 = \frac{A_0}{8}(A_0^2 + B_0^2 - 4), \qquad B_0'' - g_1 A_0 = \frac{B_0}{8}(A_0^2 + B_0^2 - 4).
$$

Multiplying the first equation by B_0 and the second by A_0, subtracting and integrating from 0 to l, and taking into account (3.5.47), we obtain

$$
g_1 \int_0^l (A_0^2(x) + B_0^2(x))dx = 0.
$$

Thus, $g_1 = 0$, and, from (3.5.56), $A_0''/A_0 = B_0''/B_0$. Therefore

$$
\int_0^x (A_0'' B_0 - B_0'' A_0)dx = c = \text{const}, \qquad \text{i.e.,}
$$

$$
(A_0' B_0 - B_0' A_0)|_0^x - \int_0^x (A_0' B_0' - B_0' A_0')dx = A_0'(x) B_0(x) - A_0(x) B_0'(x) = c.
$$

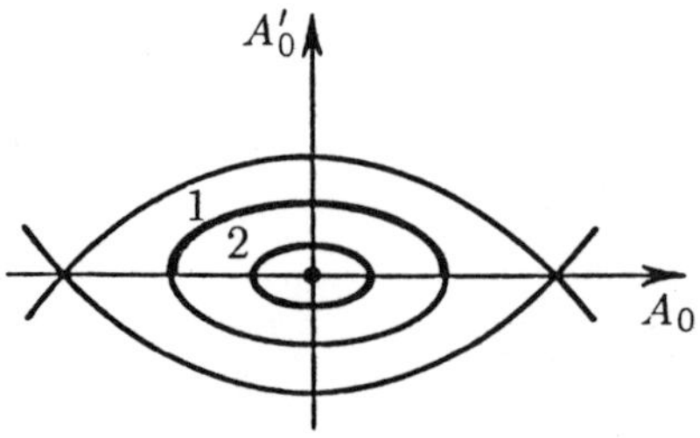

Figure 3.2: *The* cell (*homoclinic orbit formed by two heteroclinic orbits*) *in the phase-plane of the equation for* A_0.

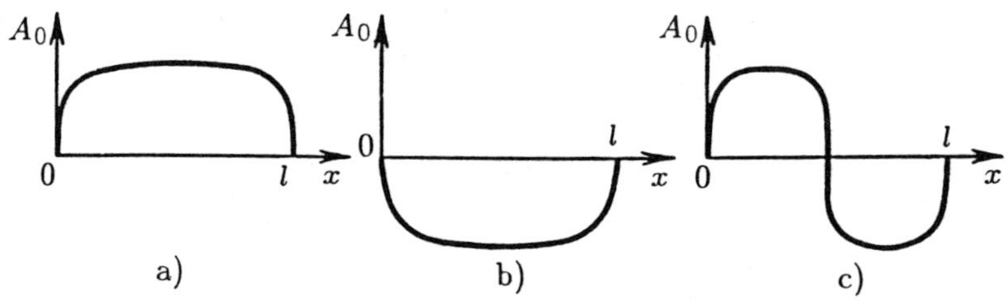

Figure 3.3: *Different possible solutions of the problem for* A_0.

From the condition at $x = l$, we obtain $c = 0$. Therefore $A_0'(x)B_0(x) - B_0'(x)A_0(x) = 0$, and hence, $B_0(x) = pA_0(x)$ where $p =$ const. Condition (3.5.52) implies that $p = 0$. Therefore $B_0(x) = 0$, and $A_0(x)$ is the solution of the boundary value problem

$$A_0'' = \frac{A_0}{8}(A_0^2 - 4), \quad A_0(0) = 0, \quad A_0(l) = 0. \tag{3.5.57}$$

In the phase-plane of (3.5.57), there exists a homoclinic cycle formed by the two heteroclinic orbits (the *cell*) filled by the closed curves (see Fig. 3.2). The boundary conditions for a given l define, generally speaking, infinitely many solutions of the type shown in Fig. 3.3. In this figure, solutions which do not have zeros in the interval $(0, l)$ (Fig. 3.3(a),(b)) and the solution that has one zero in this interval (Fig. 3.3(c)) are presented. Solutions with a greater number of zeros are also possible (but are not shown in the figure). The solution of type (a) corresponds to the upper half of orbit 1. The solution of type (c) corresponds to the phase-trajectory 2. Therefore the solution is not, in general, unique. In a sufficiently small neighborhood of each of the solutions, there are no other solutions of the problem (3.5.57) for a fixed l. Hence, the corresponding homogeneous variational equation has only trivial solution.

Let us take one of the solutions of (3.5.57) that corresponds, e.g., to case (a).

Then u_0 and v_0 are completely determined:

$$\begin{pmatrix} u_0 \\ v_0 \end{pmatrix} = A_0(x) \begin{pmatrix} \cos\tau \\ -\sin\tau \end{pmatrix}.$$

System (3.5.48) is solvable as

$$\begin{pmatrix} u_1 \\ v_1 \end{pmatrix} = A_1(x) \begin{pmatrix} \cos\tau \\ -\sin\tau \end{pmatrix} + B_1(x) \begin{pmatrix} \sin\tau \\ \cos\tau \end{pmatrix} + \begin{pmatrix} \tilde{u}_1 \\ \tilde{v}_1 \end{pmatrix},$$

where $A_1(x)$ and $B_1(x)$ are arbitrary, and $\tilde{u}_1$, $\tilde{v}_1$ is the particular solution of (3.5.48), which can be easily constructed since the right-hand side of (3.5.48) is

$$\begin{pmatrix} A_0''(x)\cos\tau \\ -A_0''(x)\sin\tau + A_0(x)\sin\tau(1 - A_0^2(x)\cos^2\tau) \end{pmatrix}.$$

By virtue of (3.5.57), this expression becomes zero when $x = 0$ and $x = l$. Therefore $\tilde{u}_1(0,\tau) = \tilde{v}_1(0,\tau) = \tilde{u}_1(l,\tau) = \tilde{v}_1(l,\tau) = 0$ as well.

Next, the system (3.5.53) has the form

$$A_1'' = \frac{A_1}{8}(3A_0^2 - 4),$$

(3.5.58)

$$B_1'' = \frac{B_1}{8}(A_0^2 - 4) + g_2 A_0 + \frac{1}{2}(A_0^{(4)} + A_0'') - \frac{3}{8}A_0^2 A_0'' - \frac{A_0^5}{256}.$$

The solution of this system should satisfy the conditions

(3.5.59) $$A_1(0) = A_1(l) = B_1(0) = B_1(l) = B_1(x_0) + \tilde{v}_1(x_0, 0) = 0.$$

From the first equation of (3.5.58), we have $A_1 = 0$, since this is the variational equation for (3.5.57), which, in turn, has only the trivial solution. Regarding the second equation of (3.5.58), the corresponding homogeneous equation with conditions $B_1(0) = B_1(l) = 0$ has, evidently, a nontrivial solution $B_1 = A_0(x)$. Thus, to solve problem (3.5.58), (3.5.59) for B_1, we need the orthogonality of the right-hand side of the second equation in (3.5.58) and A_0. From this condition, we define g_2. After that, we have $B_1 = \lambda A_0 + \tilde{B}_1$, where $\tilde{B}_1(x)$ is a particular solution of the second equation in (3.5.58) and λ is an abitrary factor that can be defined from the last condition (3.5.59). The process of construction of the asymptotic solution can be continued (cf. Dvoryaninov [45]).

Introduction of an additional small parameter μ, as in § 3.5.2, allows us to write the asymptotic approximation of the solution of (3.5.57) in terms of a parameter μ. For example, for the solution shown in Fig. 3.3(a), the expansion has two boundary layers (in the vicinities of $x = 0$ and $x = l$):

$$A_0(x,\mu) = \overline{A}_0(x) + Q_0 + Q_0^* = 2 - \frac{4}{1 + \exp(x/\mu)} - \frac{4}{1 + \exp((l - x)/\mu)} + O(\mu).$$

Parameter μ^2 will enter the the second equation of (3.5.58) as a factor multiplying B_1'' and A_0'', the factor μ^4 will appear before $A_0^{(4)}$. This allows us to construct the asymptotic series for B_1 in powers of μ. Asymptotics in μ for higher-order approximations in ε can also be derived.

It is natural to expect that if we retain only the regular parts of the expansions in μ for coefficients of expansion in ε, this will lead to the asymptotic series in ε for the Van der Pol differential equation (3.5.55). We have $\overline{A}_0 = 2$, $\overline{B}_0 = 0$. From the second equation of (3.5.58) (by virtue of $\overline{A}_0'' = \overline{A}_0^{(4)} = 0$), we obtain $g_2 = 1/16$ and $\overline{B}_1 = 0$. As a result, taking two first terms of the expansion in ε which contains only the regular parts of expansions in μ, we obtain

$$\begin{pmatrix} \overline{u}_0 \\ \overline{v}_0 \end{pmatrix} + \varepsilon^2 \begin{pmatrix} \overline{u}_1 \\ \overline{v}_1 \end{pmatrix} = 2 \begin{pmatrix} \cos(1-\varepsilon^4/16)t \\ \sin(1-\varepsilon^4/16)t \end{pmatrix} - \varepsilon^2/4 \begin{pmatrix} \sin 3(1-\varepsilon^4/16)t \\ 3\cos 3(1-\varepsilon^4/16)t \end{pmatrix}.$$

This coincides with the expression presented in Mitropol'skii [95].

Finally, it is worthwhile to mention that the similar methods can be used to study elliptic (cf. Vasil'eva [141]) and hyperbolic (cf. Vasil'eva and Saidamatov [155]) equations. Some work has been done on stability analysis of different periodic regimes in such singularly perturbed problems (see, e.g. Vasil'eva *et al.* [153]).

3.6 Hyperbolic systems

In this section we will consider some classes of hyperbolic systems with two independent variables. The special feature of this kind of problem is the presence of curves (*characteristics*) that divide the domain of independent variables into regions, in each of which the asymptotic solution is constructed using the method of boundary functions. A typical example of a hyperbolic system is the system of telegraphic equations (cf., e.g., Tikhonov and Samarskii [132]), which will be discussed from the point of view of singular perturbations in § 3.6.4.

3.6.1 Scalar partial differential equation of the first order

Consider the equation

$$\text{(3.6.1)} \quad \varepsilon\left(\frac{\partial u}{\partial t} + \Lambda(x,t)\frac{\partial u}{\partial x}\right) = a(x,t)u + f(x,t), \qquad 0 \le x \le l, \quad 0 \le t \le T,$$

where $\varepsilon > 0$ is a small parameter, subject to the initial condition

$$\text{(3.6.2)} \qquad u(x,0,\varepsilon) = \varphi(x), \qquad 0 \le x \le l.$$

We assume that $\Lambda(x,t)$, $a(x,t)$, $f(x,t)$ *and* $\varphi(x)$ *are sufficiently smooth functions in the rectangle* $\overline{\Omega} = (0 \le x \le l) \times (0 \le t \le T)$, *and further that* $\Lambda(x,t) > 0$ and $a(x,t) < 0$ *in* $\overline{\Omega}$.

We will call the family of lines defined by the equation

$$\text{(3.6.3)} \qquad \frac{dx}{dt} = \Lambda(x,t),$$

the *characteristics* of (3.6.1) (more precisely, the lines defined by the equation (3.6.3) are the projections of characteristics onto the plane (x,t); the characteristics themselves are curves in the (u,x,t) space and are defined by the system (3.6.4) presented below). For $\Lambda(x,t) > 0$, the characteristics starting from the points of the segment $0 \le x \le l$ are located above the x-axis for $t > 0$ (see Fig. 3.4). We assume that the triangular region Ω_1, bounded by the characteristic passing through the point $(0,0)$, the x-axis, and the straight line $x = l$, belongs to $\overline{\Omega}$.

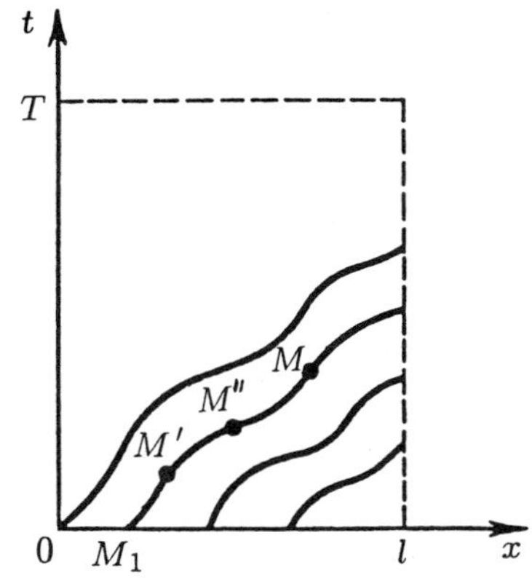

Figure 3.4: *Characteristics defined by* (3.6.3)*for* $\Lambda(x,t) > 0$.

The solution of (3.6.1), (3.6.2) can be constructed in explicit form using the *method of first integrals* (see, e.g., Tikhonov, Vasil'eva, and Sveshnikov [133]). Let us associate with (3.6.1) the following system of ordinary differential equations:

$$\frac{dx}{dt} = \Lambda(x,t), \qquad \varepsilon\frac{du}{dt} = a(x,t)u + f(x,t). \tag{3.6.4}$$

From the first equation, we derive a first integral $\Phi(x,t) = c_1$, and thus, we can obtain the general solution of (3.6.3), $x = X(t, c_1)$, describing the family of characteristics mentioned above (naturally, $X(t, \Phi(x,t)) \equiv x$). Then from the second equation of (3.6.4), we obtain

$$u = c_2 \exp\left(\frac{1}{\varepsilon}\int_0^t a(X(t',c_1),t')dt'\right)$$

$$+\frac{1}{\varepsilon}\int_0^t \exp\left(\frac{1}{\varepsilon}\int_{t'}^t a(X(t'',c_1),t'')dt''\right) f(X(t',c_1),t')dt'.$$

The initial condition (3.6.2) can be written in the form $x = \xi$, $t = 0$, $u = \varphi(\xi)$. Then we obtain $X(0, c_1) = \xi$, $c_2 = \varphi(\xi)$, so, $c_2 = \varphi(X(0,c_1))$. Thus, the solution of (3.6.1), (3.6.2) has the form

$$\begin{aligned} u(x,t) &= \varphi(X(0,\Phi(x,t))\exp\left(\frac{1}{\varepsilon}\int_0^t a(X(t',\Phi(x,t)),t')dt'\right) \\ &+\frac{1}{\varepsilon}\int_0^t \exp\left(\frac{1}{\varepsilon}\int_{t'}^t a(X(t'',\Phi(x,t)),t'')dt''\right) f(X(t',\Phi(x,t)),t')dt'. \end{aligned} \tag{3.6.5}$$

Let us clarify the geometrical meaning of this representation. It indicates that to obtain the solution at the point $M(x,t)$, we must integrate from the point on the x-axis along the characteristic passing through the point $M(x,t)$. For what follows, it is convenient to rewrite the expression for u in a form that is geometrically more clear:

$$
\begin{aligned}
u(M) = \;& u(M_1)\exp\left(\frac{1}{\varepsilon}\int_0^{t_M} a(M')dt_{M'}\right) \\
& +\frac{1}{\varepsilon}\int_0^{t_M}\exp\left(\frac{1}{\varepsilon}\int_{t_{M'}}^{t_M} a(M'')dt_{M''}\right) f(M')dt_{M'}.
\end{aligned}
\tag{3.6.6}
$$

The points $M,\ M_1,\ M',\ M''$ are shown in Fig. 3.4. Formula (3.6.5) (or (3.6.6)) defines the solution of (3.6.1), (3.6.2) in the triangular region Ω_1.

The asymptotic expansion of the solution of (3.6.1), (3.6.2) with a boundary layer character can be derived directly from the formula (3.6.5). However, we will choose another approach, taking into account the possibility of generalizing the method to more complicated cases when the solution cannot be found explicitly.

We will construct the asymptotic solution of (3.6.1), (3.6.2) in the form

$$
u = \overline{u}(x,t,\varepsilon) + \Pi(x,\tau,\varepsilon),
\tag{3.6.7}
$$

where

$$
\overline{u}(x,t,\varepsilon) = \overline{u}_0(x,t) + \varepsilon\overline{u}_1(x,t) + \cdots + \varepsilon^k\overline{u}_k(x,t) + \cdots
$$

is a regular series, and

$$
\Pi(x,\tau,\varepsilon) = \Pi_0(x,\tau) + \varepsilon\Pi_1(x,\tau) + \cdots + \varepsilon^k\Pi_k(x,\tau) + \cdots
$$

is a boundary function series in the vicinity of $t = 0$ $(\tau = t/\varepsilon)$. Substituting (3.6.7) into (3.6.1), (3.6.2), we obtain for $\overline{u}_0$ and Π_0

$$
a(x,t)\overline{u}_0 + f(x,t) = 0,
$$

$$
\frac{\partial\Pi_0}{\partial\tau} = a(x,0)\Pi_0, \qquad \Pi_0(x,0) = \varphi(x) - \overline{u}_0(x,0).
$$

For $\overline{u}_1$ and Π_1, we obtain

$$
a(x,t)\overline{u}_1 = \frac{\partial\overline{u}_0}{\partial t} + \Lambda(x,t)\frac{\partial\overline{u}_0}{\partial x},
$$

$$
\frac{\partial\Pi_1}{\partial\tau} = a(x,0)\Pi_1 + \frac{\partial a}{\partial t}(x,0)\tau\Pi_0 - \Lambda(x,0)\frac{\partial\Pi_0}{\partial x},
$$

$$
\Pi_1(x,0) = -\overline{u}_1(x,0),
$$

i.e. the terms of the expansion (3.6.7) are determined according to the usual algorithm. For the boundary functions, we have ordinary differential equations depending on x as a parameter. Since $a(x,t) < 0$, the boundary functions decay exponentially to zero as $\tau \to \infty$. The construction of higher-order terms in (3.6.7) can be done in an absolutely similar way.

Let us introduce the notation $U_n(x,t,\varepsilon) = \sum_{k=0}^{n} \varepsilon^k (\overline{u}_k + \Pi_k)$ and $w = u - U_n$ for the remainder term. Function w satisfies

(3.6.8)
$$\varepsilon \left(\frac{\partial w}{\partial t} + \Lambda(x,t)\frac{\partial w}{\partial x} \right) = a(x,t)w + R,$$

where $R = O(\varepsilon^{n+1})$ uniformly in Ω_1. The initial condition is

$$w(x,0,\varepsilon) = 0.$$

Applying formula (3.6.6) to the problem for w and since $a(x,t) < 0$, we can obtain the estimate

(3.6.9)
$$\max_{\Omega_1} |w(x,t,\varepsilon)| = O(\varepsilon^{n+1}).$$

Now consider the more general equation

(3.6.10)
$$\varepsilon \left(\frac{\partial u}{\partial t} + \Lambda(x,t)\frac{\partial u}{\partial x} \right) = F(u,x,t,\varepsilon).$$

Let $\Lambda(x,t)$ and $F(u,x,t,\varepsilon)$ be sufficiently smooth functions in the domain $\overline{\Omega} \times \{|u| \le H\}$, $H = \text{const} > 0$.

We will also seek the asymptotic solution of (3.6.10), (3.6.2) in the form of the series (3.6.7). For $\overline{u}_0$ and Π_0 we obtain

$$F(\overline{u}_0, x, t, 0) = 0,$$

(3.6.11)
$$\frac{\partial \Pi_0}{\partial \tau} = F(\overline{u}_0(x,0) + \Pi_0, x, 0, 0), \qquad \Pi_0(x,0) = \varphi(x) - \overline{u}_0(x,0).$$

From the equation for $\overline{u}_0$, we obtain $\overline{u}_0 = \overline{u}_0(x,t)$ (we assume that $\overline{u}_0(x,t)$ is an isolated root). Therefore the quantity $\overline{u}_0(x,0)$, which enters the equation for Π_0, is known.

Let the stability condition $F_u(\overline{u}_0(x,t),x,t,0) < 0$ be satisfied in Ω_1, and suppose the initial condition $\varphi(x) - \overline{u}_0(x,0)$ belongs to the domain of attraction of the asymptotically stable rest point $\Pi_0 = 0$ of (3.6.11).

Then $\Pi_0(x,\tau)$ decays exponentially to zero as $\tau \to \infty$.

For $\overline{u}_1$ and Π_1, we obtain

$$F_u(\overline{u}_0(x,t),x,t,0)\overline{u}_1 = \frac{\partial \overline{u}_0}{\partial t} + \Lambda(x,t)\frac{\partial \overline{u}_0}{\partial x},$$

$$\begin{aligned}\frac{\partial \Pi_1}{\partial \tau} &= F_u(x,\tau)\Pi_1 + (F_u(x,\tau) - \overline{F}_u(x,0))\left(\frac{\partial \overline{u}_0}{\partial t}(x,0)\tau + \overline{u}_1(x,0)\right) \\ &\quad + (F_t(x,\tau) - \overline{F}_t(x,0))\tau - \Lambda(x,0)\frac{\partial \Pi_0}{\partial x}, \qquad \Pi_1(x,0) = -\overline{u}_1(x,0).\end{aligned}$$

Here we use the same notation as in § 2.1.2. Functions $\overline{u}_1$ and Π_1 are uniquely defined from these equations. Higher-order terms of the expansion (3.6.7) are constructed according to the algorithm described in § 2.1.2.

For the remainder term $w = u - U_n$ we obtain the problem

$$\varepsilon\left(\frac{\partial w}{\partial t} + \Lambda(x,t)\frac{\partial w}{\partial x}\right) = F_u(\overline{u}_0 + \Pi_0, x, t, 0)w + g(w, x, t, \varepsilon),$$
(3.6.12)
$$w(x, 0, \varepsilon) = 0,$$

where the function g possesses the same two properties as in § 1.3.1.

The solution of (3.6.12) cannot now be represented in an explicit form, but formula (3.6.6) provides an integral equation for w:

$$w(M) = \frac{1}{\varepsilon}\int_0^{t_M} \exp\left(\frac{1}{\varepsilon}\int_{t_{M'}}^{t_M} F_u(\overline{u}_0(M'') + \Pi_0(M''), M'', 0)dt_{M''}\right)$$

$$\times g(w(M'), M', \varepsilon)dt_{M'}.$$

The method of successive approximations can be used to solve this equation (recall that a similar approach was used in § 1.3.1). Using the stability condition and the properties of the function g, we can prove the existence and uniqueness of solution of (3.6.12) and the estimate (3.6.9).

A similar result can also be obtained for an even more general equation, where Λ depends on u:

(3.6.13)
$$\varepsilon\left(\frac{\partial u}{\partial t} + \Lambda(u,x,t)\frac{\partial u}{\partial x}\right) = F(u, x, t, \varepsilon),$$

(3.6.14)
$$u(x, 0, \varepsilon) = \varphi(x).$$

In this case the asymptotic solution is again sought in the form of a series (3.6.7) with the terms determined by the usual algorithm.

Note that now, instead of the decoupled system (3.6.4), we will have the system

(3.6.15)
$$\frac{dx}{dt} = \Lambda(u, x, t), \qquad \varepsilon\frac{du}{dt} = F(u, x, t, \varepsilon).$$

This system defines the characteristics of the quasilinear equation (3.6.13). The domain of definition of the solution is the triangle bounded by x-axis, straight line $x = l$, and by the projection onto the (x,t)-plane of the characteristic, which satisfies the initial conditions $x|_{t=0} = 0$, $u|_{t=0} = \varphi(0)$. In the cases considered above, we called the *characteristics* their projections. They were defined as solutions of (3.6.3) and did not intersect. However, although the characteristics defined by (3.6.15) do not intersect in three-dimensional space, their projections may intersect and this might lead to the nonexistence of a classical solution in some subregion of $\overline{\Omega}$ (see, e.g., Tikhonov, Vasil'eva, and Sveshnikov [133]). It turns out that the presence of a small parameter ε in (3.6.13) prevents such intersections from taking place since the projections of characteristics for sufficiently small ε are described by the reduced equation

$$\frac{dx}{dt} = \Lambda(\overline{u}(x,t), x, t),$$

where $\overline{u}(x,t)$ is a root of the equation $F(\overline{u},x,t,0) = 0$, and the solutions of this equation form a one-parameter family of nonintersecting curves.

Let us show (without making precise estimates) how the main term of the asymptotic solution of (3.6.13), (3.6.14) can be obtained using the system (3.6.15). Let us construct the family of characteristics or, equivalently, let us find two first integrals of the system (3.6.15) taking for c_1 and c_2 the values of x and u at $t = 0$. The solution of (3.6.15) with these initial conditions has (to order $O(\varepsilon)$) the following representation (see § 2.1.2)

$$x = \overline{x}_0(t, c_1), \qquad u = \overline{u}(\overline{x}_0(t,c_1),t) + \Pi_0(\tau, c_1, c_2),$$

where $\overline{x}_0$ and Π_0 are the solutions of the problems

$$\frac{d\overline{x}_0}{dt} = \Lambda(\overline{u}(\overline{x}_0,t),\overline{x}_0,t), \qquad \overline{x}_0|_{t=0} = c_1,$$

$$\frac{d\Pi_0}{d\tau} = F(\overline{u}(c_1,0) + \Pi_0, c_1, 0, 0), \qquad \Pi_0|_{\tau=0} = c_2 - \overline{u}(c_1,0). \tag{3.6.16}$$

From the initial condition, we have $c_2 = \varphi(c_1)$, so with accuracy $O(\varepsilon)$

$$u = \overline{u}(\overline{x}_0(t,c_1),t) + \Pi_0(\tau, c_1, \varphi(c_1)).$$

Here $c_1 = \Phi(x,t)$ is defined by the equation $x = \overline{x}_0(t,c_1)$ (note that $\overline{x}_0(t,\Phi(x,t)) \equiv x$). Finally, with accuracy $O(\varepsilon)$, we obtain

$$u = \overline{u}(x,t) + \Pi_0(\tau, \Phi(x,t), \varphi(\Phi(x,t))). \tag{3.6.17}$$

Note that if we apply the standard construction algorithm for an asymptotic series in the form (3.6.7) to (3.6.13), (3.6.14), we obtain in the zeroth-order approximation

$$u = \overline{u}(x,t) + \Pi_0(\tau, x, \varphi(x)). \tag{3.6.18}$$

Here $\Pi_0(\tau, x, \varphi(x))$ is the solution of (3.6.16), where we set $c_1 = x$ and $c_2 = \varphi(x)$. In (3.6.17) Π_0 is the solution of the same problem but with $c_1 = \Phi(x,t)$ and $c_2 = \varphi(\Phi(x,t))$ (where x and t are parameters). Since $\Phi(x,0) = x$, the second and third arguments of the functions Π_0 in (3.6.17) and (3.6.18) differ by a quantity of order t. It can be shown that in this case the functions Π_0 differ by order ε. Thus, both formulae (3.6.17) and (3.6.18) give the solution of (3.6.13), (3.6.14) with an accuracy of order $O(\varepsilon)$.

For (3.6.1), the *mixed* problem, which involves both initial and boundary conditions

$$u(x,0,\varepsilon) = \varphi(x), \quad 0 \le x \le l; \qquad u(0,t,\varepsilon) = \psi(t), \quad 0 \le t \le T \tag{3.6.19}$$

can be posed. In this case the solution will be defined in the entire rectangle $\overline{\Omega} = (0 \le x \le l) \times (0 \le t \le T)$. In fact, we may solve two separate problems. One of them has been already studied in this subsection. Its solution is given by formula (3.6.6) and is defined in the region Ω_1 (see Fig. 3.5). In the other problem, the only difference is that x and t exchange roles. The solution of the second problem is defined in the region Ω_2. For such a *composite* solution to be classical in the entire rectangle $\overline{\Omega}$, i.e.,

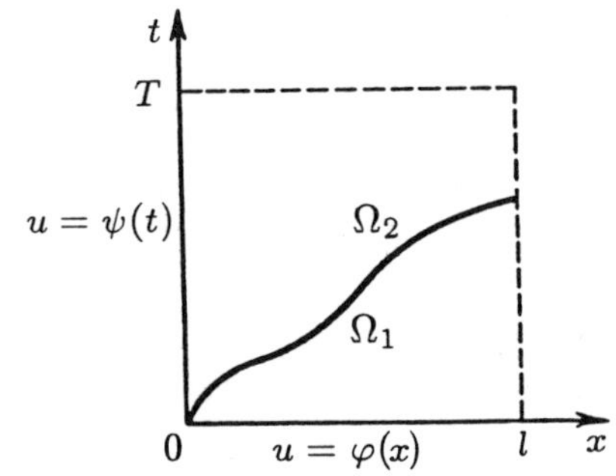

Figure 3.5: *Solution of* mixed *problem is defined in* Ω_1 *by* (3.6.6)*and in* Ω_2 *by a similar formula, where* x *is switched to* t *and vice versa.*

continuous together with its first-order derivatives, it is necessary and sufficient that the following *matching conditions* hold.

For u to be continuous along the characteristic that passes through the corner point $(0, 0)$, it is necessary and sufficient that equality

$$\varphi(0) = \psi(0) \tag{3.6.20}$$

holds. To verify that, we write the formula analogous to (3.6.6) for the solution in region Ω_2, and compare it with the formula (3.6.6) for region Ω_1. Equality (3.6.20) is called the *matching condition of the zeroth order.* To obtain the condition under which the first derivatives of u are continuous, we must differentiate (3.6.1) with respect to corresponding variable and write an expression of type (3.6.20). For example, for $v = \partial u/\partial t$,

$$\varepsilon\left(\frac{\partial v}{\partial t} + \Lambda(x,t)\frac{\partial v}{\partial x} + \frac{\partial \Lambda}{\partial t}\frac{\partial u}{\partial x}\right) = av + \frac{\partial a}{\partial t}u + \frac{\partial f}{\partial t},$$

$$v(0, t, \varepsilon) = \psi(t),$$

$$v(x, 0, \varepsilon) = \frac{1}{\varepsilon}(a(x,0)\varphi(x) + f(x,0)) - \Lambda(x,0)\varphi'(x).$$

The last relation is obtained directly from (3.6.1). For v, the function

$$\frac{1}{\varepsilon}(a(x,0)\varphi(x) + f(x,0)) - \Lambda(x,0)\varphi'(x)$$

plays the role of $\varphi(x)$, and $\psi'(t)$ plays that of $\psi(t)$. Equating these functions at $x = 0$ and $t = 0$, we obtain

$$\varepsilon(\psi'(0) + \Lambda(0,0)\varphi'(0)) = a(0,0)\varphi(0) + f(0,0).$$

Note the special structure of this relation. It has the form of (3.6.1), where the initial and boundary functions $\varphi(x)$ and $\psi(t)$ as well as their derivatives at $x = 0$ and $t = 0$

are substituted for u and its derivatives. For this equality to hold for any ε, it is necessary and sufficient that

$$\psi'(0) + \Lambda(0,0)\varphi'(0) = 0, \qquad a(0,0)\varphi(0) + f(0,0) = 0. \tag{3.6.21}$$

As can be easily verified, the condition that the derivative $\partial u/\partial x$ is continuous leads to the same relations (3.6.21). In particular, conditions (3.6.21) mean that $\varphi(0)$ (and also $\psi(0) = \varphi(0)$) cannot be taken arbitrarily, but has to equal $-f(0,0)/a(0,0)$.

Equalities (3.6.20) and (3.6.21) are called the *matching conditions of the first order.* They guarantee the existence of a classical solution to (3.6.1), (3.6.19) in the rectangle $\overline{\Omega}$. Regarding the asymptotic solution, it is already constructed in region Ω_1 in the form of expansion (3.6.7), and in the region Ω_2, it can be obtained in an absolutely similar way in the form

$$u = \overline{u}(x,t,\varepsilon) + Q(\xi,t,\varepsilon), \tag{3.6.22}$$

where Q is a boundary function series in the vicinity of $x = 0$ (so $\xi = x/\varepsilon$).

Exercise
Consider a problem

$$\varepsilon\left(\frac{\partial u}{\partial t} + \frac{\partial u}{\partial x}\right) = -(1 + x^2 + t^2)u + 1,$$

$$(x,t) \in \Omega = (0 < x \leq 1) \times (0 < t \leq 1),$$

$$u(x,0,\varepsilon) = \varphi(x) = 1 + x, \qquad v(0,t,\varepsilon) = \psi(t) = 1 - t.$$

a). Check that for $\varphi(x)$ and $\psi(t)$ the matching conditions of the zeroth and the first orders are satisfied.

b). Find the asymptotic solution up to the terms of the first order.

3.6.2 System of two first-order equations

Consider the system

$$\begin{aligned}
\varepsilon\left(\frac{\partial u}{\partial t} + \Lambda_1\frac{\partial u}{\partial x}\right) &= a_{11}u + a_{12}v + f, \\
\varepsilon\left(\frac{\partial v}{\partial t} + \Lambda_2\frac{\partial v}{\partial x}\right) &= a_{21}u + a_{22}v + g.
\end{aligned} \tag{3.6.23}$$

Here $a_{ik} = a_{ik}(x,t)$, $\Lambda_i = \Lambda_i(x,t)$, $f(x,t)$ *and* $g(x,t)$ *are sufficiently smooth functions. Further, let* $\Lambda_1 < \Lambda_2$ *in* $\overline{\Omega} = (0 \leq x \leq l) \times (0 \leq t \leq T)$. We impose the initial conditions

$$u(x,0,\varepsilon) = \varphi_1(x), \qquad v(x,0,\varepsilon) = \varphi_2(x). \tag{3.6.24}$$

For $\Lambda_1 > 0$ and $\Lambda_2 > 0$ the solution is defined in a the triangle G_1 depicted in Fig. 3.6(a); for $\Lambda_1 < 0$ and $\Lambda_2 > 0$, the solution is defined in a triangle G_2 shown in Fig. 3.6(b). In these figures, both triangles are covered by two families of characteris-

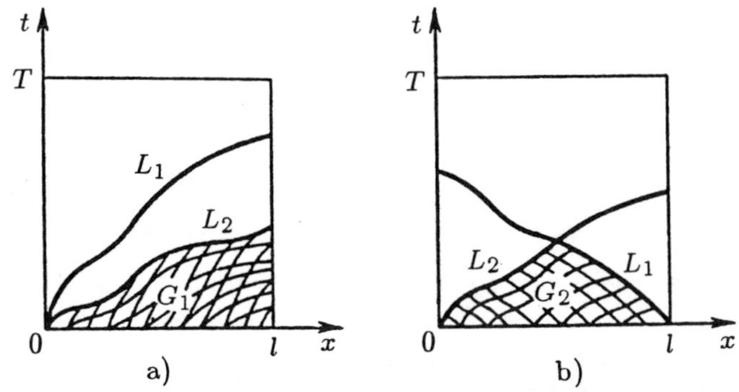

Figure 3.6: *The solution of* (3.6.23), (3.6.24)*is defined* (a)*in* G_1 *for* $\Lambda_1 > 0$, $\Lambda_2 > 0$; (b)*in* G_2 *for* $\Lambda_1 < 0$, $\Lambda_2 > 0$.

tics defined by the equations

$$\frac{dx}{dt} = \Lambda_1(x,t) \quad \text{and} \quad \frac{dx}{dt} = \Lambda_2(x,t).$$

Let the eigenvalues λ_1 *and* λ_2 *of the matrix*

$$a(x,t) = \begin{pmatrix} a_{11} & a_{12} \\ a_{21} & a_{22} \end{pmatrix}$$

noindent satisfy the conditions $\mathrm{Re}\lambda_i < 0$ *in* G.

Then the asymptotic solution is constructed in the standard way as a series of the type (3.6.7). In the zeroth-order approximation, we have the problems

$$a_{11}\overline{u}_0 + a_{12}\overline{v}_0 + f = 0, \qquad a_{21}\overline{u}_0 + a_{22}\overline{v}_0 + g = 0, \tag{3.6.25}$$

$$\frac{d\Pi_0 u}{d\tau} = a_{11}(x,0)\Pi_0 u + a_{12}(x,0)\Pi_0 v,$$

$$\frac{d\Pi_0 v}{d\tau} = a_{21}(x,0)\Pi_0 u + a_{22}(x,0)\Pi_0 v,$$

$$\Pi_0 u(x,0) = \varphi_1(x) - \overline{u}_0(x,0), \qquad \Pi_0 v(x,0) = \varphi_2(x) - \overline{v}_0(x,0),$$

from which $\overline{u}_0$, $\overline{v}_0$, $\Pi_0 u$, $\Pi_0 v$ are determined uniquely. By virtue of condition $\mathrm{Re}\lambda_i < 0$, the boundary functions $\Pi_0 u$ and $\Pi_0 v$ decay exponentially to zero as $\tau \to \infty$. Higher-order approximations can also be constructed using the standard scheme.

For the remainder term

$$w = \begin{pmatrix} w_1 \\ w_2 \end{pmatrix} = \begin{pmatrix} u - U_n \\ v - V_n \end{pmatrix}$$

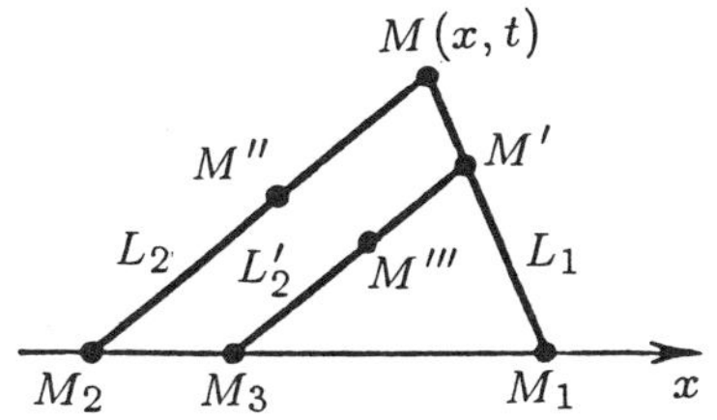

Figure 3.7: *The paths of integration in* (3.6.26).

we obtain a system of equations that can be written in the same form as (3.6.8), but now Λ is the diagonal matrix: $\Lambda = \operatorname{diag}(\Lambda_1, \Lambda_2)$.

Certain difficulties arise during the estimation of the remainder term from the vector system (3.6.8) since (for the vector case) an inversion formula analogous to (3.6.6) does not exist. A comparatively easy estimate for w can be obtained if we will make an additional assumption that $a_{11} < 0$, $a_{22} < 0$, *and* a_{12} *and* a_{21} *have sufficiently small absolute values.*

Let us demonstrate the derivation under this assumption for the case $a_{ik} = \text{const}$. We set $R = \begin{pmatrix} R_1 \\ R_2 \end{pmatrix}$. Applying formula (3.6.6) separately to each of the equations of vector system (3.6.8), we obtain a system of integral equations

$$
\begin{aligned}
w_1(M) &= \frac{1}{\varepsilon}\int_{M_1M} \exp\left(\frac{a_{11}}{\varepsilon}(t_M - t_{M'})\right)(a_{12}w_2(M') + R_1(M'))dt_{M'},\\
w_2(M) &= \frac{1}{\varepsilon}\int_{M_2M} \exp\left(\frac{a_{22}}{\varepsilon}(t_M - t_{M''})\right)(a_{21}w_1(M'') + R_2(M''))dt_{M''}.
\end{aligned}
\tag{3.6.26}
$$

This notation has the following geometrical interpretation: In the first equation, integration is performed along the characteristic L_1 from the point M_1 to the point $M(x,t)$. In the second equation we integrate from the point M_2 to the point $M(x,t)$ along the characteristic L_2. (See Fig. 3.7.)

Substituting w_2 from the second equation of (3.6.26) into the first, we obtain an integral equation for w_1. To find w_2 at the point M' on L_1, we have to integrate along the characteristic L_2' (M''' denotes the points of this characteristic) from the point M_3 to the point M' (see Fig. 3.7). As a result, we arrive at the equation

$$
\begin{aligned}
w_1(M) &= \frac{1}{\varepsilon}\int_{M_1M} \exp\left(\frac{a_{11}}{\varepsilon}(t_M - t_{M'})\right)\\
&\times \left[a_{12}\frac{1}{\varepsilon}\int_{M_3M'} \exp\left(\frac{a_{22}}{\varepsilon}(t_{M'} - t_{M'''})\right)a_{21}w_1(M''')dt_{M'''}\right]dt_{M'} + O(\varepsilon^{n+1}).
\end{aligned}
\tag{3.6.27}
$$

The $O(\varepsilon^{n+1})$ term in this equation appears due to the functions R_1 and R_2. From (3.6.27) we obtain the estimate

$$\max_G |w_1| \le \frac{\delta^2}{\kappa^2} \max_G |w_1| + O(\varepsilon^{n+1}),$$

where $\delta = \max(|a_{12}|, |a_{21}|)$ and $\kappa = \min(|a_{11}|, |a_{22}|)$. For $\delta < \kappa$, it follows from the inequality above that $\max_G |w_1| = O(\varepsilon^{n+1})$. A similar estimate holds for w_2.

Remark 1. The described algorithm (as well as the method of proof) can be extended to the case when the system (3.6.23) consists of N equations ($N > 2$), u is an N-dimensional vector function, and the right-hand side is nonlinear in u, i.e.,

$$\varepsilon\left(\frac{\partial u}{\partial t} + \Lambda(x,t)\frac{\partial u}{\partial x}\right) = F(u,x,t,\varepsilon), \qquad u(x,0,\varepsilon) = \varphi(x).$$

Here $\Lambda = \operatorname{diag}(\Lambda_1, \dots, \Lambda_N)$ and $\Lambda_1 \le \Lambda_2 \le \dots \le \Lambda_N$. The solution of this problem is defined in a triangular domain bounded from above by two characteristics L_1 and L_N corresponding to Λ_1 and Λ_N, if they have opposite signs, and by one characteristic corresponding to Λ_N, if $\Lambda_1 > 0$, or corresponding to Λ_1, if $\Lambda_N < 0$.

Remark 2. For the two-dimensional system (3.6.23) with constant coefficients, we can use another method for estimating the remainder term (cf. Vasil'eva [143]), which is based on the transformation of the system to one differential equation of the second order and a subsequent application of Riemann's formula. This method works without the assumption that a_{12} and a_{21} are small, but we still have to keep $a_{11} < 0$, $a_{22} < 0$.

Now consider the initial boundary value problem for (3.6.23):

$$\begin{aligned} &u(x,0,\varepsilon) = \varphi_1(x), \qquad v(x,0,\varepsilon) = \varphi_2(x), \\ &u(0,t,\varepsilon) = \psi_1(t), \qquad v(0,t,\varepsilon) = \psi_2(t). \end{aligned} \tag{3.6.28}$$

Let $\Lambda_2 > \Lambda_1 > 0$ (*see* Fig. 3.6(a))*and suppose the eigenvalues of matrix*

$$a = \begin{pmatrix} a_{11} & a_{12} \\ a_{21} & a_{22} \end{pmatrix}$$

satisfy the conditions $\operatorname{Re}\lambda_i < 0$.

For the existence of a classical solution in the rectangle $\overline{\Omega}$, it is necessary and sufficient that *matching conditions* similar to (3.6.20), (3.6.21) hold:

$$\varphi_i(0) = \psi_i(0), \qquad i = 1,2,$$

$$\begin{aligned} &\psi_1'(0) + \Lambda_1(0,0)\varphi_1'(0) = 0, \qquad \psi_2'(0) + \Lambda_2(0,0)\varphi_2'(0) = 0, \\ &a_{11}(0,0)\varphi_1(0) + a_{12}(0,0)\varphi_2(0) + f(0,0) = 0, \\ &a_{21}(0,0)\varphi_1(0) + a_{22}(0,0)\varphi_2(0) + g(0,0) = 0. \end{aligned} \tag{3.6.29}$$

In the region Ω_1 (see Fig. 3.8) the asymptotic series is constructed, as earlier, in the form (3.6.7). In the region $\overline{\Omega}_2$, the asymptotic series is constructed analogously, with x and t exchanging their roles (see (3.6.22)). Thus, the solution on characteristics L_1 and L_2 can be obtained to any degree of accuracy. Let us consider the region $\tilde{\Omega}_3$

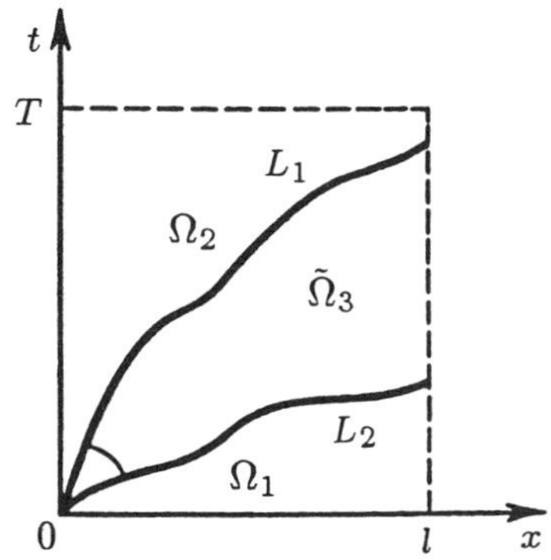

Figure 3.8: *For $\Lambda_2 > \Lambda_1 > 0$ the asymptotic solution of* (3.6.23), (3.6.28)*has different forms in different regions Ω_1, Ω_2, and $\tilde{\Omega}_3$.*

that lies between the characteristics L_1 and L_2 and does not contain some arbitrarily small, but fixed as $\varepsilon \to 0$, neighborhood of the point $(0,0)$. In the region $\tilde{\Omega}_3$, we will construct the asymptotic series in the form of regular series

$$y = \overline{y}(x,t,\varepsilon), \qquad y = \begin{pmatrix} u \\ v \end{pmatrix}$$

only. The terms of this series at the zeroth order are defined by system (3.6.25), and at higher orders by recurrent algebraic systems, as in Ω_1 and Ω_2. Let us introduce in $\tilde{\Omega}_3$ the remainder term w through the formula

$$w = y(x,t,\varepsilon) - \sum_{k=0}^{n} \varepsilon^k \overline{y}_k(x,t).$$

On the lines $\tilde{L}_1$ and $\tilde{L}_2$ (these are the parts of characteristics L_1 and L_2 lying outside a small neighborhood of $(0,0)$), we have that $\max_{\tilde{L}_i} \| w \| = O(\varepsilon^{n+1})$.

We can write the equations for $w = \begin{pmatrix} w_1 \\ w_2 \end{pmatrix}$ in $\tilde{\Omega}_3$ similar to (3.6.26). As before, we consider only constant coefficients to simplify the presentation. In contrast with (3.6.26), these equations will also contain some additional terms (see Fig. 3.9):

(3.6.30)
$$w_1(M) = w_1(M_1)\exp\left(\frac{a_{11}}{\varepsilon}(t_M - t_{M_1})\right)$$
$$+\frac{1}{\varepsilon}\int_{M_1M} \exp\left(\frac{a_{11}}{\varepsilon}(t_M - t_{M'})\right)(a_{12}w_2(M') + R_1(M'))dt_{M'},$$
$$w_2(M) = w_2(M_2)\exp\left(\frac{a_{22}}{\varepsilon}(t_M - t_{M_2})\right)$$
$$+\frac{1}{\varepsilon}\int_{M_2M} \exp\left(\frac{a_{22}}{\varepsilon}(t_M - t_{M''})\right)(a_{21}w_1(M'') + R_2(M''))dt_{M''}.$$

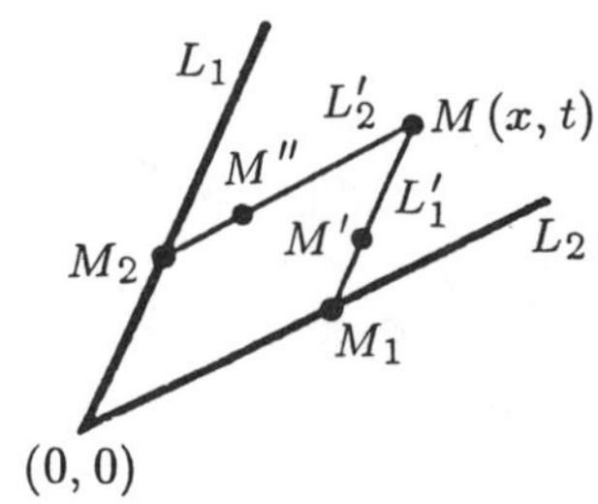

Figure 3.9: *The paths of integration in* (3.6.30) *in the case* $\Lambda_2 > \Lambda_1 > 0$.

Since

$$\max_{\tilde{L}_i} |w_i(M_i)| = O(\varepsilon^{n+1}) \quad \text{and} \quad \max_{\tilde{\Omega}_3} |R_i| = O(\varepsilon^{n+1}),$$

the system (3.6.30) can be transformed to an equation of the type (3.6.27) from which we find $\max_{\tilde{\Omega}_3} |w_1| = O(\varepsilon^{n+1})$, and therefore

$$\max_{\tilde{\Omega}_3} |w_2| = O(\varepsilon^{n+1})$$

and

$$\max_{\tilde{\Omega}_3} \| w \| = O(\varepsilon^{n+1}).$$

Let us formulate this result as follows.

Theorem 3.7 *For the solution* $y = \begin{pmatrix} u \\ v \end{pmatrix}$ *of problem* (3.6.23), (3.6.28),*the following asymptotic representation holds:*

in region Ω_1

$$y = \sum_{k=0}^{n} \varepsilon^k (\overline{y}_k(x,t) + \Pi_k y(x,\tau)) + O(\varepsilon^{n+1});$$

in region $\tilde{\Omega}_3$

$$y = \sum_{k=0}^{n} \varepsilon^k \overline{y}_k(x,t) + O(\varepsilon^{n+1});$$

and in region Ω_2

$$y = \sum_{k=0}^{n} \varepsilon^k (\overline{y}_k(x,t) + Q_k y(\xi,t)) + O(\varepsilon^{n+1}).$$

Exercise

Find the zeroth-order terms of the asymptotic solution for

$$\varepsilon\left(\frac{\partial u}{\partial t}+\frac{\partial u}{\partial x}\right)=-u+\frac{1}{2}v+x,$$

$$\varepsilon\left(\frac{\partial v}{\partial t}+2\frac{\partial v}{\partial x}\right)=\frac{1}{2}u-v+t,$$

$$(x,t)\in\Omega=(0<x\le 1)\times(0<t\le 1),$$

$$u(x,0)=v(x,0)=u(0,t)=v(0,t)=0.$$

Now consider a different case.

Let $\Lambda_1<0$ *and* $\Lambda_2>0$ (*see* Fig. 3.6(b))*and suppose the eigenvalues of*

$$a=\begin{pmatrix} a_{11} & a_{12}\\ a_{21} & a_{22}\end{pmatrix}$$

satisfy $\mathrm{Re}\lambda_i<0$.

We impose additional conditions for (3.6.23) as

$$u(x,0,\varepsilon)=\varphi_1(x),\qquad v(x,0,\varepsilon)=\varphi_2(x),$$

(3.6.31)

$$v(0,t,\varepsilon)=\psi_1(t),\qquad u(l,t,\varepsilon)=\psi_2(t).$$

We assume that the following *matching conditions* hold:

$$\varphi_2(0)=\psi_1(0),\qquad \varphi_1(l)=\psi_2(0),$$

$$\psi_2'(0)+\Lambda_1(l,0)\varphi_1'(l)=0,\qquad \psi_1'(0)+\Lambda_2(0,0)\varphi_2'(0)=0,$$

$$a_{11}(l,0)\varphi_1(l)+a_{12}(l,0)\varphi_2(l)+f(l,0)=0,$$

$$a_{21}(0,0)\varphi_1(0)+a_{22}(0,0)\varphi_2(0)+g(0,0)=0.$$

Let $a_{11}<0$, $a_{22}<0$ *and* $a_{12}a_{21}<a_{11}a_{22}$.

These conditions guarantee that the real parts of the eigenvalues of a have negative sign. Further, under these conditions

$$b(x,t)=\begin{pmatrix} a_{11}/\Lambda_1 & a_{12}/\Lambda_1\\ a_{21}/\Lambda_2 & a_{22}/\Lambda_2\end{pmatrix}$$

has real eigenvalues of opposite signs.

In the region Ω_1 (cf. Fig. 3.10) the asymptotic solution of (3.6.23), (3.6.31) is constructed in the form (3.6.7). Consider region Ω_2. The regular series there is similar to that in Ω_1. The boundary functions of the zeroth-order approximation

$$Q_0y(\xi,t)=\begin{pmatrix} Q_0u\\ Q_0v\end{pmatrix}$$

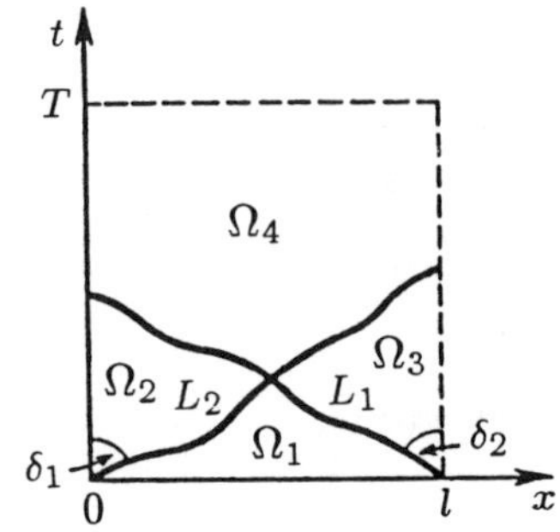

Figure 3.10: *Mixed problem for $\Lambda_1 < 0$ and $\Lambda_2 > 0$: the asymptotic solution has different representations in regions $\overline{\Omega}_1$ and $\{\overline{\Omega}/(\Omega_1 \cup \delta_1 \cup \delta_2)\}$.*

in the vicinity of $x = 0$ satisfy the following system of equations:

$$\Lambda_1(0,t)\frac{\partial Q_0 u}{\partial \xi} = a_{11}(0,t)Q_0 u + a_{12}(0,t)Q_0 v,$$

(3.6.32)

$$\Lambda_2(0,t)\frac{\partial Q_0 v}{\partial \xi} = a_{21}(0,t)Q_0 u + a_{22}(0,t)Q_0 v,$$

for $\xi = x/\varepsilon$. Since $u(0,t,\varepsilon)$ is not prescribed, there is only one additional condition for this system:

(3.6.33) $$Q_0 v(0,t) = \psi_1(t) - \overline{v}_0(0,t).$$

If we take into account the fact that the matrix of (3.6.32) coincides with the previously introduced $b(0,t)$, which has the eigenvalues of opposite signs, it becomes clear that conditions (3.6.33) and $Q_0 y(\infty,t) = 0$ uniquely define the solution of (3.6.32), and $Q_0 y$ exponentially approaches zero as $\xi \to \infty$. Thus, we have an analogy with the conditionally stable case. The Q-functions of the higher-order approximations are defined similarly.

In region Ω_3, the asymptotic expansion is constructed as in Ω_2 with the boundary functions $Q^* y(\xi_*,t)$ (for $\xi_* = (l-x)/\varepsilon$) in the vicinity of the line $x = l$. These functions are determined from conditionally stable systems analogous to (3.6.32).

Thus in region Ω_2, the asymptotic solution has the form

$$y = \overline{y}(x,t,\varepsilon) + Qy(\xi,t,\varepsilon),$$

and in region Ω_3, it has the form

$$y = \overline{y}(x,t,\varepsilon) + Q^* y(\xi_*,t,\varepsilon).$$

We will estimate the remainder term

$$w = y - \sum_{k=0}^{n} \varepsilon^k (\overline{y}_k + Q_k y)$$

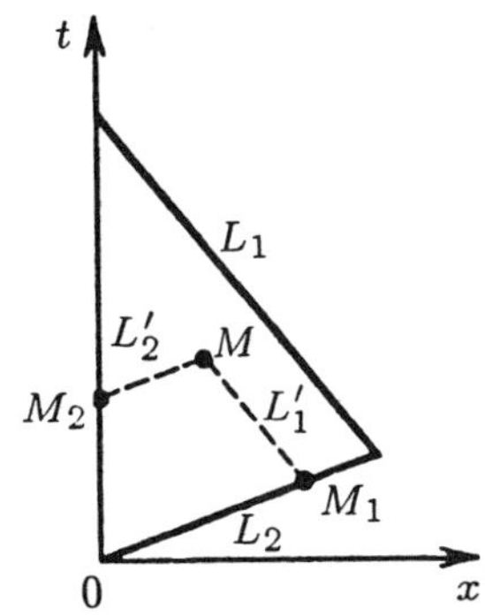

Figure 3.11: *The paths of integration in* (3.6.30)*in the case* $\Lambda_1 < 0,\ \Lambda_2 > 0$.

in the region $\tilde{\Omega}_2$ (this is the region Ω_2 without an arbitrarily small, but fixed as $\varepsilon \to 0$, vicinity δ_1 of the point $(0,0)$; see Fig. 3.10). The vector function $w = \begin{pmatrix} w_1 \\ w_2 \end{pmatrix}$ satisfies the system (3.6.8) and the additional conditions

$$w_2|_{x=0} = 0, \qquad \max_{\tilde{L}_2} |w_1| = O(\varepsilon^{n+1}), \tag{3.6.34}$$

($\tilde{L}_2$ is the characteristic L_2 without the part belonging to the previously mentioned neighborhood of $(0,0)$). The second condition in (3.6.34) is obtained as follows: Using the representation of the solution in Ω_1 derived previously in this subsection, we have

$$\left| u - \sum_{k=0}^{n} \varepsilon^k (\overline{u}_k + \Pi_k u) \right| = O(\varepsilon^{n+1})$$

on $\tilde{L}_2$. Consequently,

$$\left| u - \sum_{k=0}^{n} \varepsilon^k \overline{u}_k \right| = O(\varepsilon^{n+1}),$$

and therefore

$$|w_1| = \left| u - \sum_{k=0}^{n} \varepsilon^k (\overline{u}_k + Q_k u) \right| = O(\varepsilon^{n+1}).$$

Next, we write the integral equations for w_1, w_2. In the case of constant coefficients, these equations have the same form as (3.6.30), with a simplification since $w_2(M_2) = 0$ (the location of points M, M_1 and M_2 is shown in Fig. 3.11). After that, as for (3.6.30), the estimate

$$\max_{\tilde{\Omega}_2} \| w \| = O(\varepsilon^{n+1})$$

can be obtained. The estimation for the remainder in $\tilde{\Omega}_3$ can be proved analogously. As a result, the asymptotic expansion $y = \overline{y} + Qy + Q^* y$ is valid in $\tilde{\Omega}_2 \cup \tilde{\Omega}_3$. The same method of integral equations can be used to prove the correctness of this expansion in the region Ω_4 as well (see Figure 3.10). Thus, we have the following result.

Theorem 3.8 *The asymptotic solution of* (3.6.23), (3.6.31)*has the following form:*

$$y = \sum_{k=0}^{n} \varepsilon^k (\overline{y}_k + \Pi_k y) + O(\varepsilon^{n+1})$$

in the region $\overline{\Omega}_1$ *and*

$$y = \sum_{k=0}^{n} \varepsilon^k (\overline{y}_k + Q_k y + Q_k^* y) + O(\varepsilon^{n+1})$$

in the region $\{\overline{\Omega}/(\Omega_1 \cup \delta_1 \cup \delta_2)\}$

Remark 1. The results can be generalized to the case when the right-hand sides of (3.6.23) depend nonlinearly on u and v.

Remark 2. The above presentation is based on results of Vasil'eva and Kuchik [154], Vasil'eva and Sisoeva [156], and Kadikenov and Kasimov [74]. Similar problems were studied by Medeuov [91], Mel'nik and Tsymbal [92], Sisoeva [127], Flyud and Tsymbal [52], and Tsymbal [135]. In some of these papers the estimates were obtained in integral norms.

3.6.3 Critical case

Here we will demonstrate that the phenomenon of an internal transition layer can be observed in the system (3.6.23) when the matrix

$$a = \begin{pmatrix} a_{11} & a_{12} \\ a_{21} & a_{22} \end{pmatrix}$$

is singular.

Consider the constant coefficient system of equations

(3.6.35) $$\varepsilon\left(\frac{\partial u}{\partial t} + \Lambda_1 \frac{\partial u}{\partial x}\right) = -au + bv, \qquad \varepsilon\left(\frac{\partial v}{\partial t} + \Lambda_2 \frac{\partial v}{\partial x}\right) = \gamma(au - bv)$$

subject to conditions (3.6.28). *Let* $\Lambda_2 > \Lambda_1 > 0$, $a > 0$ *and* $\gamma b > 0$. Then the matrix

$$\begin{pmatrix} -a & b \\ \gamma a & -\gamma b \end{pmatrix}$$

has eigenvalues $\lambda_1 = 0$ and $\lambda_2 = -a - \gamma b < 0$. Thus, we have the critical case, i.e., the reduced system has a family of solutions . We will consider the solution of (3.6.35) in the rectangle $\overline{\Omega} = (0 \le x \le l) \times (0 \le t \le T)$. *We assume that the functions* $\varphi_i(x)$ *and* $\psi_i(t)$ $(i = 1, 2)$ *in* (3.6.28)*are continuous together with their first and second derivatives* (we will construct only the leading terms of the asymptotic solution). *We also assume that the matching conditions* (3.6.29)*hold.*

The asymptotic series can be constructed using the usual algorithm for problems in critical cases. The system of equations for the leading terms of the regular part of the asymptotic approximation

$$-a\overline{u}_0 + b\overline{v}_0 = 0, \qquad \gamma(a\overline{u}_0 - b\overline{v}_0) = 0$$

has the family of solutions

(3.6.36) $$\overline{u}_0 = \alpha_0, \qquad \overline{v}_0 = \frac{a}{b}\alpha_0,$$

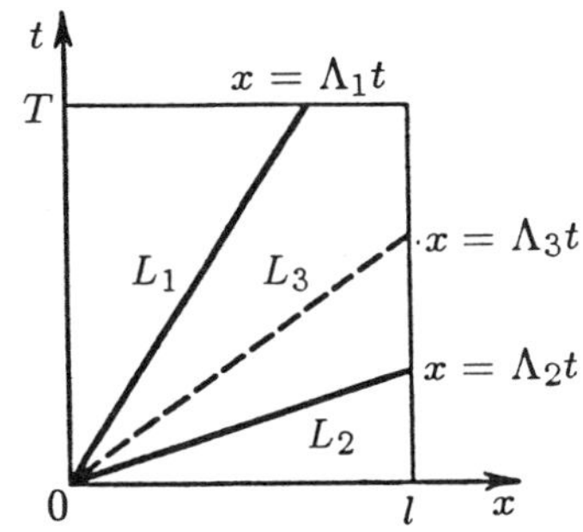

Figure 3.12: *Characteristic L_3 of the equation* (3.6.37).

where $\alpha_0 = \alpha_0(x,t)$ is a yet unknown function. The equation for α_0 will be obtained in the next order approximation from the solvability condition of the system for $\overline{u}_1$ and $\overline{v}_1$:

$$-a\overline{u}_1 + b\overline{v}_1 = \frac{\partial \alpha_0}{\partial t} + \Lambda_1 \frac{\partial \alpha_0}{\partial x}, \qquad \gamma(a\overline{u}_1 - b\overline{v}_1) = \frac{a}{b}\left(\frac{\partial \alpha_0}{\partial t} + \Lambda_2 \frac{\partial \alpha_0}{\partial x}\right).$$

Multiplying the first equation by γ and adding the two equations, we obtain the solvability condition

(3.6.37) $$(a+\gamma b)\frac{\partial \alpha_0}{\partial t} + (a\Lambda_2 + \gamma b\Lambda_1)\frac{\partial \alpha_0}{\partial x} = 0,$$

which is the equation for $\alpha_0(x,t)$. Hence,

(3.6.38) $$\alpha_0 = \Phi((a+\gamma b)x - (a\Lambda_2 + \gamma b\Lambda_1)t),$$

where Φ is still unknown.

By virtue of the conditions imposed above on coefficients of (3.6.35), the characteristic of (3.6.37) passing through the point $(0,0)$ (see Fig. 3.12, line L_3), i.e., the straight line $x = \Lambda_3 t$, for $\Lambda_3 = (a\Lambda_2 + \gamma b\Lambda_1)/(a+\gamma b)$, lies between the straight lines L_1 and L_2. The equations of L_1 and L_2 are $x = \Lambda_1 t$ and $x = \Lambda_2 t$, respectively. Below L_3, the function Φ is defined during the construction of the boundary functions $\Pi_0 u(x,\tau)$ and $\Pi_0 v(x,\tau)$ in the vicinity of $t = 0$ $(\tau = t/\varepsilon)$, for which we obtain the system

(3.6.39) $$\frac{\partial \Pi_0 u}{\partial \tau} = -a\Pi_0 u + b\Pi_0 v, \qquad \frac{\partial \Pi_0 v}{\partial \tau} = \gamma(a\Pi_0 u - b\Pi_0 v)$$

with initial conditions

(3.6.40) $$\Pi_0 u(x,0) = \varphi_1(x) - \overline{u}_0(x,0), \qquad \Pi_0 v(x,0) = \varphi_2(x) - \overline{v}_0(x,0).$$

Solving (3.6.39) with decay conditions as $\tau \to \infty$, we obtain a family of solutions depending on an arbitrary function $A(x)$:

$$\Pi_0 u = A(x)\exp(\lambda_2 \tau), \qquad \Pi_0 v = -\gamma A(x)\exp(\lambda_2 \tau).$$

Substituting these expressions into (3.6.40) and using (3.6.36), (3.6.38), we obtain a system for $A(x)$ and $\Phi((a+\gamma b)x)$:

$$\Phi((a+\gamma b)x) + A(x) = \varphi_1(x), \qquad \frac{a}{b}\Phi((a+\gamma b)x) - \gamma A(x) = \varphi_2(x).$$

Finally, obtaining Φ and A from these relations, we define $\overline{u}_0$ and $\overline{v}_0$ below the straight line L_3, as well as $\Pi_0 u$ and $\Pi_0 v$:

$$\text{(3.6.41)} \quad \overline{u}_0 = -\frac{b}{\lambda_2}(\gamma\varphi_1(x-\Lambda_3 t) + \varphi_2(x-\Lambda_3 t)) \equiv \overline{u}_0^*, \qquad \overline{v}_0 = \frac{a}{b}\overline{u}_0^* \equiv \overline{v}_0^*,$$

$$\Pi_0 u = -\frac{a}{\lambda_2}\left(\varphi_1(x) - \frac{b}{a}\varphi_2(x)\right)\exp(\lambda_2\tau), \qquad \Pi_0 v = -\gamma\Pi_0 u.$$

Here $\overline{u}_0^*$ and $\overline{v}_0^*$ mean that the functions are defined in a subdomain lying below L_3.

Similarly, during the construction of the boundary functions $Q_0 u(\xi,t)$ and $Q_0 v(\xi,t)$ in the vicinity of $x=0$ ($\xi = x/\varepsilon$), we will obtain expressions for $\overline{u}_0$ and $\overline{v}_0$ above L_3 (we will use notation $\overline{u}_0^{**}$ and $\overline{u}_0^{**}$ for the functions defined there):

$$\text{(3.6.42)} \quad \begin{aligned} \overline{u}_0 &= -\frac{b}{\Lambda_1\lambda_3}\left(\gamma\frac{\Lambda_1}{\Lambda_2}\psi_1\left(t-\frac{x}{\Lambda_3}\right) + \psi_2\left(t-\frac{x}{\Lambda_3}\right)\right) \equiv \overline{u}_0^{**}, \\ \overline{v}_0 &= \frac{a}{b}\overline{u}_0^{**} \equiv \overline{v}_0^{**}. \end{aligned}$$

For $Q_0 u(\xi,t)$ and $Q_0 v(\xi,t)$ we have

$$Q_0 u = -\frac{\Lambda_2 a}{\lambda_2\Lambda_3}\left(\psi_1(t) - \frac{b}{a}\psi_2(t)\right)\exp(\lambda_3\xi), \quad \text{and} \quad Q_0 v = -\gamma\frac{\Lambda_1}{\Lambda_2}Q_0 u,$$

where $\lambda_3 = -(a\Lambda_2 + \gamma b\Lambda_1)/(\Lambda_1\Lambda_2) < 0$.

Analyzing (3.6.41) and (3.6.42), we come to the conclusion that the leading term of the regular part of the asymptotic solution has a discontinuity on the straight line L_3. In particular, for the u- component of the solution

$$\text{(3.6.43)} \quad \triangle\,\overline{u}_0 \equiv (\overline{u}_0^{**} - \overline{u}_0^*)|_{x=\Lambda_3 t} = \varphi_1(0)\frac{\gamma b}{\lambda_2}\left(1-\frac{\Lambda_1}{\Lambda_3}\right) + \varphi_2(0)\frac{b}{\lambda_2}\left(1-\frac{\Lambda_1}{\Lambda_2}\right)$$

It turns out that the solution of (3.6.35), (3.6.28) tends to a discontinuous function as $\varepsilon \to 0$. In particular, u approaches

$$\overline{u}_0 = \begin{cases} \overline{u}_0^{**} & \text{above } L_3, \\ \overline{u}_0^* & \text{below } L_3. \end{cases}$$

Therefore, along with the ordinary boundary layers in the vicinities of $t=0$ and $x=0$, an internal transition layer appears near L_3.

Theorem 3.9 *For the solution $u(x,t,\varepsilon)$, $v(x,t,\varepsilon)$ of* (3.6.35), (3.6.28),*the following relations hold:*

$$u - (\overline{u}_0^* + \Pi_0 u) \to 0, \qquad v - (\overline{v}_0^* + \Pi_0 v) \to 0 \quad \text{for } 0 \le t \le \frac{x}{\Lambda_2};$$

$$u - \overline{u}_0^* \to 0, \qquad v - \overline{v}_0^* \to 0 \quad \text{for } \frac{x}{\Lambda_2} < t < \frac{x}{\Lambda_3};$$

$$u - \overline{u}_0^{**} \to 0, \qquad v - \overline{v}_0^{**} \to 0 \quad \text{for } \Lambda_1 t < x < \Lambda_3 t;$$

$$u - (\overline{u}_0^{**} + Q_0 u) \to 0, \qquad v - (\overline{v}_0^{**} + Q_0 v) \to 0 \quad \text{for } 0 \le x \le \Lambda_1 t$$

as $\varepsilon \to 0$.

This theorem only states the existence of an internal transition layer near L_3, and does not give its analytical description. It is shown in Nesterov [109] that this internal layer might be described by an equation of parabolic type.

To prove the theorem, we introduce in (3.6.35) the new independent variables $\alpha = x - \Lambda_1 t$, $\beta = \Lambda_2 t - x$. This will transform the characteristics L_1 and L_2 into the new coordinate axes. Then we must convert the system into a second-order differential equation (for example, for u) of the form

$$\varepsilon(\Lambda_2 - \Lambda_1)\frac{\partial^2 u}{\partial\alpha\partial\beta} + a\frac{\partial u}{\partial\alpha} + \gamma b\frac{\partial u}{\partial\beta} = 0.$$

The solution of this equation can be expressed using a Riemann function (see, e.g., Tikhonov and Samarskii [132]), and the asymptotic solution can be obtained by the method of stationary phase. We will not present the proof since it is very cumbersome (cf. Vasil'eva [143]).

Let us consider an example that allows the construction of an exact solution and gives an opportunity to illustrate the theorem formulated above:

$$\varepsilon\left(\frac{\partial u}{\partial t} + \frac{\partial u}{\partial x}\right) = v, \qquad \varepsilon\left(\frac{\partial v}{\partial t} + 2\frac{\partial v}{\partial x}\right) = -v. \tag{3.6.44}$$

This system is decoupled, and therefore we can first solve the second equation before the first.

In this example $a = 0$, $b = \gamma = 1$, $\lambda_1 = 0$, $\lambda_2 = -1$, $\Lambda_1 = 1$, $\Lambda_2 = 2$, $\Lambda_3 = 1$, $\lambda_3 = -\frac{1}{2}$. Since $\Lambda_1 = \Lambda_3$, the characteristics L_1 and L_3 coincide, so the discontinuity of the limiting solution will take place on L_1. Its magnitude, according to (3.6.43), is

$$\triangle\, \overline{u}_0 = \varphi_2(0). \tag{3.6.45}$$

Let us now obtain the exact solution of (3.6.44), (3.6.28). First, we consider the second equation of (3.6.44). In region Ω_1 (see Fig. 3.13), the function v is defined entirely by condition $v(x, 0, \varepsilon) = \varphi_2(x)$. Solving this equation by the method of first integrals, we obtain $x - 2t = c_1$, $v = c_2 \exp(-t/\varepsilon)$. At $t = 0$, we have $x = c_1$ and $\varphi_2(x) = c_2$, i.e., $c_2 = \varphi_2(c_1)$. Thus, in Ω_1

$$v = \varphi_2(x - 2t)\exp\left(-\frac{t}{\varepsilon}\right).$$

In an absolutely similar way, the function v in the region $\Omega_2 \cup \Omega_3$ is defined by the condition $v(0, t, \varepsilon) = \psi_2(t)$ and has the form

$$v = \psi_2(t - x/2)\exp\left(-\frac{x}{2\varepsilon}\right).$$

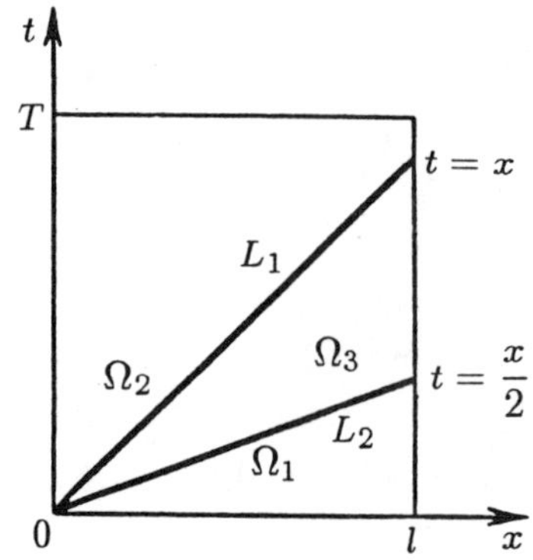

Figure 3.13: *Characteristics for the system* (3.6.44).

Let us now consider the first equation of (3.6.44). We will also solve it by the method of the first integrals. In region Ω_1 we have

$$x - t = c_1, \qquad u = c_2 + \int_0^t \frac{1}{\varepsilon}\varphi_2(c_1 - t')\exp(-t'/\varepsilon)dt'$$

$$= c_2 - \varphi_2(c_1 - t)\exp(-t/\varepsilon) + \varphi_2(c_1) + O(\varepsilon).$$

The last expression was derived using integration by parts. At $t = 0$, we obtain $x = c_1$ and $\varphi_1(x) = c_2$, so $c_2 = \varphi_1(c_1)$. Thus, in Ω_1

$$u = \varphi_1(x - t) - \varphi_2(x - 2t)\exp(-t/\varepsilon) + \varphi_2(x - t) + O(\varepsilon).$$

We seek the function u in the region Ω_3 by the same method. Taking into account the continuity of u on L_2, we obtain

$$x - t = c_1, \qquad u = c_2 + \int_0^x \frac{1}{\varepsilon}\,\psi_2(x'/2 - c_1)\exp\left(-\frac{x'}{2\varepsilon}\right)dx'$$

$$= c_2 - 2\psi_2(x/2 - c_1)\exp\left(-\frac{x}{2\varepsilon}\right) + 2\psi_2(-c_1) + O(\varepsilon).$$

For $x = 2t$ (on L_2), we have

$$t = c_1, \qquad u = c_2 - 2\psi_2(0)\exp\left(-\frac{t}{\varepsilon}\right) + 2\psi_2(-t) + O(\varepsilon)$$

$$= \varphi_1(t) - \varphi_2(0)\exp\left(-\frac{t}{\varepsilon}\right) + \varphi_2(t) + O(\varepsilon),$$

so

$$c_2 = \varphi_1(c_1) - \varphi_2(0)\exp\left(-\frac{c_1}{\varepsilon}\right) + \varphi_2(c_1) + 2\psi_2(0)\exp\left(-\frac{c_1}{\varepsilon}\right) - 2\psi_2(-c_1) + O(\varepsilon).$$

Hence, using the matching condition $\varphi_2(0) = \psi_2(0)$, we obtain

$$u = \varphi_1(x - t) + \varphi_2(0)\exp\left(-\frac{x - t}{\varepsilon}\right) + \varphi_2(x - t) - 2\psi_2\left(t - \frac{x}{2}\right)\exp\left(-\frac{x}{2\varepsilon}\right) + O(\varepsilon).$$

in Ω_3, where $t < x$ holds, and therefore $t - x/2 < x/2$. Thus, we can substitute zero for the argument $t - x/2$ in the last term (then this term will change to a quantity of order $O(\varepsilon)$). Thus in Ω_3,

$$(3.6.46)\qquad \begin{aligned} u = \varphi_1(x-t) + \varphi_2(x-t) + \varphi_2(0)\exp\left(-\tfrac{x-t}{\varepsilon}\right) \\ -2\psi_2(0)\exp\left(-\tfrac{x}{2\varepsilon}\right) + O(\varepsilon). \end{aligned}$$

In the region Ω_2, function u is defined by the condition at $x = 0$ and can be constructed as in Ω_1. As a result, in Ω_2 we obtain

$$(3.6.47)\qquad u = \psi_1(t-x) + 2\psi_2(t-x) - 2\psi_2\left(t - \frac{x}{2}\right)\exp\left(-\frac{x}{2\varepsilon}\right) + O(\varepsilon).$$

It can be seen from (3.6.46), (3.6.47) that when approaching L_1 from above, u has the limit $\overline{u}^{**} = \psi_1(t-x) + 2\psi_2(t-x)$, and when approaching L_1 from below, it has the limit $\overline{u}^{*} = \varphi_1(x-t) + \varphi_2(x-t)$. These limits are not equal on the straight line L_1 $(x = t)$, and their difference (with matching conditions taken into account) is equal to $\Delta\overline{u} = \varphi_2(0)$, which coincides with (3.6.45).

3.6.4 A physical problems leading to hyperbolic systems

Telegraphic equations

The telegraphic equations have the form

$$(3.6.48)\qquad -\frac{\partial j}{\partial x} = C\frac{\partial v}{\partial t} + Gv, \qquad -\frac{\partial v}{\partial x} = L\frac{\partial j}{\partial t} + Rj.$$

They describe the propagation of electric impulses in wires (see, e.g., Tikhonov and Samarskii [132]). Here j and v are the current and the voltage (the difference of potentials between the wire and the earth); C, L, G, and R are physical parameters: capacitance, inductance, leakage, and resistance, respectively.

Let us make the change of variables $j = \sqrt{C}(p+q)$, $v = \sqrt{L}(-p+q)$ to transform (3.6.48) into a system of the form (3.6.23) (without a small parameter):

$$(3.6.49)\qquad \begin{aligned} \frac{\partial p}{\partial t} - \frac{1}{\sqrt{LC}}\frac{\partial p}{\partial x} &= -\frac{GL+RC}{2LC}p + \frac{GL-RC}{2LC}q, \\ \frac{\partial q}{\partial t} + \frac{1}{\sqrt{LC}}\frac{\partial q}{\partial x} &= \frac{GL-RC}{2LC}p - \frac{GL+RC}{2LC}q. \end{aligned}$$

Suppose that the resistance and leakage are large and set $R = r/\varepsilon$ and $G = g/\varepsilon$, with $\varepsilon > 0$ being a small parameter. Then (3.6.49) finally can be written in the form of (3.6.23) with the small parameter ε. The matrix

$$a = \frac{1}{2LC}\begin{pmatrix} -(gL+rC) & gL-rC \\ gL-rC & -(gL+rC) \end{pmatrix}$$

has eigenvalues λ_1 and λ_2 with negative real parts: $\lambda_{1,2} = -(gL+rC) \pm i(gL-rC)$. Note that $\Lambda_1 = -1/\sqrt{LC}$ and $\Lambda_2 = 1/\sqrt{LC}$ have opposite signs. For system (3.6.48), we impose the conditions $j(x,0) = \varphi_1(x)$, $v(x,0) = \varphi_2(x)$ and $v(0,t) = v(l,t) = v^0$ (the ends of the wire are kept under the voltage v^0). Then for p and q we have

$$(3.6.50)\quad p(x,0) = \frac{1}{2}\left(\frac{\varphi_1(x)}{\sqrt{C}} - \frac{\varphi_2(x)}{\sqrt{L}}\right), \qquad q(x,0) = \frac{1}{2}\left(\frac{\varphi_1(x)}{\sqrt{C}} + \frac{\varphi_2(x)}{\sqrt{L}}\right),$$

$$(3.6.51) \qquad -p(0,t)+q(0,t)=v^0/\sqrt{L}, \qquad -p(l,t)+q(l,t)=v^0/\sqrt{L}.$$

The asymptotic solution of this problem can be constructed as in Theorem 3.8. The regular part of the asymptotics is trivial, and the Π-functions are defined by the algorithm described in § 3.6.2. Concerning Q_k and Q_k^*, we should mention that they are constructed in a slightly different way, since conditions (3.6.51) are different from (3.6.31). However, this difference is not important. For example, for the Q_0-functions we have

$$(3.6.52) \qquad \begin{aligned} \frac{\partial Q_0 p}{\partial \xi} &= \frac{gL+rC}{2\sqrt{LC}} Q_0 p - \frac{gL-rC}{2\sqrt{LC}} Q_0 q, \\ \frac{\partial Q_0 q}{\partial \xi} &= \frac{gL-rC}{2\sqrt{LC}} Q_0 p - \frac{gL+rC}{2\sqrt{LC}} Q_0 q, \end{aligned}$$

$$(3.6.53) \qquad \begin{aligned} -Q_0 p(0,t)+Q_0 q(0,t) &= v^0/L, \\ Q_0 p(\infty,t) = Q_0 q(\infty,t) &= 0. \end{aligned}$$

The eigenvalues of the matrix in system (3.6.52) equal $\pm\sqrt{gr}$. The solution corresponding to $\lambda=-\sqrt{gr}$ has the form

$$\begin{pmatrix} Q_0 p \\ Q_0 q \end{pmatrix} = A \begin{pmatrix} 1 \\ \dfrac{gL+rC+2\sqrt{grLC}}{gL-rC} \end{pmatrix} \exp(-\sqrt{gr}\,\xi),$$

where A is defined from (3.6.53) as

$$A = \frac{v^0}{L}\frac{2(rC+\sqrt{grLC})}{gL-rC} = \frac{2\sqrt{rC}v^0}{L(\sqrt{gL}-\sqrt{rC})}.$$

(Note that for $gL=rC$, we have $Q_0 p=0$ and $Q_0 q=(v^0/L)\exp(-\sqrt{gr}\,\xi)$.)

Let us now consider the case when only the leakage is large, i.e., $G=g/\varepsilon$, or only the resistance is large, i.e., $R=r/\varepsilon$. Both cases will be critical ones. Let us consider the case of large R in more detail. The system (3.6.49) then has the form

$$(3.6.54) \qquad \begin{aligned} \varepsilon\left(\frac{\partial p}{\partial t} - \frac{1}{\sqrt{LC}}\frac{\partial p}{\partial x}\right) &= -\frac{r}{2L}p - \frac{r}{2L}q - \varepsilon\frac{G}{2C}p + \varepsilon\frac{G}{2C}q, \\ \varepsilon\left(\frac{\partial q}{\partial t} + \frac{1}{\sqrt{LC}}\frac{\partial q}{\partial x}\right) &= -\frac{r}{2L}p - \frac{r}{2L}q + \varepsilon\frac{G}{2C}p - \varepsilon\frac{G}{2C}q. \end{aligned}$$

Let conditions (3.6.50), (3.6.51) be imposed. Only the difference in signs of Λ_1 and Λ_2 and the presence of terms of the order ε in the right-hand side distinguishes system (3.6.54) from (3.6.35). This difference in signs, however, leads to some changes in the construction algorithm. In the zeroth order for the regular terms, we obtain $\overline{p}_0=\alpha_0$ and $\overline{q}_0=-\alpha_0$, where $\alpha_0=\alpha_0(x,t)$ is arbitrary. At the next order approximation, we obtain the equation

$$\frac{\partial \alpha_0}{\partial t} = -\frac{G}{C}\alpha_0$$

for α_0. Unlike the case considered in § 3.6.3, this equation is an ordinary differential equation, so the characteristic L_3 coincides with the t-axis ($\Lambda_3 = 0$). Solving for α_0, we obtain $\alpha_0 = \Phi(x)\exp\left(-\frac{G}{C}t\right)$ where $\Phi(x)$ is yet unknown. The boundary functions $\Pi_0 p$ and $\Pi_0 q$ near $t = 0$ are constructed as before, and together with them, $\Phi(x)$ is defined. Therefore $\overline{p}_0$ and $\overline{q}_0$ are also defined. The boundary layers near $x = 0$ and $x = l$ cannot now be described by introducing Q-functions, as for the noncritical case, since the matrix of the system, analogous to (3.6.52), now has two zero eigenvalues. This boundary layer has a more complicated structure. It is similar to the internal transition layer in the vicinity of L_3 discussed in § 3.6.3. (In our case, L_3 coincides with t- axis.)

To describe the boundary layer near $x = 0$, we transform (3.6.48) to a second-order equation (see Tikhonov and Samarskii [132]). In the case $R = r/\varepsilon$, it has the form

$$\varepsilon\frac{\partial^2 v}{\partial x^2} = \varepsilon LC\frac{\partial^2 v}{\partial t^2} + (rC + \varepsilon LG)\frac{\partial v}{\partial t} + rGv.$$

It is not difficult to show that the boundary layer near $x = 0$ is described by the parabolic equation

$$\frac{\partial^2 Qv}{\partial \xi^2} = rC\frac{\partial Qv}{\partial t} + rG\,Qv, \qquad \xi = x/\sqrt{\varepsilon}.$$

If $R = r/\varepsilon$ and $G = g/\varepsilon$, the corresponding equation has the form

$$\varepsilon^2\frac{\partial^2 v}{\partial x^2} = \varepsilon^2 LC\frac{\partial^2 v}{\partial t^2} + \varepsilon(rC + Lg)\frac{\partial v}{\partial t} + rg\,v,$$

and the boundary layer in the vicinity of $x = 0$ is described by the ordinary differential equation

$$\frac{\partial^2 Qv}{\partial \xi^2} = rg\,Qv \quad \text{for} \qquad \xi = x/\varepsilon.$$

Exercise

Find the zeroth-order terms of the asymptotic solution of (3.6.49)–(3.6.51) in the case when the leakage is large ($G = g/\varepsilon$, $0 < \varepsilon \ll 1$ is a small parameter).

The problem of gas absorption in the presence of mass transfer

A problem of dynamics of gas absorption leads to the following system

$$\text{(3.6.55)} \qquad \varepsilon\frac{\partial u}{\partial t} = -a(x)u + b(x)v, \qquad \varepsilon\left(\frac{\partial v}{\partial t} + \frac{\partial v}{\partial x}\right) = a(x)u - b(x)v.$$

(cf. Tikhonov and Samarskii [132]). Here v is the concentration of gas moving along the tube filled by a porous absorbent, u is the quantity of gas absorbed by a unit volume of absorbent, ε is inversely proportional to the so-called kinetic coefficient that characterizes the rate of absorption.

The system (3.6.55) is again an example in the critical case, as discussed in § 3.6.3. Here $\Lambda_1 = 0$, $\Lambda_2 = 1$ and $\Lambda_3 = a/(a+b)$. The internal transition layer appears in the vicinity of the characteristic $x = \Lambda_3 t$. The asymptotic solution is similar to one constructed in § 3.6.3. For this problem the structure of the internal transition layer was studied in Nesterov [109], where the parabolic equation describing this layer was obtained and a uniform asymptotic solution was constructed.

Chapter 4

Applied Problems

Methods and ideas used to solve applied problems considered in this chapter complement the ones presented in previous chapters of the book. In §§ 4.1 and 4.2, problems in the critical case, where the reduced equations are not algebraic, but differential, are presented (compare with § 3.4). For a combustion problem (§ 4.1), the asymptotic series consists of only a regular part and initial layer (ordinary) boundary functions, whereas in the asymptotic solution of the heat conduction problem (§ 4.2), corner boundary functions are also present. A problem from the theory of semiconductor devices (§ 4.3) in one spatial dimension can be posed as a boundary value problem in the critical case (in § 2.2 only initial value problems were considered). A generalization to the two-dimensional case is also discussed. Section 4.4 illustrates how the boundary function method can be applied to construct the relaxation wave solution of the FitzHugh–Nagumo equations (similar ideas can be used to construct moving pulse and moving front solutions of more general reaction-diffusion type systems). And finally, a brief survey of other results on applied problems, solved by the boundary function method and its modifications, is given in § 4.5.

4.1 Mathematical model of combustion process in the case of autocatalytic reaction

4.1.1 Statement of the problem

The system of equations describing combustion process in the case of a first-order autocatalytic reaction has the form

$$
\begin{aligned}
&c\rho\frac{\partial T}{\partial t} - \lambda \,\triangle\, T = q(v_0 + v)(1 - v)k_0 \exp\left(-\frac{E}{RT}\right),\\
(4.1.1)\qquad &\frac{\partial v}{\partial t} - D \,\triangle\, v = (v_0 + v)(1 - v)k_0 \exp\left(-\frac{E}{RT}\right),
\end{aligned}
$$

(see, e.g., Vol'pert and Khudyaev [164]). Here T is the absolute temperature, v is the depth of conversion of the combustible component that characterizes the relative amount of substance that has been burnt ($v = 1 - u$, where u is the relative concentration of the combustible component, $0 \le u \le 1$, $0 \le v \le 1$), c, ρ, λ are physical characteristics of the medium: the specific heat capacity, density of the medium in which reaction occurs and thermal conductivity, respectively, q is the thermal effect of reaction, v_0 is the criterion of autocatalytic behavior (the ratio of the initial rate of reaction to the autocatalytic constant, $0 < v_0 < 1$), k_0 is the reaction constant,

E is the activation energy, R is the universal gas constant, and D is the diffusion coefficient of the combustible component. We assume that all these quantities, with the exception of T and v, are constant.

The functions v and $u = 1 - v$ are already dimensionless. Let us introduce the nondimensional temperature θ, time t', coordinates x', y', z' and constants β, ε, a_0, b_0 by the formulae

$$\theta = \frac{E}{RT_*^2}(T - T_*), \quad \beta = RT_*/E, \quad t' = tk_0 \exp(-1/\beta),$$

$$x' = \frac{x}{r}, \quad y' = \frac{y}{r}, \quad z' = \frac{z}{r}, \quad \varepsilon = \frac{c\rho}{q}\frac{RT_*^2}{E},$$

$$a_0 = \frac{\lambda}{c\rho r^2 k_0 \exp(-1/\beta)}, \qquad b_0 = \frac{D}{r^2 k_0 \exp(-1/\beta)}.$$

Here T_* is a characteristic temperature, e.g., the temperature of the surrounding medium, and r is the characteristic size of the domain in which the reaction takes place. In the new variables, system (4.1.1) has the form (we again use the notation t, x, y, z instead of t', x', y', z'):

(4.1.2)
$$\varepsilon\left(\frac{\partial \theta}{\partial t} - a_0 \triangle \theta\right) = (v_0 + v)(1 - v)\exp\left(\frac{\theta}{1 + \beta\theta}\right),$$
$$\frac{\partial v}{\partial t} - b_0 \triangle v = (v_0 + v)(1 - v)\exp\left(\frac{\theta}{1 + \beta\theta}\right).$$

For combustion processes, the parameters ε and β are usually small. We will investigate the case of a fast reaction (large k_0) such that $k_0 \exp(-1/\beta) = O(1)$. Since the dependence on β in the right-hand side of (4.1.2) is regular, we can simplify the system by setting there $\beta = 0$ (this will not have much impact on the characteristics of the combustion process). Further, we will consider only the one-dimensional case, when θ and v do not depend on the variables y and z (and therefore $\triangle = \partial^2/\partial x^2$). We will also assume that the nondimensional coefficients of heat transfer a_0 and diffusion b_0 are large: $a_0 \sim b_0 \sim 1/\varepsilon$. From the physical point of view, this means that either the dimensional coefficients of heat transfer and diffusion are large, or the combustion process occurs in a thin layer (the characteristic length r is small). Introducing the notation $a_0 = a/\varepsilon$ and $b_0 = b/\varepsilon$, we arrive at a system of singularly perturbed equations

(4.1.3)
$$\varepsilon\frac{\partial \theta}{\partial t} - a\frac{\partial^2 \theta}{\partial x^2} = (v_0 + v)(1 - v)\exp\theta,$$
$$\varepsilon\frac{\partial v}{\partial t} - b\frac{\partial^2 v}{\partial x^2} = \varepsilon(v_0 + v)(1 - v)\exp\theta.$$

Let us consider this system in the rectangle $\Omega = (0 < x < 1) \times (0 < t \leq T)$. We impose the natural boundary conditions

(4.1.4) $$\theta|_{t=0} = v|_{t=0} = 0, \quad \theta|_{x=0} = \theta|_{x=1} = 0, \quad \left.\frac{\partial v}{\partial x}\right|_{x=0} = \left.\frac{\partial v}{\partial x}\right|_{x=1} = 0.$$

The conditions for θ mean that at the initial time $t = 0$, and on the boundary of the domain ($x = 0$ and $x = 1$), the temperature is that of the surrounding medium: $\theta = 0$, i.e. $T = T_*$. The initial condition for v shows that at $t = 0$, the amount of burnt substance is zero (the reaction has not yet started). The boundary conditions for v describe the absence of the flow of combustible component through the boundary (the impermeability condition).

4.1.2 Construction of the asymptotic solution

We will construct the asymptotic expansion of the solution of (4.1.3), (4.1.4) in the form of a sum of regular and boundary layer parts:

$$\theta = \overline{\theta} + \Pi\theta = \sum_{k=0}^{\infty} \varepsilon^k (\overline{\theta}_k(x,t) + \Pi_k\theta(x,\tau)),$$

(4.1.5)

$$v = \overline{v} + \Pi v = \sum_{k=0}^{\infty} \varepsilon^k (\overline{v}_k(x,t) + \Pi_k v(x,\tau)),$$

where $\tau = t/\varepsilon$. Substituting (4.1.5) into (4.1.3), (4.1.4), and representing the right-hand sides of (4.1.3) in the form $f = \overline{f} + \Pi f$, in the standard way, we obtain the equations for the terms.

For $\overline{v}_0$ and $\overline{\theta}_0$, we have the system

$$\frac{\partial^2 \overline{v}_0}{\partial x^2} = 0, \qquad \frac{\partial^2 \overline{\theta}_0}{\partial x^2} = -\frac{1}{a}(v_0 + \overline{v}_0)(1 - \overline{v}_0) \exp \overline{\theta}_0$$

with boundary conditions

$$\frac{\partial \overline{v}_0}{\partial x}(0,t) = \frac{\partial \overline{v}_0}{\partial x}(1,t) = 0, \qquad \overline{\theta}_0(0,t) = \overline{\theta}_0(1,t) = 0. \tag{4.1.6}$$

This implies that

$$\overline{v}_0 = \alpha_0(t),$$

where $\alpha_0(t)$ is an arbitrary function. We can rewrite the equation for $\overline{\theta}_0$ as

$$\frac{\partial^2 \overline{\theta}_0}{\partial x^2} = -\delta(\alpha_0(t), a) \exp \overline{\theta}_0, \tag{4.1.7}$$

where $\delta(\alpha_0(t), a) = (v_0 + \alpha_0)(1 - \alpha_0)/a$. Such an equation has been considered in Gel'fand [54] and Frank-Kamenetskii [53]. Let us use the results of [54], where the boundary value problem

$$\frac{\partial^2 \theta_0}{\partial x^2} = -\delta \exp \theta, \qquad \theta(-1) = \theta(1) = 0 \tag{4.1.8}$$

was discussed. It is shown in Gel'fand [54] that for $\delta < \delta_{cr} \approx 0.878$, there exist two positive solutions of this problem, the smaller of which is stable. Applying this result to the problem (4.1.7), (4.1.6) for $\overline{\theta}_0$, whose solution depends on $\alpha_0(t)$ and a as parameters, we obtain that when

$$a > a_1 = \frac{\max\limits_{\alpha_0}(v_0 + \alpha_0)(1 - \alpha_0)}{4\delta_{cr}} = \frac{(1 + v_0)^2}{16\,\delta_{cr}}, \tag{4.1.9}$$

there exist two positive solutions. Let us take the smaller solution and denote it by $\overline{\theta}_0 = \tilde{\theta}_0(x, \alpha_0(t), a)$. It is known (see [54]) that the solution of (4.1.8) which we have chosen is an increasing function of δ for $\delta < \delta_{cr}$. Therefore $\tilde{\theta}_0$ decreases with growing a, and $\tilde{\theta}_0 \to 0$ for $a \to \infty$. It should be also noted that

$$\max_{0 \le x \le 1} \tilde{\theta}_0(x, \alpha_0(t), a) = \tilde{\theta}_0\left(\frac{1}{2}, \alpha_0(t), a\right).$$

Thus, the leading terms $\overline{v}_0$ and $\overline{\theta}_0$ of the regular part of the asymptotic solution depend on the yet unknown function $\alpha_0(t)$. This means that (4.1.3), (4.1.4) is a problem in the *critical case*: the corresponding reduced system has a family of solutions.

Let us now define the boundary functions of the zeroth order. For $\Pi_0 v$, we have

$$\frac{\partial \Pi_0 v}{\partial \tau} - b\frac{\partial^2 \Pi_0 v}{\partial x^2} = 0,$$

$$\Pi_0 v(x, 0) = -\overline{v}_0(x, 0) = -\alpha_0(0), \qquad \frac{\partial \Pi_0 v}{\partial x}(0, \tau) = \frac{\partial \Pi_0 v}{\partial x}(1, \tau) = 0.$$

This implies that $\Pi_0 v(x, \tau) = -\alpha_0(0)$. It is natural to demand that Π-functions approach zero as $\tau \to \infty$, so $\Pi_0 v(x, \tau) = 0$, and the initial condition $\alpha_0(0) = 0$ for the yet unknown $\alpha_0(t)$ has been found. This function will be completely determined at the next step of the asymptotic algorithm (during the study of equation for $\overline{v}_1(x, t)$).

For $\Pi_0 \theta$, we obtain

$$\begin{aligned} &\frac{\partial \Pi_0 \theta}{\partial \tau} - a\frac{\partial^2 \Pi_0 \theta}{\partial x^2} = v_0[\exp(\overline{\theta}_0(x, 0) + \Pi_0 \theta) - \exp(\overline{\theta}_0(x, 0))], \\ &\Pi_0 \theta(x, 0) = -\overline{\theta}_0(x, 0), \qquad \Pi_0 \theta(0, \tau) = \Pi_0 \theta(1, \tau) = 0. \end{aligned} \tag{4.1.10}$$

Note that initial and boundary conditions are matched so as to be continuous since $\overline{\theta}_0(0, 0) = \overline{\theta}_0(1, 0) = 0$. We should also mention that, although the function $\overline{\theta}_0(x, t) = \tilde{\theta}_0(x, \alpha_0(t), a)$ is still not defined (it depends on the unknown $\alpha_0(t)$), its value at $t = 0$ is known: $\overline{\theta}_0(x, 0) = \tilde{\theta}_0(x, \alpha_0(0), a) = \tilde{\theta}_0(x, 0, a)$.

Let us make the change of variables $w(x, \tau) = \overline{\theta}_0(x, 0) + \Pi_0 \theta(x, \tau)$. For $w(x, \tau)$, we obtain

$$\frac{\partial w}{\partial \tau} - a\frac{\partial^2 w}{\partial x^2} = v_0 \exp w, \quad w(x, 0) = 0, \quad w(0, \tau) = w(1, \tau) = 0.$$

It has been shown in Vol'pert and Khudyaev [164] that under condition (4.1.9), a unique solution of this problem exists and $w(x, \tau) \to \overline{\theta}_0(x, 0)$ monotonically as $\tau \to \infty$. Consequently, $\Pi_0 \theta(x, \tau) \to 0$ when $\tau \to \infty$, $\Pi_0 \theta \le 0$.

Let us prove that for sufficiently large a the function $\Pi_0 \theta(x, \tau)$ exponentially approaches zero for growing τ. We introduce $\tilde{\Pi} = -\Pi_0 \theta$. From (4.1.10) we obtain

$$\frac{\partial \tilde{\Pi}}{\partial \tau} - a\frac{\partial^2 \tilde{\Pi}}{\partial x^2} = v_0 \exp(\overline{\theta}_0(x, 0))(1 - \exp(-\tilde{\Pi})),$$

$$\tilde{\Pi}(x, 0) = \overline{\theta}_0(x, 0), \qquad \tilde{\Pi}(0, \tau) = \tilde{\Pi}(1, \tau) = 0.$$

Let us introduce the notation

$$k = v_0 \max_{0\le x\le 1} \exp(\overline{\theta}_0(x,0)) = v_0 \exp\left(\tilde{\theta}_0\left(\frac{1}{2}, 0, a\right)\right),$$

and consider the problem

$$\frac{\partial Q}{\partial \tau} - a\frac{\partial^2 Q}{\partial x^2} = kQ, \quad Q(x,0) = \overline{\theta}_0(x,0), \quad Q(0,\tau) = Q(1,\tau) = 0.$$

Using a comparison theorem for parabolic equations (cf. Kolesov [82]), and taking into account the inequality $\tilde{\Pi} \ge 0$, we obtain

$$0 \le \tilde{\Pi}(x,\tau) \le Q(x,\tau),$$

where $Q(x,\tau)$ can be represented by the Fourier series:

$$Q(x,\tau) = \sum_{n=1}^{\infty} q_n \exp(-(a\lambda_n - k)\tau)\sin(\pi n x),$$

for

$$\lambda_n = \pi^2 n^2, \qquad q_n = 2\int_0^1 \overline{\theta}_0(x,0)\sin(\pi n x)dx.$$

Hence,

$$Q(x,\tau) \le c\exp(-(a\lambda_1 - k)\tau) \le c\exp(-\kappa\tau)$$

if

$$a > \frac{k}{\lambda_1} = \frac{v_0 \exp\left(\tilde{\theta}_0\left(\frac{1}{2}, 0, a\right)\right)}{\pi^2}. \tag{4.1.11}$$

Since $\tilde{\theta}_0\left(\frac{1}{2}, 0, a\right) \to 0$ as $a \to \infty$, there exists an a_2 such that for $a > a_2$ inequality (4.1.11) holds. Taking into account (4.1.9), we can write the condition on a as

$$a > \max(a_1, a_2). \tag{4.1.12}$$

Then for a satisfying (4.1.12), the function $\tilde{\Pi}$ (and therefore, $\Pi_0\theta$) possesses the exponential estimate

$$|\Pi_0\theta(x,\tau)| \le c\exp(-\kappa\tau), \qquad 0 \le x \le 1,\ \tau \ge 0. \tag{4.1.13}$$

For $\overline{v}_1$ and $\overline{\theta}_1$ we have the system

$$\begin{aligned} \frac{\partial^2 \overline{v}_1}{\partial x^2} &= \frac{1}{b}[\alpha_0'(t) - (v_0 + \alpha_0(t))(1 - \alpha_0(t))\exp(\tilde{\theta}_0(x, \alpha_0(t), a)] \\ &\equiv \psi_0(x, \alpha_0), \end{aligned} \tag{4.1.14}$$

$$\begin{aligned} \frac{\partial^2 \overline{\theta}_1}{\partial x^2} = \frac{1}{a} \Bigg\{ \frac{\partial \overline{\theta}_0}{\partial t} &- [(v_0 + \alpha_0(t))(1 - \alpha_0(t))\overline{\theta}_1 \\ &+ (1 - v_0 - 2\alpha_0(t))\overline{v}_1]\exp(\tilde{\theta}_0(x, \alpha_0(t), a))\Bigg\} \end{aligned}$$

with boundary conditions

$$\frac{\partial \overline{v}_1}{\partial x}(0,t) = 0, \quad \frac{\partial \overline{v}_1}{\partial x}(1,t) = 0, \quad \overline{\theta}_1(0,t) = \overline{\theta}_1(1,t) = 0. \tag{4.1.15}$$

Integrating (4.1.14) with the first condition, we obtain

$$\frac{\partial \overline{v}_1}{\partial x}(x,t) = \int_0^x \psi_0(x_0, \alpha_0)\,dx_0. \tag{4.1.16}$$

The second condition of (4.1.15) is satisfied only if

$$\int_0^1 \psi_0(x_0, \alpha_0)\,dx_0 = 0. \tag{4.1.17}$$

This provides an equation for the unknown function $\alpha_0(t)$. Taking into account the explicit form of $\psi_0(x, \alpha_0)$, we can write this equation as

$$\alpha_0' = (v_0 + \alpha_0)(1 - \alpha_0)g(\alpha_0), \tag{4.1.18}$$

where

$$g(\alpha_0) = \int_0^1 \exp(\tilde{\theta}_0(x_0, \alpha_0(t), a)\,dx_0.$$

Since $g(\alpha_0) > 0$, the solution of (4.1.18) with the initial condition $\alpha_0(0) = 0$ monotonically approaches the rest point $\alpha_0 = 1$ as $t \to \infty$. Hence,

$$0 \le \alpha_0(t) < 1 \quad \text{for} \quad 0 \le t \le T.$$

It should be mentioned that (4.1.8) can be integrated by quadratures.

Thus, $\alpha_0(t)$ is constructed, and hence all the terms of the zeroth-order approximation are defined. Integrating (4.1.16), we will obtain $\overline{v}_1(x,t) = \alpha_1(t) + \tilde{v}_1(x,t)$ where

$$\tilde{v}_1(x,t) = \int_0^x \left(\int_0^{x_1} \psi_0(x_0, \alpha_0(t), a)\,dx_0 \right) dx_1$$

is known, and $\alpha_1(t)$ is arbitrary.

Let us rewrite the equation for $\overline{\theta}_1$ in the form

$$\frac{\partial^2 \overline{\theta}_1}{\partial x^2} + \beta(x,t)\overline{\theta}_1 = \gamma(x,t)\alpha_1(t) + g(x,t),$$

where

$$\beta(x,t) = \frac{1}{a}(v_0 + \alpha_0(t))(1 - \alpha_0(t)) \exp(\tilde{\theta}_0(x, \alpha_0(t), a)),$$

$$\gamma(x,t) = -\frac{1}{a}(1 - v_0 - 2\alpha_0(t)) \exp(\tilde{\theta}_0(x, \alpha_0(t), a)),$$

$$g(x,t) = \frac{1}{a}\left(\frac{\partial \overline{\theta}_0}{\partial t} - (1 - v_0 - 2\alpha_0(t))\tilde{v}_1 \exp(\tilde{\theta}_0(x, \alpha_0(t), a)) \right)$$

are known.

We seek $\overline{\theta}_1$ in the form

$$\overline{\theta}_1 = A(x,t) + B(x,t)\alpha_1(t). \tag{4.1.19}$$

For A and B we obtain linear equations

$$\frac{\partial^2 A}{\partial x^2} + \beta(x,t)A = g(x,t), \qquad \frac{\partial^2 B}{\partial x^2} + \beta(x,t)B = \gamma(x,t) \tag{4.1.20}$$

(where t enters as a parameter) with boundary conditions

$$A(0,t) = A(1,t) = 0, \qquad B(0,t) = B(1,t) = 0.$$

The question of solvability arises. It is well known that for existence of a unique solution of the boundary value problem

$$u'' + d(x)u = f(x), \qquad u(0) = u(1) = 0,$$

it is necessary and sufficient that the corresponding homogeneous problem has only the trivial solution. In turn, the homogeneous problem has only the trivial solution if and only if there exists a nonnegative function $w(x)$ such that

$$w'' + d(x)w \leq 0 \quad \text{for} \quad 0 < x < 1,$$

$$w(0) \geq 0, \qquad w(1) \geq 0,$$

and at least one of these inequalities is strict (cf. Protter and Weinberger [120]). For (4.1.20) such a function w can be easily found for sufficiently large a. Let us take $w(x) = \sin \pi x$. Then $w(x) \geq 0$ for $0 \leq x \leq 1$ and, furthermore,

$$w'' + \beta(x,t)w = (-\pi^2 + \beta(x,t)) \sin \pi x < 0 \quad \text{for} \quad 0 < x < 1,$$

if $\beta(x,t) < \pi^2$. It follows from the expression for $\beta(x,t)$ that there exists an a_3 such that for $a > a_3$, $\beta(x,t) < \pi^2$ holds. Thus, the condition (4.1.12) on a should be changed to

$$a > \max(a_1, a_2, a_3). \tag{4.1.21}$$

Then the boundary value problems for A and B are uniquely solvable, and we obtain $\overline{\theta}_1$ in the form (4.1.19). Therefore $\overline{v}_1$ and $\overline{\theta}_1$ depend linearly on the yet unknown function $\alpha_1(t)$.

Let us now construct the boundary functions of the first order. For $\Pi_1 v$ we have

$$\frac{\partial \Pi_1 v}{\partial \tau} - b\frac{\partial^2 \Pi_1 v}{\partial x^2} = v_0[\exp(\overline{\theta}_0(x,0) + \Pi_0\theta(x,\tau)) - \exp(\overline{\theta}_0(x,0))] \equiv h(x,\tau),$$

$$\Pi_1 v(x,0) = -\overline{v}_1(x,0) = -\tilde{v}_1(x,0) - \alpha_1(0), \qquad \frac{\partial \Pi_1 v}{\partial x}(0,\tau) = \frac{\partial \Pi_1 v}{\partial x}(1,\tau) = 0.$$

Note that initial and boundary conditions are matched since

$$\frac{\partial \overline{v}_1}{\partial x}(0,0) = \frac{\partial \overline{v}_1}{\partial x}(1,0) = 0.$$

Function $h(x,\tau)$ has an estimate of the type (4.1.13). The solution of this problem can be found as a Fourier series:

$$\Pi_1 v(x,\tau) = \sum_{n=0}^{\infty} c_n(\tau)\cos(\pi n x). \tag{4.1.22}$$

For the coefficients $c_n(\tau)$, we obtain equations

$$c_n{}' = -b\pi^2 n^2 c_n + h_n(\tau), \qquad n = 0,1,2,\dots, \tag{4.1.23}$$

with initial conditions

$$c_0(0) = -\int_0^1 \tilde{v}_1(x,0)dx - \alpha_1(0) = \varphi_0 - \alpha_1(0),$$

$$c_n(0) = \varphi_n, \qquad n = 1,2,\dots\ .$$

Here $h_n(\tau)$ and φ_n are coefficients of the Fourier series expansions of $h(x,\tau)$ and $(-\tilde{v}_1(x,0))$ in $\cos(\pi n x)$, respectively; the $h_n(\tau)$ evidently have an estimate of the type (4.1.13).

Solving (4.1.23) for $n = 0$, we obtain

$$c_0(\tau) = \varphi_0 - \alpha_1(0) + \int_0^\tau h_0(\tau_0)d\tau_0.$$

For $\Pi_1 v(x,\tau)$ to approach zero as $\tau \to \infty$, it is necessary that $c_0(\infty) = 0$. This condition allows us to find the initial value

$$\alpha_1(0) = \varphi_0 + \int_0^\infty h_0(\tau)d\tau. \tag{4.1.24}$$

The function $\alpha_1(t)$ will be completely defined during the study of the problem for $\overline{v}_2(x,t)$.

Solutions of (4.1.23) for $n = 1,2,\dots$ can be written as

$$c_n(\tau) = \varphi_n \exp(-b\pi^2 n^2 \tau) + \int_0^\tau \exp(-b\pi^2 n^2(\tau - \tau_0))h_n(\tau_0)d\tau_0.$$

The exponential estimate for $c_n(\tau)$ follows in an elementary fashion from this formula. It can be easily verified that the $c_n(\tau)$ decay like $1/n^4$ when $n \to \infty$. Therefore the series (4.1.22) converges. It can be differentiated term wise, once with respect to τ and twice with respect to x, and it satisfies an estimate of the type (4.1.13).

For the $\Pi_1\theta$ we have

$$\frac{\partial \Pi_1\theta}{\partial \tau} - a\frac{\partial^2 \Pi_1\theta}{\partial x^2} = v_0 \exp(\overline{\theta}_0(x,0) + \Pi_0\theta(x,\tau))\Pi_1\theta + \sigma(x,\tau),$$

$$\Pi_1\theta(x,0) = -\overline{\theta}_1(x,0), \qquad \Pi_1\theta(0,\tau) = \Pi_1\theta(1,\tau) = 0.$$

Here $\sigma(x,\tau)$ is known and satisfies an estimate of the type (4.1.13). As for $\Pi_0\theta$, the estimate like (4.1.13) for $\Pi_1\theta$ can be obtained using a comparison theorem.

Consider the problem for $\overline{v}_2(x,t)$:

$$\frac{\partial^2\overline{v}_2}{\partial x^2} = \frac{1}{b}\{\alpha_1'(t) - [(v_0+\alpha_0(t))(1-\alpha_0(t))\overline{\theta}_1$$

$$+(1-v_0-2\alpha_0(t))\alpha_1(t)]\exp(\tilde{\theta}_0(x,t)) + F(x,t)\} \equiv \psi_1(x,\alpha_1(t),t),$$

$$\frac{\partial\overline{v}_2}{\partial x}(0,t) = \frac{\partial\overline{v}_2}{\partial x}(1,t) = 0,$$

where $F(x,t)$ is known, and $\overline{\theta}_1$ has the form (4.1.19). The solvability condition for this problem (analogous to (4.1.17)),

$$\int_0^1 \psi_1(x,\alpha_1(t),t)dx = 0,$$

provides a linear differential equation for the yet unknown $\alpha_1(t)$:

$$\alpha_1' = k_1(t)\alpha_1 + k_2(t),$$

where k_1 and k_2 are known. Solving this equation with the initial condition (4.1.24), we obtain $\alpha_1(t)$. Thus, all the terms of the first-order approximation have been completely determined.

Higher-order terms of the series (4.1.5) can be obtained analogously.

Let Θ_n and V_n denote the nth partial sums of the series (4.1.5)

Theorem 4.1 *For sufficiently small ε, and sufficiently large a, there exists a unique solution θ, v of (4.1.3), (4.1.4), and the series (4.1.5) is the asymptotic series for this solution in the rectangle $\overline{\Omega}$ as $\varepsilon \to 0$, i.e., the the following estimates hold:*

$$\max_{\overline{\Omega}} |\theta - \Theta_n| = O(\varepsilon^{n+1}), \qquad \max_{\overline{\Omega}} |v - V_n| = O(\varepsilon^{n+1}).$$

The detailed proof is presented in Butuzov and Kalachev [21]. It is important to mention that during the proof of the theorem one more condition on a appears:

$$a > \frac{(1+v_0)^2}{32} \max_{0\le t\le T} \exp\left(\tilde{\theta}_0\left(\frac{1}{2},\alpha_0(t),a\right)\right). \tag{4.1.25}$$

By virtue of behavior of $\tilde{\theta}_0$ for $a \to \infty$, it is clear that for sufficiently large a this inequality holds. Thus, we add (4.1.25) to the condition (4.1.21) on a. It can be shown that (4.1.21) is a consequence of (4.1.25). We should also mention that of all the conditions on a, only (4.1.9) is the necessary one which provides the solvability of the problem for $\tilde{\theta}_0(x,t)$. The other conditions are sufficient. They are associated with obtaining the estimates and with the method of proof. The condition (4.1.25) on a could possibly be somewhat relaxed.

4.1.3 Physical interpretation of the asymptotic solution

Note that the quantity a has quite an important practical meaning. If the conditions on a are satisfied, a nonexplosive reaction occurs.

The constructed asymptotics allow us to analyze the behavior of the temperature θ and the depth of conversion of the combustible component v. Since Π-functions have an exponential estimate of the type (4.1.13), the temperature θ during a short period of time changes rapidly from zero to values close to $\overline{\theta}_0(x,t)$. This corresponds to the fast stage of reaction. During this stage, the value of v changes very little since $\Pi_0 v = 0$. Then the slow stage of reaction begins: $\theta = \overline{\theta}_0(x,t) + O(\varepsilon)$ and $v = \alpha_0(t) + O(\varepsilon)$. As we mentioned earlier, the function $\alpha_0(t)$ grows monotonically from zero at $t = 0$ to one as $t \to \infty$. Thus, during a sufficiently long time, the combustible component burns out almost completely. The process of combustion is uniform in the domain $0 \le x \le 1$, since the leading term $\alpha_0(t)$ of the asymptotic expansion for v does not depend on x. The nonuniformity appears in the process only starting with the terms of order ε.

As $\alpha_0(t)$ grows from 0 to 1, the function $\delta(\alpha_0(t), a) = (v_0+\alpha_0(t))(1-\alpha_0(t))/a$ first increases (for $0 \le \alpha_0(t) \le (1+v_0)/2$), and then decreases (for $(1+v_0)/2 \le \alpha_0(t) \le 1$). The fact that $\tilde{\theta}_0(x,t)$ is an increasing function of δ implies that the temperature θ during the slow stage of reaction first grows, and then, after $\alpha_0(t)$ reaches $(1+v_0)/2$, starts to decrease.

We considered a spatially one-dimensional case. We can construct analogously the asymptotic solution and estimate the remainder for the combustion problem (4.1.1) in a cylindrical domain. See also Sattinger [124] where a problem for the nonautocatalytic reaction is discussed and the theorem on passage to the limit is proved.

Exercises

1. Consider a nondimensionalized problem describing a reaction-diffusion process which, in the case of large ($\sim O(1/\varepsilon)$) diffusion, can be written as

$$\varepsilon\frac{\partial u}{\partial t} - \frac{\partial^2 u}{\partial x^2} = -\varepsilon[(1+x)u + 1],$$

$$(x,t) \in \Omega = (0 < x < 1) \times (0 < t < T),$$

$$\frac{\partial u}{\partial x}(0,t,\varepsilon) - \varepsilon u(0,t,\varepsilon) = \varepsilon, \qquad \frac{\partial u}{\partial x}(1,t,\varepsilon) + 2\varepsilon u(1,t,\varepsilon) = 0.$$

$$u(x,0,\varepsilon) = 1 + \cos \pi x.$$

Find the zeroth- and the first-order approximations (in small parameter $0 < \varepsilon \ll 1$) to the solution of this problem.

2. Consider the system

$$\varepsilon\frac{\partial u}{\partial t} - \frac{\partial^2 u}{\partial x^2} = -u + v,$$

$$\varepsilon\frac{\partial u}{\partial t} - b\frac{\partial^2 u}{\partial x^2} = \varepsilon(u - 2v),$$

$$(x,t) \in \Omega = (0 < x < 1) \times (0 < t < T),$$

subject to conditions

$$u(x,0) = x, \quad u(0,t) = 0, \quad u(1,t) = 1,$$

$$v(x,0) = 0, \qquad \frac{\partial v}{\partial x}(0,t) = \frac{\partial v}{\partial x}(1,t) = 0,$$

where $b > 0$ is a constant and $0 < \varepsilon \ll 1$ is a small parameter.

(a) Construct the zeroth-order terms of the asymptotic solution. What is the $\lim_{t\to\infty} \overline{v}_0$?

(b) What will change in the zeroth-order approximation if $v(x,0) = \cos \pi x$?

4.2 Heat conduction in thin bodies

4.2.1 Statement of the problem

We will say that a body is thin if one (or more) of its characteristic dimensions is much smaller than the others. A thin rod or a thin plate are examples of such bodies. We will consider a boundary value problem describing a heat conduction process in the thin rod, where the ratio ε of the thickness of the rod to its length is a small parameter. To simplify the presentation, we consider the problem for a planar rod, i.e., in the thin rectangle $(0 \le x \le 1) \times (0 \le z \le \varepsilon)$:

$$\frac{\partial u}{\partial t} - a(x)\left(\frac{\partial^2 u}{\partial x^2} + \frac{\partial^2 u}{\partial z^2}\right) = f(u,x,t), \tag{4.2.1}$$

$$u|_{t=0} = \varphi(x,z), \qquad u|_{x=0} = \psi_1(z,t), \qquad u|_{x=1} = \psi_2(z,t), \tag{4.2.2}$$

$$\left.\frac{\partial u}{\partial z} - A\varepsilon u\right|_{z=0} = 0, \qquad \left.\frac{\partial u}{\partial z} + A\varepsilon u\right|_{z=\varepsilon} = 0. \tag{4.2.3}$$

The heat conduction equation (4.2.1) contains the term $f(u,x,t)$, which describes the heat sources within the rod and depends in general on temperature u, coordinate x, and time t. Conditions (4.2.2) prescribe the initial temperature, as well as that of the surfaces $x = 0$ and $x = 1$ of the rod. Conditions (4.2.3) mean that, on the lateral surfaces $z = 0$ and $z = \varepsilon$, a weak heat exchange with the surrounding medium occurs.

Note that in (4.2.1)–(4.2.3) not only the rod is thin ($0 \le z \le \varepsilon$), but also the heat exchange through its lateral surfaces is small (the corresponding heat exchange coefficient is $A\varepsilon$). At first glance, it might seem that to obtain the approximate solution of the problem, we should omit the second derivative with respect to z in (4.2.1), as well as the conditions (4.2.3), and solve the resulting *truncated* (one-dimensional) equation with additional conditions (4.2.2). However, as the asymptotic analysis will show, the distribution of temperature within the rod is described by a spatially one-dimensional equation which is different from the *truncated* equation mentioned above. This asymptotically correct equation will be obtained during the construction of the asymptotic solution.

Let us make the change of variable $z = \varepsilon y$. Then (4.2.1)–(4.2.3) will have the form of a singularly perturbed problem (which we consider on the time interval $0 \le t \le T$)

$$\varepsilon^2 \frac{\partial u}{\partial t} - a(x)\left(\varepsilon^2 \frac{\partial^2 u}{\partial x^2} + \frac{\partial^2 u}{\partial y^2}\right) = \varepsilon^2 f(u,x,t), \tag{4.2.4}$$

$$(x, y, t) \in \Omega = (0 < x < 1) \times (0 < y < 1) \times (0 < t \leq T),$$

(4.2.5) $$u|_{t=0} = \varphi(x, y, \varepsilon), \qquad u|_{x=0} = \psi_1(y, t, \varepsilon), \qquad u|_{x=1} = \psi_2(y, t, \varepsilon),$$

(4.2.6) $$\left.\frac{\partial u}{\partial y} - A\varepsilon^2 u\right|_{y=0} = 0, \qquad \left.\frac{\partial u}{\partial y} + A\varepsilon^2 u\right|_{y=1} = 0.$$

We assume that all the given functions are sufficiently smooth, that $a(x) > 0$, and that the boundary values (4.2.5) are matched at the corner points, i.e.,

$$\varphi(0, y, \varepsilon) = \psi_1(y, 0, \varepsilon), \qquad \varphi(1, y, \varepsilon) = \psi_2(y, 0, \varepsilon).$$

Note that in what follows we consider in (4.2.5) a general dependence of φ and ψ_i $(i = 1, 2)$ on y and ε, and not only in the form of a product εy that appears after the change of variable $z = \varepsilon y$.

Let us describe in detail the construction of the zeroth-order approximation.

4.2.2 Construction of the asymptotic solution

We will construct the asymptotic solution of (4.2.4)–(4.2.6) in the form

(4.2.7) $$u(x, y, t, \varepsilon) = \sum_{k=0}^{\infty} \varepsilon^k (\overline{u}_k(x, y, t) + \Pi_k(x, y, \tau) + Q_k(\xi, y, t) + Q_k^*(\xi_*, y, t) + P_k(\xi, y, \tau) + P_k^*(\xi_*, y, \tau)),$$

where the $\overline{u}_k$ are terms of a regular expansion, Π_k, Q_k and Q_k^* are boundary functions describing the boundary layers near $t = 0$, $x = 0$ and $x = 1$, respectively; P_k and P_k^* are corner boundary functions; $\tau = t/\varepsilon^2$, $\xi = x/\varepsilon$ and $\xi_* = (1 - x)/\varepsilon$ are boundary layer variables. In the standard way, substituting the series (4.2.7) into (4.2.4)–(4.2.6), and representing f in the form $\overline{f} + \Pi f + Qf + Q^*f + Pf + P^*f$, we obtain equations for successive terms (see § 3.2.3).

Setting $\varepsilon = 0$ in (4.2.4) and (4.2.6), we obtain the equation

$$a(x)\frac{\partial^2 \overline{u}_0}{\partial y^2} = 0$$

for $\overline{u}_0(x, y, t)$, with boundary conditions

$$\left.\frac{\partial \overline{u}_0}{\partial y}\right|_{y=0} = \left.\frac{\partial \overline{u}_0}{\partial y}\right|_{y=1} = 0.$$

The solution of this problem is an arbitrary function of x and t: $\overline{u}_0 = \alpha_0(x, t)$. Therefore, (4.2.4)–(4.2.6) is a problem in the *critical case*. Analogously, we obtain $\overline{u}_1 = \alpha_1(x, t)$, an arbitrary function. For $\overline{u}_2(x, t)$ we have

$$a(x)\frac{\partial^2 \overline{u}_2}{\partial y^2} = \frac{\partial \alpha_0}{\partial t} - a(x)\frac{\partial^2 \alpha_0}{\partial x^2} - f(\alpha_0, x, t) \equiv \psi_0(x, t, \alpha_0),$$

$$\left.\frac{\partial \overline{u}_2}{\partial y}\right|_{y=0} = A\alpha_0(x, t), \qquad \left.\frac{\partial \overline{u}_2}{\partial y}\right|_{y=1} = -A\alpha_0(x, t).$$

Integration of the this equation with the first boundary condition implies

$$\frac{\partial \overline{u}_2}{\partial y} = A\alpha_0(x,t) + \frac{\psi_0(x,t,\alpha_0)y}{a(x)}. \tag{4.2.8}$$

Substituting this expression into the second boundary condition, we obtain

$$\psi_0(x,t,\alpha_0) = -2Aa(x)\alpha_0.$$

Thus, $\overline{u}_2$ is obtainable only if the above equality holds. This provides an equation for the unknown function $\alpha_0(x,t)$, i.e., for the leading term of the asymptotic expansion. Taking into account the form of $\psi_0(x,t,\alpha_0)$, this relation can by rewritten as

$$\frac{\partial \alpha_0}{\partial t} - a(x)\frac{\partial^2 \alpha_0}{\partial x^2} = g(\alpha_0, x, t), \tag{4.2.9}$$

where $g(\alpha_0,x,t) = f(\alpha_0,x,t) - 2Aa(x)\alpha_0$. The initial and boundary conditions for α_0 will be obtained during construction of the boundary functions. Then we will be able to define $\alpha_0(x,t)$ completely.

For $\Pi_0(x,y,\tau)$ we have the problem

$$\frac{\partial \Pi_0}{\partial \tau} - a(x)\frac{\partial^2 \Pi_0}{\partial y^2} = 0, \qquad 0 < y < 1, \quad \tau > 0,$$

$$\Pi_0(x,y,0) = \varphi_0(x,y) - \alpha_0(x,0), \qquad \frac{\partial \Pi_0}{\partial y}(x,0,\tau) = \frac{\partial \Pi_0}{\partial y}(x,1,\tau) = 0,$$

where x enters as a parameter ($0 \le x \le 1$), and $\varphi_0(x,t)$ is the leading term of the expansion for $\varphi(x,t,\varepsilon)$ into a power series in ε. Note that $\alpha_0(x,0)$ (still unknown) enters the initial condition. Solving this problem by Fourier series, we obtain

$$\Pi_0 = \sum_{n=0}^{\infty} b_n(x)\exp(-a(x)\pi^2 n^2 \tau)\cos(\pi n y),$$

where

$$b_0(x) = \int_0^1 (\varphi_0(x,y) - \alpha_0(x,0))dy, \qquad b_n(x) = 2\int_0^1 \varphi_0(x,y)\cos(\pi n y)dy, \qquad n \ge 1.$$

The standard decay condition for Π-functions (as $\tau \to \infty$) leads to $b_0 = 0$. This allows us to determine $\alpha_0(x,0)$:

$$\alpha_0(x,0) = \int_0^1 \varphi_0(x,y)dy = \alpha_{00}(x). \tag{4.2.10}$$

Thus, Π_0 is completely determined and it satisfies the estimate

$$|\Pi_0(x,y,\tau)| \le c\exp(-\kappa\tau).$$

For $Q_0(\xi,y,t)$, we obtain the problem

$$\frac{\partial^2 Q_0}{\partial \xi^2} + \frac{\partial^2 Q_0}{\partial y^2} = 0, \qquad \xi > 0, \quad 0 < y < 1,$$

$$Q_0(0,y,t) = \psi_{10}(y,t) - \alpha_0(0,t), \qquad \frac{\partial Q_0}{\partial y}(\xi,0,t) = \frac{\partial Q_0}{\partial y}(\xi,1,t) = 0.$$

Notice that the unknown $\alpha_0(0,t)$ enters the boundary condition at $\xi = 0$. Using the Fourier method, we obtain

$$Q_0 = \sum_{n=0}^{\infty} d_n(t) \exp(-\pi n \xi) \cos(\pi n y),$$

where

$$d_0(t) = \int_0^1 (\psi_{10}(y,t) - \alpha_0(0,t))dy, \quad \text{and} \quad d_n(t) = 2\int_0^1 \psi_{10}(y,t)\cos(\pi n y)dy, \qquad n \geq 1.$$

The standard assumption $Q_0(\infty, y, t) = 0$ leads to the condition $d_0 = 0$ and this allows us to find

$$\alpha_0(0,t) = \int_0^1 \psi_{10}(y,t)dy = \alpha_0^0(t). \tag{4.2.11}$$

Thus, Q_0 is completely determined and

$$|Q_0(\xi,y,t)| \leq c\exp(-\kappa\xi).$$

Functions $Q_0^*(\xi_*, y, t)$ is defined analogously to the Q_0 and has an exponential estimate in the variable ξ_*. During its construction, we obtain boundary value

$$\alpha_0(1,t) = \int_0^1 \psi_{20}(y,t)dy = \alpha_0^1(t). \tag{4.2.12}$$

For the unknown $\alpha_0(x,t)$, we obtained equation (4.2.9), as well as the initial condition (4.2.10) and boundary conditions (4.2.11), (4.2.12). It can be easily verified that these conditions are matched to be continuous at the corner points, i.e. $\alpha_{00}(0) = \alpha_0^0(0)$, $\alpha_{00}(1) = \alpha_0^1(0)$. These equalities directly follow from (4.2.10), (4.2.11), (4.2.12), and conditions $\varphi_0(0,y) = \psi_{10}(y,0)$, $\varphi_0(1,y) = \psi_{20}(y,0)$.

Let the equation (4.2.9) (*which might be nonlinear*) *with conditions* (4.2.10), (4.2.11), (4.2.12) *have a solution for* $0 \leq t \leq T$.

The corner boundary functions $P_0(\xi,y,\tau)$ and $P_0^*(\xi_*,y,\tau)$ are needed to compensate for the discrepancies introduced by the Π_0 in the boundary conditions at $x = 0$ and $x = 1$, and by the Q_0 and Q_0^* in the initial condition. For $P_0(\xi,y,\tau)$, we obtain

$$\frac{\partial P_0}{\partial \tau} - a\left(\frac{\partial^2 P_0}{\partial \xi^2} + \frac{\partial^2 P_0}{\partial y^2}\right) = 0, \qquad \xi > 0, \quad 0 < y < 1, \quad \tau > 0,$$

$$P_0(\xi,y,0) = -Q_0(\xi,y,0) = -\sum_{n=1}^{\infty} d_n(0)\exp(-\pi n\xi)\cos(\pi n y),$$

$$P_0(0,y,\tau) = -\Pi_0(0,y,\tau) = -\sum_{n=1}^{\infty} b_n(0)\exp(-a\pi^2 n^2 \tau)\cos(\pi n y),$$

$$\frac{\partial P_0}{\partial y}(\xi, 0, \tau) = \frac{\partial P_0}{\partial y}(\xi, 1, \tau) = 0,$$

where $a = a(0)$. From the expressions for $d_n(t)$ and $b_n(x)$ together with the equality $\varphi_0(0, y) = \psi_{10}(y, 0)$, it follows that $d_n(0) = b_n(0)$. Thus, the initial and boundary (at $\xi = 0$) conditions for P_0 are matched so as to be continuous at $(0, y, 0)$. We will seek the solution of this problem in the form

$$P_0 = \sum_{n=1}^{\infty}(v_n(\xi, \tau) - d_n(0)\exp(-\pi n\xi - a\pi^2 n^2\tau))\cos(\pi ny).$$

Then for $v_n(\xi, \tau)$, we have the problem

$$\frac{\partial v_n}{\partial \tau} - a\frac{\partial^2 v_n}{\partial \xi^2} + a\pi^2 n^2 v_n = h_n(\xi, \tau), \qquad \xi > 0, \quad \tau > 0,$$

$$v_n(\xi, 0) = 0, \qquad v_n(0, \tau) = 0,$$

$$h_n = -ad_n(0)\pi^2 n^2 \exp(-\pi n\xi - a\pi^2 n^2\tau),$$

whose solution can be represented as

$$v_n(\xi, \tau) = \int_0^\tau d\tau_0 \int_0^\infty G_n(\xi, \xi_0, \tau - \tau_0)h_n(\xi_0, \tau_0)d\xi_0,$$

where G_n is the Green's function:

$$G_n(\xi, \xi_0, \tau - \tau_0) = \frac{1}{2\sqrt{a\pi(\tau - \tau_0)}}\exp(-a\pi^2 n^2(\tau - \tau_0))$$

$$\times\left[\exp\left(-\frac{(\xi - \xi_0)^2}{4a(\tau - \tau_0)}\right) - \exp\left(-\frac{(\xi + \xi_0)^2}{4a(\tau - \tau_0)}\right)\right].$$

Using this representation for $v_n(\xi, \tau)$, it is not difficult to prove that

$$|v_n| \le c|d_n(0)|\exp(-\kappa(n\xi + n^2\tau)).$$

Hence, $P_0(\xi, y, \tau)$ satisfies

$$|P_0(\xi, y, \tau)| \le c\exp(-\kappa(\xi + \tau)).$$

Corner boundary function $P_0^*(\xi_*, y, \tau)$ is constructed in a similar way, and has analogous exponential estimate.

Let

$$U_0(x, y, t, \varepsilon) = \alpha_0(x, t) + \Pi_0(x, y, \tau) + Q_0(\xi, y, t)$$

$$+Q_0^*(\xi_*, y, t) + P_0(\xi, y, \tau) + P_0^*(\xi_*, y, \tau).$$

Theorem 4.2 *For sufficiently small ε, the problem (4.2.1)–(4.2.3)has a unique solution $u(x,y,t,\varepsilon)$, and $U_0(x,y,t,\varepsilon)$ is the asymptotic approximation for this solution in the parallelepiped $\overline{\Omega}$ as $\varepsilon \to 0$ with accuracy of order $O(\varepsilon)$, i.e., the following estimate holds:*

$$\max_{\overline{\Omega}} |u - U_0| = O(\varepsilon).$$

This theorem can be proved by applying the method of successive approximations to the equation for the remainder and using the maximum principle to estimate these successive approximations. A detailed proof is presented in Butuzov and Urazgil'dina [37].

Under more restrictive matching conditions at the corner points $(0,y,0)$ and $(1,y,0)$, higher-order terms of the asymptotic series (4.2.7) can be constructed.

4.2.3 Concluding remarks

By virtue of corresponding exponential estimates, the boundary functions Π_0, P_0 and P_0^* become arbitrarily small for $t \geq \delta > 0$, and the approximation of the solution with an accuracy of order ε (such approximation is often sufficient for practical purposes) is given by the sum

$$\alpha_0(x,t) + Q_0(\xi,y,t) + Q_0^*(\xi_*,y,t).$$

The boundary functions Q_0 and Q_0^* describe fast change of temperature near the ends $x = 0$ and $x = 1$ of the rod. However, inside the rod (for $\delta \leq x \leq 1-\delta$) they are arbitrarily small, and the distribution of temperature there is approximated by the function $\alpha_0(x,t)$, which is the solution of the one-dimensional heat conduction equation (4.2.9) with additional conditions (4.2.10), (4.2.11), (4.2.12). As we can see, this equation differs substantially from the *truncated* one-dimensional equation, obtained from the original equation (4.2.1) by omitting the second derivative with respect to z (such truncated version of (4.2.1) is often inappropriately used in practical calculations). In comparison with the truncated equation, there is an additional term $-2Aa(x)\alpha_0$ on the right hand-side of (4.2.9), whose influence is large for a large thermal diffusion coefficient $a(x)$, and for a large coefficient A in the heat exchange term. Therefore the asymptotic method allows us to construct the correct one-dimensional model which provides a good approximation for the two-dimensional problem.

Exercise

Find the zeroth-order terms of the asymptotic approximation of the solution for

$$\varepsilon^2 \frac{\partial u}{\partial t} - \varepsilon^2 \frac{\partial^2 u}{\partial x^2} - \frac{\partial^2 u}{\partial y^2} = -\varepsilon^2 u,$$

$$(x,y,t) \in \Omega = (0 < x < 1) \times (0 < y < 1) \times (0 < t < T),$$

$$u|_{t=0} = x, \qquad u|_{x=0} = 0, \qquad u|_{x=1} = 1,$$

$$\left.\frac{\partial u}{\partial y} - \varepsilon^2 u\right|_{y=0} = 0, \qquad \left.\frac{\partial u}{\partial y} + \varepsilon^2 u\right|_{y=1} = 0.$$

We considered the heat conduction problem in the thin rectangle. The asymptotic solution can be constructed similarly for the problem in a thin three-dimensional rod of a constant cross-section S. In this case, by stretching the variables y and z by the coefficient ε, we obtain

$$\varepsilon^2\frac{\partial u}{\partial t} - a(x)\left(\varepsilon^2\frac{\partial^2 u}{\partial x^2} + \frac{\partial^2 u}{\partial y^2} + \frac{\partial^2 u}{\partial z^2}\right) = \varepsilon^2 f(u, x, t),$$

$$0 < x < 1, \qquad (y, z) \in S, \qquad 0 < t \le T,$$

$$u|_{t=0} = \varphi(x, y, z, \varepsilon), \qquad u|_{x=0} = \psi_1(y, z, t, \varepsilon), \qquad u|_{x=1} = \psi_2(y, z, t, \varepsilon),$$

$$\left.\frac{\partial u}{\partial n} + A\varepsilon^2 u\right|_{\Gamma} = 0.$$

Here Γ is the lateral surface of the bar and $\partial/\partial n$ is the derivative along the outer normal to Γ.

The asymptotic solution in this case can be constructed analogously to the case of a rectangle. In particular, equation for α_0 has the same form (4.2.9), with the only difference being that the coefficient 2 in the term $2Aa(x)\alpha_0$ should be replaced by l/s, where l is the length of the boundary of the cross-section S and s is its area.

The initial and boundary conditions for the $\alpha_0(x, t)$ are defined, as before, during the construction of the boundary functions. For example, for $Q_0(\xi, y, z, t)$, we obtain the problem

$$\frac{\partial^2 Q_0}{\partial \xi^2} + \frac{\partial^2 Q_0}{\partial y^2} + \frac{\partial^2 Q_0}{\partial z^2} = 0, \qquad \xi > 0, \quad (y, z) \in S,$$

$$Q_0(0, y, z, t) = \psi_{10}(y, z, t) - \alpha_0(0, t), \qquad \left.\frac{\partial Q_0}{\partial n}\right|_{\Gamma} = 0.$$

We can find the solution of this problem as

$$Q_0 = \sum_{n=0}^{\infty} d_n(t)\exp(-\lambda_n \xi) F_n(y, z),$$

where the λ_n $(0 = \lambda_0 < \lambda_1 \le \ldots \le \lambda_n \le \ldots)$ and the $F_n(y, z)$ are the eigenvalues and eigenfunctions, respectively, of the problem

$$\frac{\partial^2 F}{\partial y^2} + \frac{\partial^2 F}{\partial z^2} + \lambda^2 F = 0, \quad (y, z) \in S; \qquad \left.\frac{\partial F}{\partial n}\right|_{\Gamma} = 0.$$

In particular, $F_0 = 1$, and

$$d_n(t) = \frac{1}{\| F_n \|^2}\iint_S (\psi_{10}(y, z, t) - \alpha_0(0, t)) F_n(y, z)\,dy\,dz.$$

The condition $Q_0(\infty, y, z, t) = 0$ leads to $d_0(t) = 0$, which, in turn, allows us to find the boundary condition for the yet unknown $\alpha_0(x, t)$ at $x = 0$:

$$\alpha_0(0, t) = \frac{1}{s}\iint_S \psi_{10}(y, z, t)\,dy\,dz.$$

The initial condition $\alpha_0(x,0)$ and the second boundary condition $\alpha_0(1,t)$ are defined analogously. Thus, the process of constructing the boundary functions is similar to that described in § 4.2.2 (here we use the expansions in a Fourier series in eigenfunctions $F_n(y,z)$ instead of $\cos(\pi ny)$).

Detailed discussion of the heat conduction problem in a thin plate can be found in Urazgil'dina [137].

The above scheme can also be applied to some more complicated problems, e.g., problems of thermoelasticity in thin bodies. The corresponding system of equations for the displacement vector $\mathbf{u}(x,y,z,t)$ and temperature $\Theta(x,y,z,t)$ in a domain G in the linear approximation has the form (cf. Nowacki [111])

$$\mu \triangle \mathbf{u} + (\lambda+\mu)\operatorname{grad}\operatorname{div}\mathbf{u} + \mathbf{F} = \gamma \operatorname{grad}\Theta + \rho_0 \frac{\partial^2 \mathbf{u}}{\partial t^2},$$

$$\triangle\Theta - \frac{1}{\kappa}\frac{\partial\Theta}{\partial t} - \eta \operatorname{div}\frac{\partial \mathbf{u}}{\partial t} = -\frac{f}{\kappa},$$

where μ and λ are the elastic moduli, $\mathbf{F}$ is a vector of mass forces, $\gamma = (3\lambda+2\mu)/\alpha$, α is the coefficient of thermal expansion, $\rho_0(x,y,z)$ is the apparent density, κ is the coefficient of thermal diffusivity, $\eta = \gamma\Theta_0/\lambda_0$, Θ_0 is the average temperature of the body, λ_0 is the coefficient of heat conduction, and f represents thermal sources. In the case of a thin body (for a thin rod or plate), this system can be transformed to a singularly perturbed one by an appropriate stretching of variables. This allows us to apply the boundary function method to construct the asymptotic solution. In Butuzov and Urazgil'dina [37] the asymptotic solution for the thin rod is constructed; the discussion of the case of the thin plate is pesented in Urazgil'dina [138]. Asymptotic analysis allows us to choose the correct simplified model of lower dimension, which provides a good approximation to the solution of the original problem.

4.3 Application of the boundary function method in the theory of semiconductor devices

4.3.1 Statement of the problem and asymptotic algorithm for a one-dimensional model

Although our main goal will be the investigation of a two-dimensional model of a diode, let us start the discussion with comments on a problem for a one-dimensional (p-n)-junction (cf. Vasil'eva and Butuzov [149] and Vasil'eva and Stel'makh [157]); this problem plays an important role in analyzing one-dimensional models of semiconductor devices (e.g., diodes, transistors, etc.). Consider a (p-n)-junction, i.e., a contact, located at the point $x=0$, between semiconductors of p- and n-types, characterized by hole and electron conductivities, respectively. A semiconductor of p-type is placed to the left of the contact ($-x_0 \le x \le 0$), and a semiconductor of n-type is placed to the right of the contact ($0 \le x \le x_0$). Such a semiconductor scheme, in a stationary case with no externally generated sources, can be described by a system of equations (cf. Pol'skii [118], Markowich [88], and Markowich, Ringhofer, and Schmeiser [89]) consisting of Poisson's equation

$$(4.3.1) \qquad \varepsilon\frac{dE}{dx} = -q(n-p-c(x))$$

(here E is the polar electric voltage, n and p are the respective concentrations of electrons and holes, $c(x,y)$ (dopant concentration) is the difference between concentrations of donors and acceptors in a semiconductor material: $c(x,y) > 0$ holds in the n-region, and $c(x,y) < 0$ in the p-region; q is the charge of the electron, ε is the dielectric permeability), and the equations for the densities of electron and hole currents, J_n and J_p, respectively,

$$J_n = q\mu_n En + qD_n\frac{dn}{dx}, \qquad J_p = q\mu_p Ep - qD_p p\frac{dp}{dx}. \tag{4.3.2}$$

(Here D_n and D_p are the diffusion coefficients for electrons and holes, respectively, and μ_n and μ_p are their mobilities.) It is known that $D_n/\mu_n = D_p/\mu_p = q/(kT) \equiv U_T = \text{const}$, where k is Boltzmann's constant and T is the absolute temperature.

From the continuity equations it follows that in the one-dimensional problem the scalars J_n and J_p are constant. We will consider here a special case, when $c(x) = -\bar{c} < 0$ to the left of the contact, while to the right $c(x) = \bar{c} > 0$, $\bar{c} = \text{const}$. Let us introduce the dimensionless variables

$$\tilde{x} = x/x_0, \quad \tilde{n} = n/\bar{c}, \quad \tilde{p} = p/\bar{c}, \quad I_n = J_n x_0/(q\bar{c}D_n), \quad I_p = J_p x_0/(q\bar{c}D_p),$$

$$\tilde{E} = E/(x_0 U_T), \quad \mu^2 = \frac{\varepsilon U_T}{q\bar{c}x_0^2}, \quad N(x) = c/\bar{c} = \begin{cases} -1, & \tilde{x} \in [-1,0), \\ 1, & \tilde{x} \in (0,1]. \end{cases}$$

Note that for many semiconductor devices μ^2 is a small parameter ($\mu^2 \le 10^{-7}$). Omitting tildes in the new variables, we rewrite (4.3.1), (4.3.2) as

$$\begin{gathered} \mu^2\frac{dE}{dx} = p - n + N(x), \\ \frac{dn}{dx} = -nE + I_n, \qquad \frac{dp}{dx} = pE - I_p. \end{gathered} \tag{4.3.3}$$

Various kinds of boundary conditions can be considered for (4.3.3). We will restrict ourselves to one of the simplest, the so-called *symmetric case*, when $I_n = I_p = I = \text{const}$ is given, so the boundary conditions are prescribed separately for the intervals $[-1,0]$ and $[0,1]$; namely,

$$n(-1) = 0, \quad p(-1) = 1, \quad n(0) = p(0), \tag{4.3.4}$$

and

$$n(0) = p(0), \quad n(1) = 1, \quad p(1) = 0. \tag{4.3.5}$$

It can be easily seen that the problem (4.3.3), (4.3.4) reduces to the problem (4.3.3), (4.3.5) under the change of variables $x \to -x$, $n \to p$, and $p \to n$. Therefore it suffices to consider only one of these problems, e.g., (4.3.3), (4.3.5) for $0 \le x \le 1$.

By introducing the new variables $u = p + n$ and $v = p - n$, we obtain the system

$$\begin{gathered} \mu^2\frac{dE}{dx} = v + 1, \\ \frac{du}{dx} = vE, \qquad \frac{dv}{dx} = uE - 2I. \end{gathered} \tag{4.3.6}$$

Differentiating the first equation and introducing the new variables

$$\varphi = \mu E, \qquad \chi = \mu \frac{d\varphi}{dx},$$

we obtain

$$(4.3.7) \qquad \begin{aligned} &\mu \frac{d\chi}{dx} = u\varphi - 2\mu I, \qquad \mu \frac{d\varphi}{dx} = \chi, \\ &\mu \frac{du}{dx} = v\varphi, \qquad \mu \frac{dv}{dx} = u\varphi - 2\mu I. \end{aligned}$$

The boundary conditions (4.3.5) in the new variables have the form

$$(4.3.8) \qquad v(0) = 0, \quad u(1) = 1, \quad v(1) = -1.$$

In addition, it is necessary to supply one more condition for (4.3.7), which is obtained from the first equation of (4.3.6) in the two forms:

$$(4.3.9) \qquad \chi(0) = 1, \qquad \chi(1) = 0.$$

It is easy to see that of the five conditions (4.3.8) and (4.3.9), we need only consider four

$$(4.3.10) \qquad \chi(0) = 1, \quad \chi(1) = 0, \quad u(1) = 1, \quad v(1) = -1,$$

since the condition $v(0) = 0$ is automatically satisfied. Indeed, from the first and last equations of (4.3.7), it follows that $d\chi/dx = dv/dx$, and hence, $\chi = v + \text{const}$. Noting the boundary conditions $\chi(1) = 0$ and $v(1) = -1$, we obtain $\chi = v + 1$. And, then, by virtue of condition $\chi(0) = 1$, it follows that $v(0) = 0$.

Let us construct the asymptotic solution of the problem (4.3.7), (4.3.10) in the form

$$z = \bar{z}(x, \mu) + \Pi z(\xi, \mu) + Qz(\xi_*, \mu),$$

where below z stands for all functions φ, χ, u and v, and $\xi = x/\mu$ and $\xi_* = (x-1)/\mu$; $\bar{z}$ is the regular part of the solution, and Πz and Qz represent the boundary functions near $x = 0$ and $x = 1$, respectively. The right-hand sides of (4.3.7) should be written in a similar form (see formula (2.3.7)).

Note that (4.3.7), (4.3.10) is a boundary value problem in the *critical case* since the *reduced system* , obtained from (4.3.7) for $\mu = 0$, has a family of solutions. In § 2.2 we studied the initial value problem in the critical case. Now we are dealing with the boundary value problem. However, using the ideas of § 2.2 and § 2.3.1, we can construct the boundary layer asymptotics in this case as well. The boundary value problems in the critical case are discussed in more detail in Vasil'eva and Butuzov [149].

For the regular functions of the zeroth order, which satisfy the reduced equations, we have

$$\bar{\chi}_0 = 0, \quad \bar{\varphi}_0 = 0, \quad \bar{u}_0 = \alpha_1(x), \quad \bar{v}_0 = \alpha_2(x),$$

where α_1 and α_2 are as yet arbitrary. The equations for α_1 and α_2 are obtained in the usual way (see § 2.2) from solvability conditions for the system in the first order approximation. These equations have the form

$$\frac{d\alpha_1}{dx} = \frac{2I\alpha_2}{\alpha_1}, \qquad \frac{d\alpha_2}{dx} = 0. \tag{4.3.11}$$

To find supplementary conditions for (4.3.11), we must consider boundary functions of the zeroth order. Since u and v are prescribed at $x = 1$, we consider first the system of equations for $Q_0 z(\xi_*)$:

$$\begin{aligned} \frac{dQ_0\chi}{d\xi_*} &= (\alpha_1(1) + Q_0 u)Q_0\varphi, & \frac{dQ_0\varphi}{d\xi_*} &= Q_0\chi, \\ \frac{dQ_0 u}{d\xi_*} &= (\alpha_2(1) + Q_0 v)Q_0\varphi, & \frac{dQ_0 v}{d\xi_*} &= (\alpha_1(1) + Q_0 u)Q_0\varphi. \end{aligned} \tag{4.3.12}$$

The additional conditions for $Q_0 z$ are

$$\begin{gathered} Q_0\chi(0) = 0, \quad Q_0 u(0) = 1 - \alpha_1(1), \quad Q_0 v(0) = -1 - \alpha_2(1), \\ Q_0 z(-\infty) = 0. \end{gathered} \tag{4.3.13}$$

The solution of (4.3.12), satisfying (4.3.13), belongs to an *unstable manifold* Ω^- for which in this concrete case we can derive an analytic representation. We obtain from (4.3.12) the equations

$$\frac{dQ_0 u}{dQ_0\chi} = \frac{\alpha_2(1) + Q_0 v}{\alpha_1(1) + Q_0 u}, \qquad \frac{dQ_0 v}{dQ_0\chi} = 1.$$

The solution of this system with conditions $Q_0 u = Q_0 v = 0$ for $Q_0\chi = 0$ (following from the boundary conditions at negative infinity) is

$$Q_0 u = -\alpha_1(1) + \operatorname{sgn}(\alpha_1(1))\sqrt{\alpha_1^2(1) + 2\alpha_2(1)Q_0\chi + (Q_0\chi)^2} \tag{4.3.14}$$

$$Q_0 v = Q_0\chi. \tag{4.3.15}$$

The expression for $Q_0\varphi$ can be obtained using the first two equations of (4.3.12):

$$\frac{(Q_0\varphi)^2}{2} = \operatorname{sgn}(\alpha_1(1)) \int_0^{Q_0\chi} \frac{\eta d\eta}{\sqrt{\alpha_1^2(1) + 2\alpha_2(1)\eta + \eta^2}} . \tag{4.3.16}$$

Equations (4.3.14)–(4.3.16) therefore provide an analytic representation of the unstable manifold.

By substituting the boundary values (4.3.13) into (4.3.14), (4.3.15), we obtain additional conditions

$$\alpha_1(1) = 1, \qquad \alpha_2(1) = -1 \tag{4.3.17}$$

for $\alpha_i(x)$, $i = 1, 2$, and the functions $Q_0 z(\xi_*)$ are easily seen to be identically zero.

The solution of system (4.3.11) with conditions (4.3.17) is

$$\alpha_1(x) = \sqrt{1 + 4I(1 - x)}, \qquad \alpha_2(x) = -1.$$

For $\Pi_0 z(\xi)$ we have the system

$$\frac{d\Pi_0\chi}{d\xi} = (\alpha_1(0) + \Pi_0 u)\Pi_0\varphi, \qquad \frac{d\Pi_0\varphi}{d\xi} = \Pi_0\chi,$$

(4.3.18)

$$\frac{d\Pi_0 u}{d\xi} = (-1 + \Pi_0 v)\Pi_0\varphi, \qquad \frac{d\Pi_0 v}{d\xi} = (\alpha_1(0) + \Pi_0 u)\Pi_0\varphi,$$

where $\alpha_1(0) = \sqrt{1 + 4I}$. The additional conditions for are

(4.3.19) $$\Pi_0\chi(0) = 1, \qquad \Pi_0 z(\infty) = 0.$$

The solution of (4.3.18), satisfying (4.3.19), belongs to the *stable manifold* Ω^+, represented by

$$\Pi_0 u = -\sqrt{1 + 4I} + \sqrt{(1 + 4I)^2 - 2\Pi_0\chi + (\Pi_0\chi)^2},$$

$$\Pi_0 v = \Pi_0\chi,$$

(4.3.20)

$$\Pi_0\varphi = -\left(2\int_0^{\Pi_0\chi} \frac{\eta d\eta}{\sqrt{(1 + 4I)^2 - 2\eta + \eta^2}}\right)^{1/2} \operatorname{sgn}(\Pi_0\chi).$$

Now we must substitute these expressions for $\Pi_0 u$ and $\Pi_0\varphi$ into the first equation of (4.3.18) and solve the resulting differential equation for $\Pi_0\chi$ together with the initial condition $\Pi_0\chi(0) = 1$. After that the remainder of the Π_0-functions are determined from (4.3.20).

Construction of higher-order terms of the asymptotic solution can be executed as in Vasil'eva and Butuzov [149]. They are defined as solutions of linear equations.

A comparison with experiments shows that the asymptotic algorithm presented provides the solution which is even suitable in the zeroth approximation to describe processes involving transistors (with a high degree of accuracy). A more detailed physical analysis of the mathematical results is given in Vasil'eva and Stel'makh [157], where an estimate of the remainder is also presented. More complicated one-dimensional problems (and estimates of remainders) are considered in Belyanin [5], [6].

4.3.2 A two-dimensional model of a diode: statement of the problem

Operation of a simple semiconductor device (diode) is based on the contact of two materials with different types of conductivity. A two-dimensional model of a diode is presented in Fig. 4.1. Here materials n and p are characterized by electron and hole types of conductivity, respectively; C_1 and C_2 are the Ohmic contacts, Γ is the boundary separating regions Ω_1 and Ω_2, and $\partial\Omega_i$ are the isolated boundaries.

The system of equations describing the stationary distribution of electrostatic potential ψ and the densities of charge carriers (n and p are the densities of electrons and holes, respectively) has the form (cf. Pol'skii [118], Markowich [88], and Markowich, Ringhofer, and Schmeiser [89])

$$\nabla(\varepsilon \nabla \psi) = q(n - p - c(x, y)),$$

(4.3.21)

$$\nabla(D_n \nabla n - \mu_n n \nabla \psi) = R(n, p), \quad \nabla(D_p \nabla p + \mu_p p \nabla \psi) = R(n, p).$$

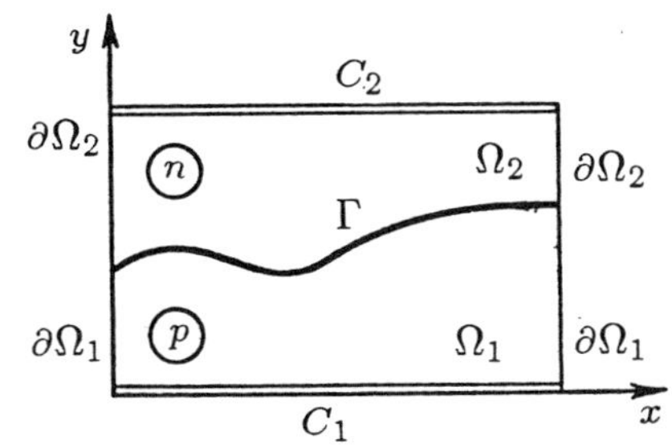

Figure 4.1: *Two-dimensional model of a diode.*

Here q is the charge of the electron, ε is the dielectric permeability, D_n and D_p are the diffusion coefficients of the charge carriers, and μ_n and μ_p are their mobilities. The function $c(x, y)$ (dopant concentration) describes the difference between concentrations of donors and acceptors in the semiconductor material: $c(x, y) > 0$ holds in the n-region (i.e in Ω_2), and $c(x, y) < 0$ in the p-region (i.e. in Ω_1); $R(n, p)$ measures recombination of electrons and holes. The first equation of (4.3.21) is the Poisson equation, the other two are continuity equations. Taking into account the expressions

$$\mathbf{J}_n = q(D_n \bigtriangledown n - \mu_n n \bigtriangledown \psi) \quad \text{and} \quad \mathbf{J}_p = -q(D_p \bigtriangledown p + \mu_p p \bigtriangledown \psi)$$

for the densities of electron and hole currents, respectively, we can rewrite the last two equations of (4.3.21) in a form similar to that of the first equation:

$$\bigtriangledown \mathbf{J}_n = qR, \qquad \bigtriangledown \mathbf{J}_p = -qR.$$

Remark. It can be easily noted that the system (4.3.1), (4.3.2) of § 4.3.1 is a one-dimensional analog of the first equation in (4.3.21) together with expressions for $\mathbf{J}_n$ and $\mathbf{J}_p$ (which are scalars in the one-dimensional case), since $E = -\bigtriangledown \psi$, where E is the polar electric voltage.

In what follows we will assume that the function $c(x, y)$ is constant in each of the regions Ω_1 and Ω_2: $c = -\bar{c} < 0$ in Ω_1, while $c = \bar{c} > 0$ in Ω_2. We will also assume that $R(n, p) = 0$, and that the curve Γ is sufficiently smooth.

The boundary conditions for (4.3.21) have the following form.

(a) On the Ohmic contacts

$$\psi|_{C_k} = \frac{1}{2} U_T \ln \left. \frac{n}{p} \right|_{C_k} + U_k, \qquad np|_{C_k} = n_k^2, \tag{4.3.22}$$

$$(n - p - c(x, y))|_{C_k} = 0, \qquad k = 1, 2.$$

Here $U_T = D_n/\mu_n = D_p/\mu_p$ and n_k are constants that characterize the semiconductor material, and U_k are applied potentials.

(b) On the isolated boundaries

$$(\mathbf{J}_n \cdot \nu)|_{\partial\Omega_k} = 0, \qquad (\mathbf{J}_p \cdot \nu)|_{\partial\Omega_k} = 0, \tag{4.3.23}$$

$$(\nabla\psi \cdot \nu)|_{\partial\Omega_k} = 0, \tag{4.3.24}$$

where ν is the normal vector to $\partial\Omega_k$, and $(\,\cdot\,)$ stands for the inner product. Using the expressions for $\mathbf{J}_n$ and $\mathbf{J}_p$, (4.3.23) can be substituted by conditions

$$(\nabla n \cdot \nu)|_{\partial\Omega_k} = 0, \qquad (\nabla p \cdot \nu)|_{\partial\Omega_k} = 0. \tag{4.3.25}$$

We demand that the functions ψ, u, v and their normal derivatives are continuous on Γ.

Let $N = c/\bar{c}$, and x_0 be a characteristic dimension of the device (along the x-axis). We introduce the dimensionless variables by the formulae

$$\tilde{x} = x/x_0, \quad \tilde{y} = y/x_0, \quad \tilde{p} = p/\bar{c}, \quad \tilde{n} = n/\bar{c}, \quad \tilde{\psi} = \psi/U_T.$$

Omitting tildes in the new variables, we arrive at the equations

$$\begin{gathered} \mu^2 \triangle \psi = n - p - N(x,y), \\ \Delta n = n \triangle \psi + (\nabla n \cdot \nabla \psi), \qquad \Delta p = -p \triangle \psi - (\nabla p \cdot \nabla \psi), \end{gathered} \tag{4.3.26}$$

where

$$\mu^2 = \frac{\varepsilon U_T}{q\bar{c}x_0^2} \quad \text{and} \quad N(x,y) = \begin{cases} N_1 = -1 & \text{in } \Omega_1, \\ N_2 = 1 & \text{in } \Omega_2. \end{cases}$$

Quantity μ^2 is a small parameter: for many semiconductor devices, $\mu^2 \le 10^{-7}$.

Let us introduce the new variables $u = p + n$ and $v = p - n$. Then (4.3.26) is transformed to the system

$$\begin{gathered} \mu^2 \triangle \psi = -v - N, \\ \Delta u = -v \triangle \psi - (\nabla v \cdot \nabla \psi), \qquad \Delta v = -u \triangle \psi - (\nabla u \cdot \nabla \psi). \end{gathered} \tag{4.3.27}$$

The boundary conditions (4.3.22) now have the form

$$\begin{gathered} \psi|_{C_k} = \ln \frac{N_k + \sqrt{N_k^2 + \delta_k^2}}{2\delta_k^2} + \frac{U_k}{U_T} \quad \text{with } \delta_k = n_k/\bar{c}, \\ u|_{C_k} = \sqrt{N_k^2 + 4\delta_k^2}, \qquad v|_{C_k} = -N_k, \end{gathered} \tag{4.3.28}$$

and conditions (4.3.24), (4.3.25) are transformed to

$$(\nabla\psi \cdot \nu)|_{\partial\Omega_k} = (\nabla u \cdot \nu)|_{\partial\Omega_k} = (\nabla v \cdot \nu)|_{\partial\Omega_k} = 0$$

or

$$\left.\frac{\partial\psi}{\partial x}\right|_{x=0} = \left.\frac{\partial\psi}{\partial x}\right|_{x=1} = \left.\frac{\partial u}{\partial x}\right|_{x=0} = \left.\frac{\partial u}{\partial x}\right|_{x=1} = \left.\frac{\partial v}{\partial x}\right|_{x=0} = \left.\frac{\partial v}{\partial x}\right|_{x=1} = 0. \tag{4.3.29}$$

We demand that the nondimensional functions ψ, n, p and their normal derivatives be continuous on the curve Γ. We will call the problem stated above problem ((4.3.27)–(4.3.29),Γ).

Remark. Note that in the one-dimensional problem discussed in § 4.3.1, $n_k = 0$, and the currents $J_p = J_n$ (or $I_p = I_n = I$ in nondimensional form) were prescribed, instead of the potentials, as in (4.3.28).

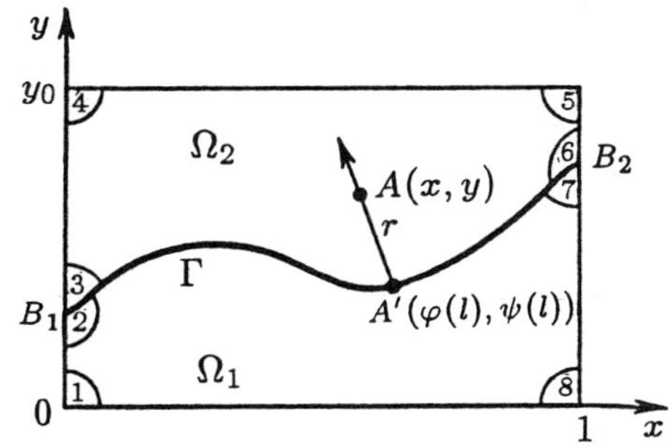

Figure 4.2: *Local coordinates (r, l) in the vicinity of* Γ.

4.3.3 Construction of the asymptotic solution

Methods developed in Chapter 3 allow us to construct the asymptotic solution, which satisfies ((4.3.27)–(4.3.29),Γ) with a discrepancy that is asymptotically small for small μ. Since the system (4.3.27) is quite complicated, it is difficult to estimate the remainder. We will obtain the leading terms of the asymptotic solution.

Let us consider the problem ((4.3.27)–(4.3.29),Γ). We introduce boundary layer variables near the boundaries of the device: $\eta = y/\mu$, $\xi = x/\mu$, $\eta_* = (y_0 - y)/\mu$, $\xi_* = (1 - x)/\mu$. Let the parametric equations of the curve Γ have the form $x = \varphi(l)$, $y = \psi(l)$, $0 \le l \le l_0$. As in § 3.1.1, we introduce local coordinates (r, l) in the vicinity of Γ (see Fig. 4.2, where $A'A$ is the normal to Γ). We introduce a boundary layer variable near Γ by the formula $\rho = r/\mu$. Note that $r > 0$ for points of the region Ω_2 and $r < 0$ for points in Ω_1.

The asymptotic solution of ((4.3.27)–(4.3.29),Γ) will be sought in the form

$$z = \bar{z}(x, y, \mu) + \overset{(1)}{\Pi} z(x, \eta, \mu) + \overset{(2)}{\Pi} z(\xi, y, \mu) + \overset{(3)}{\Pi} z(x, \eta_*, \mu)$$

$$+ \overset{(4)}{\Pi} z(\xi_*, y, \mu) + \Pi z(\rho, l, \mu), \tag{4.3.30}$$

where below z stands for all functions ψ, u and v and $\bar{z}$ is the regular series; $\overset{(1)}{\Pi} z$, $\overset{(2)}{\Pi} z$, $\overset{(3)}{\Pi} z$ and $\overset{(4)}{\Pi} z$ are boundary function series in the vicinities of $y = 0$, $x = 0$, $y = y_0$ and $x = 1$, respectively; and Πz is the series describing the internal transition layer near Γ.

We should also add corner boundary functions to expansion (4.3.30) in the regions 1–8 shown in Fig. 4.2, but we will not construct them here.

For the leading term of the regular series, we obtain the system

$$\bar{v}_0 + N = 0, \qquad \Delta \bar{u}_0 = -\bar{v}_0 \,\Delta\, \bar{\psi}_0 - (\nabla \bar{v}_0 \cdot \nabla \bar{\psi}_0),$$

$$\Delta \bar{v}_0 = -\bar{u}_0 \,\Delta\, \bar{\psi}_0 - (\nabla \bar{u}_0 \cdot \nabla \bar{\psi}_0). \tag{4.3.31}$$

Since

$$N(x, y) = \begin{cases} N_1 = -1 & \text{in } \Omega_1 \\ N_2 = 1 & \text{in } \Omega_2 \end{cases},$$

we should have introduced subindexes 1 and 2 for $\overline{z}_0 = \{\overline{\psi}_0, \overline{u}_0, \overline{v}_0\}$ depending on the region in which the functions are defined. But, to simplify the notation, we will not do so.

From (4.3.31), we obtain

$$\overline{v}_0 = -N, \tag{4.3.32}$$

and, therefore, $\overline{v}_0$ satisfies the conditions (4.3.28), (4.3.29) for v. The rest of the conditions (4.3.28), (4.3.29), imposed on $\overline{\psi}_0$ and $\overline{u}_0$, have the form

$$\begin{gathered}\overline{\psi}_0|_{C_k} = \ln\frac{N_k + \sqrt{N_k^2 + 4\delta_k^2}}{2\delta_k^2} + \frac{U_k}{U_T}, \qquad \overline{u}_0|_{C_k} = \sqrt{N_k^2 + 4\delta_k^2}, \\ \left.\frac{\partial \overline{\psi}_0}{\partial x}\right|_{x=0} = \left.\frac{\partial \overline{\psi}_0}{\partial x}\right|_{x=1} = \left.\frac{\partial \overline{u}_0}{\partial x}\right|_{x=0} = \left.\frac{\partial \overline{u}_0}{\partial x}\right|_{x=1} = 0.\end{gathered} \tag{4.3.33}$$

Since $\overline{\psi}_0$, $\overline{u}_0$ and $\overline{v}_0$ satisfy all the conditions (4.3.28), (4.3.29), we have $\overset{(i)}{\Pi}_0\, z \equiv 0$ and the expansions for $\overset{(i)}{\Pi}$-functions start from terms of higher order. However, the solution of the problem (4.3.31), (4.3.33) does not, in general, satisfy the conditions on Γ, due to the discontinuity of N. Therefore a term of the zeroth order $\Pi_0 z$ will be present in the expansion for Πz.

Let us construct $\Pi_0 z$. First, we must write (4.3.27) in local coordinates (r, l). Differential operators, entering (4.3.27), have the following form in the variables r and l:

$$\Delta\psi = \frac{\partial^2\psi}{\partial r^2} + \alpha(r,l)\frac{\partial\psi}{\partial r} + \beta(r,l)\frac{\partial^2\psi}{\partial l^2} + \gamma(r,l)\frac{\partial\psi}{\partial l},$$

$$(\nabla\psi\cdot\nabla u) = \frac{\partial\psi}{\partial r}\frac{\partial u}{\partial r} + \beta(r,l)\frac{\partial\psi}{\partial l}\frac{\partial u}{\partial l}.$$

Here α, β, and γ are the same coefficients as in § 3.1.1. Using the stretched variable $\rho = r/\mu$ in the standard way, we obtain the equations for $\Pi_0 z = \{\Pi_0\psi,\ \Pi_0 u\ , \Pi_0 v\}$

$$\begin{gathered}\frac{\partial^2\Pi_0\psi}{\partial\rho^2} = -\Pi_0 v, \\ \frac{\partial^2\Pi_0 u}{\partial\rho^2} = -(\overline{v}_0(\pm 0, l) + \Pi_0 v)\frac{\partial^2\Pi_0\psi}{\partial\rho^2} - \frac{\partial\Pi_0 v}{\partial\rho}\frac{\partial\Pi_0\psi}{\partial\rho}, \\ \frac{\partial^2\Pi_0 v}{\partial\rho^2} = -(\overline{u}_0(\pm 0, l) + \Pi_0 u)\frac{\partial^2\Pi_0\psi}{\partial\rho^2} - \frac{\partial\Pi_0 u}{\partial\rho}\frac{\partial\Pi_0\psi}{\partial\rho},\end{gathered} \tag{4.3.34}$$

where l enters as a parameter. In fact, we have two systems of equations: one (with $\overline{v}_0(+0, l)$, $\overline{u}_0(+0, l)$) is defined for $\rho \geq 0$, and the other (with $\overline{v}_0(-0, l)$, $\overline{u}_0(-0, l)$) for $\rho \leq 0$. Integrating the second and the third equations with the decay conditions at infinity, we arrive at the system

$$\frac{\partial^2\Pi_0\psi}{\partial\rho^2} = -\Pi_0 v,$$

$$\frac{\partial \Pi_0 u}{\partial \rho} = -(\overline{v}_0(\pm 0, l) + \Pi_0 v)\frac{\partial \Pi_0 \psi}{\partial \rho}, \tag{4.3.35}$$

$$\frac{\partial \Pi_0 v}{\partial \rho} = -(\overline{u}_0(\pm 0, l) + \Pi_0 u)\frac{\partial \Pi_0 \psi}{\partial \rho}.$$

The functions ψ, u, v and their normal derivatives must be continuous on the curve Γ. This leads to the following additional conditions:

$$\overline{z}_0^+ + (\Pi_0 z)^+ = \overline{z}_0^- + (\Pi_0 z)^- \tag{4.3.36}$$

$$\frac{\partial (\Pi_0 z)^+}{\partial \rho} = \frac{\partial (\Pi_0 z)^-}{\partial \rho}. \tag{4.3.37}$$

Here and below, by $\overline{z}_0^+$ (or $\overline{z}_0^-$) we understand the limiting value of $\overline{z}_0$ on Γ when approaching this curve from the region Ω_2 (respectively, from Ω_1), i.e., $\overline{z}_0^+ = \overline{z}_0(+0, l)$ and $\overline{z}_0^- = \overline{z}_0(-0, l)$. The notation $(\Pi_0 z)^+$ and $(\Pi_0 z)^-$ has similar meaning. Further, the usual decay conditions for the boundary functions must hold:

$$\Pi_0 z(\pm\infty, l) = 0. \tag{4.3.38}$$

We rewrite (4.3.35) as a system of four first-order ordinary differential equations

$$\begin{gathered} \frac{\partial \Pi_0 \psi}{\partial \rho} = -\Pi_0 \varphi, \qquad \frac{\partial \Pi_0 \varphi}{\partial \rho} = \Pi_0 v, \\ \frac{\partial \Pi_0 u}{\partial \rho} = (\overline{v}_0^\pm + \Pi_0 v)\Pi_0 \varphi, \qquad \frac{\partial \Pi_0 v}{\partial \rho} = (\overline{u}_0^\pm + \Pi_0 u)\Pi_0 \varphi. \end{gathered} \tag{4.3.39}$$

We should mention that this system is of conditionally stable type and in the critical case. Such systems were investigated in Vasil'eva and Butuzov [149]. However, here we will not use the general theory. Instead, taking into account the special form of the system, we will construct and study the Π-functions directly (as we did in § 4.3.1).

Excluding $\Pi_0 \varphi$ from the last two equations of (4.3.39), we obtain a first integral of this system:

$$\overline{u}_0^\pm \Pi_0 u + \frac{1}{2}(\Pi_0 u)^2 = \overline{v}_0^\pm \Pi_0 v + \frac{1}{2}(\Pi_0 v)^2 \tag{4.3.40}$$

(the constant of integration is zero by virtue of (4.3.38)). We will find one more first integral by rewriting the third equation of (4.3.39) in the form

$$\frac{\partial \Pi_0 u}{\partial \rho} = \overline{v}_0^\pm \left(-\frac{\partial \Pi_0 \psi}{\partial \rho}\right) + \Pi_0 \varphi \frac{\partial \Pi_0 \varphi}{\partial \rho},$$

and hence,

$$\Pi_0 u = -\overline{v}_0^\pm \Pi_0 \psi + \frac{1}{2}(\Pi_0 \varphi)^2 \tag{4.3.41}$$

(the constant of integration here is also zero by virtue of (4.3.38)).

Using the first integrals derived above, we obtain matching conditions on the curve Γ for the system (4.3.31). Taking into account relations $\overline{v}_0^+ = -1$ and $\overline{v}_0^- = 1$, which follow from (4.3.32), we rewrite (4.3.40) as

$$\begin{aligned} 2\overline{u}_0^+(\Pi_0 u)^+ + ((\Pi_0 u)^+)^2 &= 2\overline{v}_0^+(\Pi_0 v)^+ + ((\Pi_0 v)^+)^2 \\ &= -2(\Pi_0 v)^+ + ((\Pi_0 v)^+)^2, \\ 2\overline{u}_0^-(\Pi_0 u)^- + ((\Pi_0 u)^-)^2 &= 2\overline{v}_0^-(\Pi_0 v)^- + ((\Pi_0 v)^-)^2 \\ &= -2(\Pi_0 v)^- + ((\Pi_0 v)^-)^2. \end{aligned} \tag{4.3.42}$$

Taking the squares of both sides of (4.3.36) for the u- and v-functions, respectively, we obtain

$$(\overline{u}_0^+)^2 + 2\overline{u}_0^+(\Pi_0 u)^+ + ((\Pi_0 u)^+)^2 = (\overline{u}_0^-)^2 + 2\overline{u}_0^-(\Pi_0 u)^- + ((\Pi_0 u)^-)^2, \tag{4.3.43}$$

and a similar expression for v. Subtracting one of the equations (4.3.42) from the other, and taking into account (4.3.43), we arrive at

$$(\overline{u}_0^+)^2 - (\overline{u}_0^-)^2 = (\overline{v}_0^+)^2 - (\overline{v}_0^-)^2 = 0.$$

Therefore,

$$\overline{u}_0^+ = \overline{u}_0^-. \tag{4.3.44}$$

In the derivation of (4.3.44) we used the fact that the physically meaningful solutions are only those for which $u > 0$. But then, by virtue of (4.3.36),

$$(\Pi_0 u)^+ = (\Pi_0 u)^-. \tag{4.3.45}$$

Next, from the relation (4.3.37) for ψ and from the first equation of (4.3.39), we obtain

$$(\Pi_0 \varphi)^+ = (\Pi_0 \varphi)^-. \tag{4.3.46}$$

Condition (4.3.44) is one of the matching conditions for the solution of (4.3.31) on Γ. However, the condition (4.3.44) is not sufficient to complete the statement of the problem for (4.3.31). We also need the some other conditions that will be obtained below. For now, assuming that $\overline{u}_0$ is known, let us proceed with the study of $\Pi_0 z$.

From (4.3.40) we can express $\Pi_0 u$ through $\Pi_0 v$:

$$\Pi_0 u = -\overline{u}_0^{\pm} + \sqrt{(\overline{u}_0^{\pm})^2 + 2\overline{v}_0^{\pm}\Pi_0 v + (\Pi_0 v)^2} \tag{4.3.47}$$

(the positive sign in front of the square root is chosen using the conditions at infinity). For $\Pi_0\varphi$ and $\Pi_0 v$, we now have the system

$$\begin{aligned} \frac{\partial \Pi_0\varphi}{\partial \rho} &= \Pi_0 v, \\ \frac{\partial \Pi_0 v}{\partial \rho} &= \Pi_0\varphi\sqrt{(\overline{u}_0^{\pm})^2 + 2\overline{v}_0^{\pm}\Pi_0 v + (\Pi_0 v)^2}, \end{aligned} \tag{4.3.48}$$

in which the rest point $\Pi_0\varphi = 0$, $\Pi_0 v = 0$ is a saddle (the corresponding eigenvalues are $\lambda_{1,2} = \pm\sqrt{\overline{u}_0^{\pm}}$). From (4.3.48),

$$\frac{1}{2}(\Pi_0\varphi)^2 = \int \frac{\Pi_0 v \, d\Pi_0 v}{\sqrt{(\overline{u}_0^{\pm})^2 + 2\overline{v}_0^{\pm}\Pi_0 v + (\Pi_0 v)^2}}.$$

We can evaluate this integral using some additional ideas that follow from the physical content of the problem. First, $u > 0$ (as already used). Second, $u^2 > v^2$ (recall that u is the sum of concentrations of electrons and holes, and v is their difference). Then we can write

$$\frac{1}{2}(\Pi_0\varphi)^2 = \Pi_0 u - \overline{v}_0^{\pm}\ln\left(\frac{\overline{v}_0^{\pm} + \Pi_0 v + \overline{u}_0^{\pm} + \Pi_0 u}{\overline{v}_0^{\pm} + \overline{u}_0^{\pm}}\right), \tag{4.3.49}$$

(the constant of integration is zero by virtue of conditions at infinity). From (4.3.46), it now follows that

$$\begin{aligned} &(\Pi_0 u)^+ - \overline{v}_0^+\ln\left(\frac{\overline{v}_0^+ + (\Pi_0 v)^+ + \overline{u}_0^+ + (\Pi_0 u)^+}{\overline{v}_0^+ + \overline{u}_0^+}\right) \\ &= (\Pi_0 u)^- - \overline{v}_0^-\ln\left(\frac{\overline{v}_0^- + (\Pi_0 v)^- + \overline{u}_0^- + (\Pi_0 u)^-}{\overline{v}_0^- + \overline{u}_0^-}\right), \end{aligned} \tag{4.3.50}$$

and from (4.3.36) for v, we have

$$(\Pi_0 v)^+ - (\Pi_0 v)^- = 2. \tag{4.3.51}$$

Using equalities (4.3.51) and (4.3.45), and substituting in (4.3.50) $\overline{u}_0^+ + \Pi_0 u$ by its expression through $\Pi_0 v$ (see (4.3.47)), we arrive at

$$\left(1 + (\Pi_0 v)^- + \sqrt{(u^0)^2 + 2(\Pi_0 v)^- + ((\Pi_0 v)^-)^2}\right)^2 = (u^0)^2 - 1,$$

where $u^0 = \overline{u}_0^+ = \overline{u}_0^-$ (see (4.3.44)). From here, after some transformations, we obtain

$$\begin{aligned} &(1 + (\Pi_0 v)^-)\left(1 + (\Pi_0 v)^- + \sqrt{(u^0)^2 + 2(\Pi_0 v)^- + ((\Pi_0 v)^-)^2}\right) \\ &= (1 + (\Pi_0 v)^-)(v + u)|_\Gamma = 0, \end{aligned}$$

and, therefore, $(\Pi_0 v)^- = -1$. Hence, by virtue of (4.3.51), $(\Pi_0 v)^+ = 1$. Thus, we obtained the boundary conditions for the conditionally stable systems (4.3.48):

$$(\Pi_0 v)^- = -1, \qquad (\Pi_0 v)^+ = 1. \tag{4.3.52}$$

Using (4.3.49), we can find $\Pi_0 v$ as a function of ρ along the separatrix of the saddle point $\Pi_0 v = 0$, $\Pi_0\varphi = 0$ of (4.3.48), both for $\rho \geq 0$ and for $\rho \leq 0$. Keeping, for now, both plus and minus signs before the radical, we write

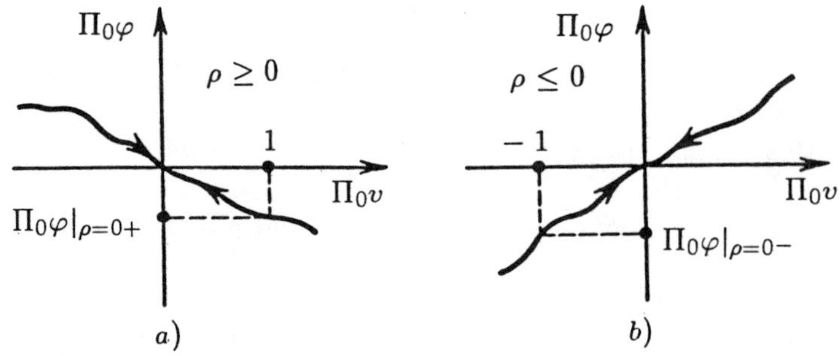

Figure 4.3: *The choice of separatrix for* (a) $\rho \geq 0$, (b) $\rho \leq 0$.

$$\begin{aligned}\frac{\partial \Pi_0 v}{\partial \rho} = {} & \pm\sqrt{2}\sqrt{(u^0)^2 + 2\overline{v}_0^{\pm}\Pi_0 v + (\Pi_0 v)^2} \\ & \times \Bigg[-u^0 + \sqrt{(u^0)^2 + 2\overline{v}_0^{\pm}\Pi_0 v + (\Pi_0 v)^2} \\ & -\overline{v}_0^{\pm}\ln\left(\frac{\overline{v}_0^{\pm} + \Pi_0 v + \sqrt{(u^0)^2 + 2\overline{v}_0^{\pm}\Pi_0 v + (\Pi_0 v)^2}}{u^0 + \overline{v}_0^{\pm}}\right)\Bigg]^{1/2}, \end{aligned} \tag{4.3.53}$$

$$\Pi_0 v|_{\rho=0} = -\overline{v}_0^{\pm}.$$

One of these equations (with $\overline{v}_0^{+}$) is formulated for the domain $\rho > 0$, and the other (with $\overline{v}_0^{-}$) is formulated for $\rho < 0$. So far, the $\pm$ sign in front of the $\sqrt{2}$ is not connected with the $\pm$ sign in $\overline{v}_0^{\pm}$. Our next goal is to study this connection.

Let us consider, e.g., $\Pi_0 v$ in Ω_2, i.e. for $\rho \geq 0$. Here the equation of the stable separatrix of the saddle $(\Pi_0 v, \Pi_0\varphi) = (0,0)$ of (4.3.48) in the linear approximation has the form $\Pi_0 v = -\sqrt{u^0}\,\Pi_0\varphi$. This equation can be obtained if we keep in (4.3.48) the terms linear in the variables $\Pi_0 v$ and $\Pi_0\varphi$, and take into account that of the two exponents $\exp(\lambda_{1,2}\rho) = \exp(\pm\sqrt{u^0}\,\rho)$, we must keep the one with the minus sign since $\rho > 0$. We should mention that (according to (4.3.49)) $\Pi_0\varphi$ as a function of $\Pi_0 v$ is nowhere zero, but at $\Pi_0 v = 0$, and the separatrix is located in the second and fourth quadrants of the phase-plane (see Fig. 4.3(a)). According to (4.3.52), we have $\Pi_0 v|_{\rho=+0} = 1$. Therefore $\Pi_0\varphi|_{\rho=+0} < 0$, as can be seen in Fig. 4.3(a). Thus, integrating the system (4.3.53) for $\rho \geq 0$, we must take the minus sign in front of $\sqrt{2}$. A similar argument for $\rho \leq 0$ leads to the same conclusion that we need the minus sign.

If $\Pi_0 v$ is known (as the solution of (4.3.53), where we take minus sign for $\rho \geq 0$, as well as for $\rho \leq 0$), the function $\Pi_0 u$ can be defined from (4.3.47), and $\Pi_0\psi$ can be obtained from (4.3.41), where we substitute expression (4.3.49) for $(\Pi_0\varphi)^2$. Therefore, all the boundary functions of the zeroth order in the vicinity of Γ (i.e., all the functions $\Pi_0 z$) have been constructed. It is important to mention, however, that this statement is conditional, since the yet unknown quantity u^0 enters (4.3.53).

It is not difficult to obtain, in addition to (4.3.44), one more condition on Γ for the solution of (4.3.31). By virtue of (4.3.41), (4.3.49),

$$(\Pi_0\psi)^{\pm} = -\ln\left(\frac{\overline{v}_0^{\pm} + \Pi_0 v + \overline{u}_0^{\pm} + \Pi_0 u}{\overline{v}_0^{\pm} + \overline{u}_0^{\pm}}\right),$$

and hence, taking into account (4.3.36), we obtain (recall that $\overline{u}_0^{\pm} = u^0$, and $\overline{v}_0^{\pm} = \mp 1$)

$$\overline{\psi}_0^{+} - \overline{\psi}_0^{-} = \ln\left(\frac{u^0 - 1}{u^0 + 1}\right). \tag{4.3.54}$$

Conditions (4.3.44) and (4.3.54) are still not sufficient to complete solution of (4.3.31). We also need some conditions on Γ for the derivatives of $\overline{u}_0$ and $\overline{\psi}_0$. These conditions will be obtained during the study of equations of the first-order approximation, i.e. the equations for $\overline{z}_1$ and $\Pi_1 z$.

The first equation of (4.3.27) implies $\overline{v}_1 = 0$. For $\overline{\psi}_1$ and $\overline{u}_1$, we obtain a linear system containing $\Delta\overline{\psi}_1$ and $\Delta\overline{u}_1$, as well as the first-order derivatives. We impose zero boundary conditions for $\overline{\psi}_1$ and $\overline{u}_1$ on C_k, and we also demand that the derivatives of the these functions with respect to x are zero on the isolated boundaries. Hence, it follows that the boundary functions $\overset{(i)}{\Pi}_1 z$, same as $\overset{(i)}{\Pi}_0 z$, are identically zero. These functions will appear only in the next order approximation, when $\overline{v}_2 = -\Delta\,\overline{\psi}_0$ will not satisfy the imposed homogeneous Dirichlet boundary conditions on C_k and the homogeneous Neumann conditions on $\partial\Omega_k$.

The following conditions will hold on Γ at the first-order approximation:

$$\overline{z}_1^{+} + (\Pi_1 z)^{+} = \overline{z}_1^{-} + (\Pi_1 z)^{-}, \tag{4.3.55}$$

$$\left(\frac{\partial}{\partial\rho}\Pi_1 z\right)^{+} + \left(\frac{\partial\overline{z}_0}{\partial r}\right)^{+} = \left(\frac{\partial}{\partial\rho}\Pi_1 z\right)^{-} + \left(\frac{\partial\overline{z}_0}{\partial r}\right)^{-}. \tag{4.3.56}$$

Since $\overline{v}_1 = 0$, the condition (4.3.55) for v implies

$$(\Pi_1 v)^{+} = (\Pi_1 v)^{-},$$

and from (4.3.56), by virtue of (4.3.32), we obtain

$$\left(\frac{\partial}{\partial\rho}\Pi_1 v\right)^{+} = \left(\frac{\partial}{\partial\rho}\Pi_1 v\right)^{-}. \tag{4.3.57}$$

The system for $\Pi_1 z$ has the form

$$\frac{\partial^2\Pi_1\psi}{\partial\rho^2} = -\Pi_1 v - \alpha(0,l)\frac{\partial\Pi_0\psi}{\partial\rho},$$

$$
\begin{aligned}
\frac{\partial^2 \Pi_1 u}{\partial \rho^2} = {} & -(\overline{v}_0^{\pm} + \Pi_0 v)\frac{\partial^2 \Pi_1 \psi}{\partial \rho^2} - \frac{\partial \Pi_0 v}{\partial \rho}\frac{\partial \Pi_1 \psi}{\partial \rho} - \frac{\partial \Pi_1 v}{\partial \rho}\frac{\partial \Pi_0 \psi}{\partial \rho} \\
& - \left(\overline{v}_1^{\pm} + \Pi_1 v + \left(\frac{\partial \overline{v}_0}{\partial r}\right)^{\pm} \rho\right)\frac{\partial^2 \Pi_0 \psi}{\partial \rho^2} - \alpha(0,l)\frac{\partial \Pi_0 u}{\partial \rho} \\
& -(\overline{v}_0^{\pm} + \Pi_0 v)\alpha(0,l)\frac{\partial \Pi_0 \psi}{\partial \rho} - \left(\frac{\partial \overline{\psi}_0}{\partial r}\right)^{\pm}\frac{\partial \Pi_0 v}{\partial \rho} - \left(\frac{\partial \overline{v}_0}{\partial r}\right)^{\pm}\frac{\partial \Pi_0 \psi}{\partial \rho},
\end{aligned}
\tag{4.3.58}
$$

while the third equation of (4.3.58) for $\Pi_1 v$ has the same structure as that for $\Pi_1 u$, where the function u is substituted by v, and the function v is substituted by u in the right-hand side. Note that, by virtue of (4.3.39), the terms containing $\alpha(0,l)$ in the second and the third equations of (4.3.58) will cancel each other.

Integrating, as in (4.3.34), we obtain

$$
\frac{\partial^2 \Pi_1 \psi}{\partial \rho^2} = -\Pi_1 v - \alpha(0,l)\frac{\partial \Pi_0 \psi}{\partial \rho},
$$

$$
\begin{aligned}
\frac{\partial \Pi_1 u}{\partial \rho} = {} & -(\overline{v}_0^{\pm} + \Pi_0 v)\frac{\partial \Pi_1 \psi}{\partial \rho} - (\overline{v}_1^{\pm} + \Pi_1 v)\frac{\partial \Pi_0 \psi}{\partial \rho} \\
& - \left(\frac{\partial \overline{\psi}_0}{\partial r}\right)^{\pm}\Pi_0 v - \left(\frac{\partial \overline{v}_0}{\partial r}\right)^{\pm}\left(\Pi_0 \psi + \int_{\pm\infty}^{\rho} \rho \frac{\partial^2 \Pi_0 \psi}{\partial \rho^2} d\rho\right),
\end{aligned}
$$

and the third differential equation for $\Pi_1 v$, with the right-hand side similar to that of the second equation of the system above, where v is switched for u. The integrals entering the system can be transformed by integration by parts. Finally, we have

$$
\begin{aligned}
\frac{\partial^2 \Pi_1 \psi}{\partial \rho^2} = {} & -\Pi_1 v - \alpha(0,l)\frac{\partial \Pi_0 \psi}{\partial \rho}, \\
\frac{\partial \Pi_1 u}{\partial \rho} = {} & -(\overline{v}_0^{\pm} + \Pi_0 v)\frac{\partial \Pi_1 \psi}{\partial \rho} - (\overline{v}_1^{\pm} + \Pi_1 v)\frac{\partial \Pi_0 \psi}{\partial \rho} \\
& - \left(\frac{\partial \overline{\psi}_0}{\partial r}\right)^{\pm}\Pi_0 v - \left(\frac{\partial \overline{v}_0}{\partial r}\right)^{\pm}\rho\frac{\partial \Pi_0 \psi}{\partial \rho}, \\
\frac{\partial \Pi_1 v}{\partial \rho} = {} & -(\overline{u}_0^{\pm} + \Pi_0 u)\frac{\partial \Pi_1 \psi}{\partial \rho} - (\overline{u}_1^{\pm} + \Pi_1 u)\frac{\partial \Pi_0 \psi}{\partial \rho} \\
& - \left(\frac{\partial \overline{\psi}_0}{\partial r}\right)^{\pm}\Pi_0 u - \left(\frac{\partial \overline{u}_0}{\partial r}\right)^{\pm}\rho\frac{\partial \Pi_0 \psi}{\partial \rho}.
\end{aligned}
\tag{4.3.59}
$$

From the last equation of (4.3.59), using (4.3.57), we get

$$
-(u_0 + (\Pi_0 u)^+)\left(\frac{\partial \Pi_1 \psi}{\partial \rho}\right)^+ - (\overline{u}_1^+ + (\Pi_1 u)^+)\left(\frac{\partial \Pi_0 \psi}{\partial \rho}\right)^+ - \left(\frac{\partial \overline{\psi}_0}{\partial r}\right)^+ (\Pi_0 u)^+
$$

$$= -(u_0 + (\Pi_0 u)^-) \left(\frac{\partial \Pi_1 \psi}{\partial \rho}\right)^- - (\overline{u}_1^- + (\Pi_1 u)^-) \left(\frac{\partial \Pi_0 \psi}{\partial \rho}\right)^- - \left(\frac{\partial \overline{\psi}_0}{\partial r}\right)^- (\Pi_0 u)^-.$$

Hence, by virtue of (4.3.45), (4.3.46), (4.3.39), we obtain

$$(4.3.60) \qquad \begin{aligned} &-(u_0 + \Pi_0 u)\left[\left(\frac{\partial \Pi_1 \psi}{\partial \rho}\right)^+ - \left(\frac{\partial \Pi_1 \psi}{\partial \rho}\right)^-\right] \\ &-\left[\left(\frac{\partial \overline{\psi}_0}{\partial r}\right)^+ - \left(\frac{\partial \overline{\psi}_0}{\partial r}\right)^-\right] \Pi_0 u = 0, \end{aligned}$$

where we use the notation $\Pi_0 u$ for $(\Pi_0 u)^+ = (\Pi_0 u)^-$. We write the condition (4.3.56) for ψ in the form

$$(4.3.61) \qquad \left[\left(\frac{\partial \Pi_1 \psi}{\partial \rho}\right)^+ - \left(\frac{\partial \Pi_1 \psi}{\partial \rho}\right)^-\right] + \left[\left(\frac{\partial \overline{\psi}_0}{\partial r}\right)^+ - \left(\frac{\partial \overline{\psi}_0}{\partial r}\right)^-\right] = 0.$$

From (4.3.60) and (4.3.61), it follows that

$$(4.3.62) \qquad \left(\frac{\partial \Pi_1 \psi}{\partial \rho}\right)^+ = \left(\frac{\partial \Pi_1 \psi}{\partial \rho}\right)^-,$$

$$(4.3.63) \qquad \left(\frac{\partial \overline{\psi}_0}{\partial r}\right)^+ = \left(\frac{\partial \overline{\psi}_0}{\partial r}\right)^-.$$

Therefore from the second equation of (4.3.59), we obtain

$$\left(\frac{\partial \Pi_1 u}{\partial \rho}\right)^+ - \left(\frac{\partial \Pi_1 u}{\partial \rho}\right)^- = -\left(\frac{\partial \overline{\psi}_0}{\partial r}\right)^+ (\Pi_0 v)^+ + \left(\frac{\partial \overline{\psi}_0}{\partial r}\right)^- (\Pi_0 v)^-,$$

and hence, by virtue of (4.3.52),

$$\left(\frac{\partial \Pi_1 u}{\partial \rho}\right)^+ - \left(\frac{\partial \Pi_1 u}{\partial \rho}\right)^- = -2\frac{\partial \overline{\psi}_0}{\partial r},$$

where $\partial \overline{\psi}_0 / \partial r$ is the notation for $(\partial \overline{\psi}_0 / \partial r)^+ = (\partial \overline{\psi}_0 / \partial r)^-$.

But then, (4.3.56) implies

$$(4.3.64) \qquad \left(\frac{\partial \overline{u}_0}{\partial r}\right)^+ - \left(\frac{\partial \overline{u}_0}{\partial r}\right)^- = 2\frac{\partial \overline{\psi}_0}{\partial r}.$$

Thus, we determined the missing additional conditions needed for defining $\overline{\psi}_0$ and $\overline{u}_0$ from (4.3.31), namely, the conditions (4.3.63) and (4.3.64). Now we can find $\overline{\psi}_0$ and $\overline{u}_0$ and, finally, determine $\Pi_0 z$ (recall that the equations for $\Pi_0 z$ depended on $u_0 = \overline{u}_0(0, l)$, which is now known).

At the same time, we obtained the additional conditions (4.3.57), (4.3.62) for system (4.3.59), which involves yet other unknown quantities $\overline{u}_1^{\pm}$. As (4.3.35), this system can be studied with the help of the first integrals. This will enable us to find part of the additional conditions on Γ for the regular terms $\overline{\psi}_1$ and $\overline{u}_1$ of the first-order approximation. Other additional conditions will be obtained during the construction of $\Pi_2 z$. Terms of higher-order approximations can be found analogously.

So, the zeroth-order approximation $\overline{z}_0 + \Pi_0 z$ is completely constructed. It satisfies the boundary conditions on C_k and $\partial\Omega_k$. This sum is continuous across Γ, and the first derivatives are continuous only in the leading order $(1/\mu)$, i.e., with an accuracy $O(1)$. To obtain matching of the normal derivatives with an accuracy of order $O(\mu)$, we must add terms $\mu\overline{z}_1 + \mu\Pi_1 z$. Higher-order terms in expansion for Πz will increase the accuracy of matching.

The boundary functions $\Pi_i z$ introduce errors in the boundary conditions on the insulated boundaries. These errors are essential in the vicinities of points B_1 and B_2 (see Fig. 4.2), and they decay exponentially as the arguments move away from these points along the boundaries. Such errors are compensated by introducing corner boundary functions. Similarly, corner boundary functions should be introduced near points where the Ohmic contacts C_k join the isolated boundaries $\partial\Omega_i$. As we have mentioned earlier, we will not construct corner boundary functions here. Thus, the partial sum of the series (4.3.30) provides a uniform approximation with respect to discrepancy everywhere in the domain, except the neighborhoods of corner points (enumerated 1–8 in Fig. 4.2).

Note that corner boundary functions of the zeroth order are absent because the Neumann boundary conditions are prescribed on $\partial\Omega_i$, and therefore $\overline{z}_0 + \Pi_0 z$ provides the approximation with respect to discrepancy with an accuracy of order $O(\mu)$ uniformly in the entire domain $\overline{\Omega} = \overline{\Omega}_1 \cup \overline{\Omega}_2$.

In conclusion, we should mention that the method of boundary functions can also be applied to solve nonstationary problems in the theory of semiconductor devices (see, e.g., Belyanin [4]).

4.4 Relaxation Waves in the FitzHugh–Nagumo System

4.4.1 Statement of the problem

Consider the *FitzHugh–Nagumo* (FHN) system of equations (cf. FitzHugh [51] and Nagumo, Arimoto, and Yoshizawa [101])

$$\varepsilon\frac{\partial u}{\partial t} - \varepsilon^2\frac{\partial^2 u}{\partial x^2} = f(u) - v + I \equiv F(u,v,I), \qquad (4.4.1)$$
$$\frac{\partial v}{\partial t} = bu - \gamma v \equiv H(u,v),$$

where $f(u) = u(a-u)(u-1)$. This system is often used as a model for a more complicated Hodgkin–Huxley system (cf. Hodgkin and Huxley [63]) that describes the propagation of excitation in a nerve axon. In the dimensionless equations (4.4.1) the function u corresponds to the nerve membrane potential, v represents ion concentrations, I is an applied current, $0 < \varepsilon \ll 1$ is a small parameter that appears in the case of fast kinetics ($\sim O(\varepsilon^{-1})$), described by the right-hand side of the first equation, and the small "diffusion" coefficient ($\sim O(\varepsilon)$) associated with the axial current in the axon (this coefficient is directly proportional to the axon diameter and inversely proportional to resistivity), b and γ are positive constants, while $0 < a < 1$.

For $0 < a < \frac{1}{2}$, $I = 0$, the FHN system represents a classical example of a so-called *excitable medium* (cf., e.g., Keener [75]): a special type of initial excitation (e.g.,

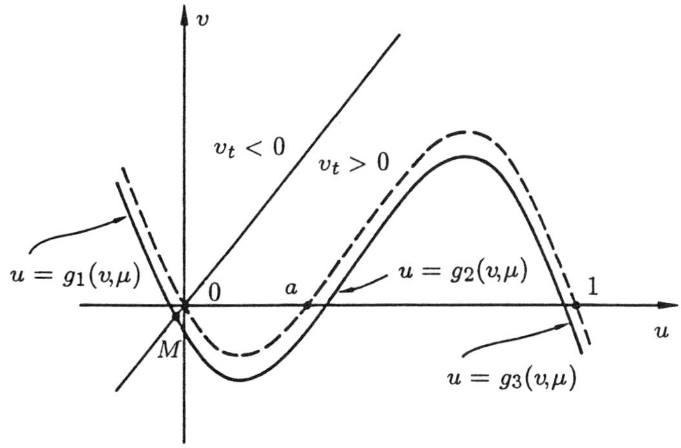

Figure 4.4: *Null clines for the FitzHugh–Nagumo system* (4.4.1)*when* $I = -\mu$, $(0 < \mu \ll 1)$:*the rest point* M *does not belong to interval* $0 \leq v \leq 1$.

$u(x,0) > a$, $v(x,0) = 0$ over a small spatial domain) initiates the solitary pulse (in the variable u) that propagates with a certain speed. Results on excitable systems, as well as on the FHN equations, can also be found in Rinzel [121], Zykov [167], Tyson and Keener [136]. See also Murray [99] for survey of related problems and numerous references.

Here we will construct the zeroth-order approximation for the relaxation wave solution of the FHN equations using the boundary function method; we will also point to the relation between the boundary function method and matching. Our presentation is based on results obtained in Kalachev [69].

Consider the case when a small current $I = -\mu$ $(0 < \mu \ll 1)$ is applied to the FHN system. The null clines for $I = -\mu$ are represented in Fig. 4.4 (the null cline of the first equation for $I = 0$ is shown by the dashed line). From the physical meaning of the dependent variable v follows the constraint: $0 \leq v \leq 1$. The rest point M of the system, which originally coincided with the origin O, now lies in the "forbidden zone," since the v-coordinate of M is negative. Instead of the pulse-like solution that can be constructed for u when $I = 0$, we now will have a *relaxation wave*: a solution of a form of a moving periodic structure (see Fig. 4.5) represented by the closed-loop in Fig. 4.6, with subintervals of slow and fast changes in u.

Let us introduce the new independent variable $z = x + ct$ associated with the moving structure (here c is the as yet unknown velocity of the relaxation wave). We look for the solution of (4.4.1) in a special form:

$$u(x,t) = u(z), \qquad v(x,t) = v(z). \tag{4.4.2}$$

Substitution of (4.4.2) into (4.4.1) implies

$$\begin{aligned} &\varepsilon^2 \frac{d^2u}{dx^2} - \varepsilon c \frac{du}{dt} + f(u) - v - \mu = 0, \\ &c\frac{dv}{dt} = bu - \gamma v. \end{aligned} \tag{4.4.3}$$

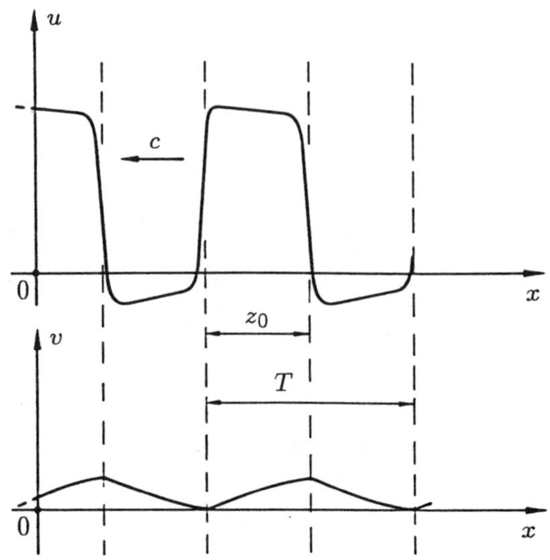

Figure 4.5: *Sketch of the asymptotic solution for the FitzHugh–Nagumo system* (4.4.1)*when* $I = -\mu$.

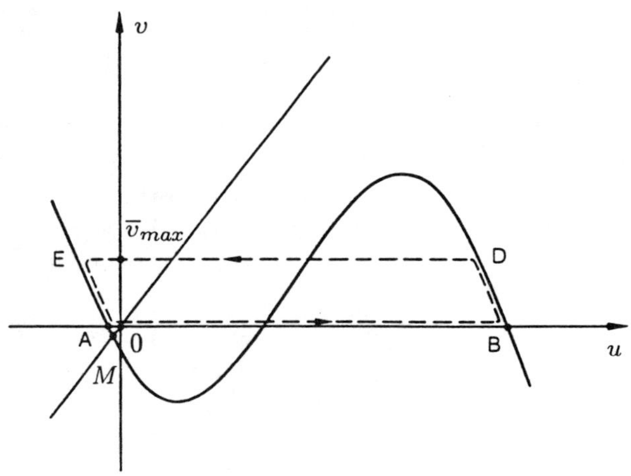

Figure 4.6: *The relaxation wave solution is represented by the closed-loop trajectory in a* (u, v)*-plane.*

Since we are looking for a periodic solution of the variable z, we must impose the periodicity conditions:

$$u(z) = u(z+T), \qquad v(z) = v(z+T). \tag{4.4.4}$$

The unknown period T of the moving structure, as well as the distance z_0 between sequential transition layers (see Fig. 4.5), will be determined during the construction of the asymptotic solution. We note that due to periodicity, it is sufficient to construct the solution for only one period. Let us begin the construction of the asymptotic expansion at $z = 0$, where the "jump up," or the so-called leading edge transition layer takes place (the initial point can be chosen arbitrarily within the period, e.g., it could be the point where the "jump down" (trailing edge transition layer) occurs, or any other point).

4.4.2 Asymptotic algorithm and zeroth-order approximation

We construct uniform (within the period) asymptotic approximation for the solution of (4.4.3), (4.4.4), with the constraint $0 \le v \le 1$, in the form

$$\begin{aligned} u(z,\varepsilon) &= \overline{u}^1(z,\varepsilon) + \Pi u(\xi,\varepsilon) + Q^*u(\xi^*,\varepsilon), \qquad z \in [0, z_0], \\ u(z,\varepsilon) &= \overline{u}^2(z,\varepsilon) + \Pi^* u(\xi^*,\varepsilon) + Qu(\xi,\varepsilon), \qquad z \in [z_0, T], \end{aligned} \tag{4.4.5}$$

with similar expressions for $v(z,\varepsilon)$. Here $\overline{u}^1$ and $\overline{u}^2$ are the regular parts of the asymptotic expansion for u within subintervals $[0, z_0]$ and $[z_0, T]$, respectively; Πu, Qu, and $\Pi^* u$, $Q^* u$ are the transition layer boundary functions depending on the stretched variables $\xi = z/\varepsilon$ for $\xi \ge 0$, $\xi = (z-T)/\varepsilon$ for $\xi \le 0$ and $\xi^* = (z - z_0)/\varepsilon$ for all ξ^*. Every term in (4.4.5) is a power series in ε. For example,

$$\overline{u}^1(z,\varepsilon) = \sum_{i=0}^{\infty} \varepsilon^i \overline{u}_i^1(z), \qquad \Pi u(\xi,\varepsilon) = \sum_{i=0}^{\infty} \varepsilon^i \Pi_i u(\xi), \qquad \text{etc.}$$

We also impose the usual decay conditions on the boundary functions: all the Π- and Π^*-functions should approach zero as the corresponding stretched variables tend to infinity, and all the Q- and Q^*-functions should decay to zero as the corresponding layer variables tend to negative infinity. Regular and boundary functions for u are represented in Fig. 4.7, where the terms $\overline{u}^1 + \Pi u$, $\overline{u}^2 + \Pi^* u$, etc., represent only the leading-order terms of the asymptotics, and all the exponentially small terms are omitted. We should also represent the function $f(u)$ in a form similar to (4.4.5):

$$\begin{aligned} f &= \overline{f}^1 + \Pi f + Q^* f, \qquad z \in [0, z_0], \\ f &= \overline{f}^2 + \Pi^* f + Qf, \qquad z \in [z_0, T], \end{aligned} \tag{4.4.6}$$

where $\overline{f}^i = f(\overline{u}^i(z,\varepsilon))$, $\Pi f = f(\overline{u}^1(\varepsilon\xi,\varepsilon) + \Pi u(\xi,\varepsilon)) - f(\overline{u}^1(\varepsilon\xi,\varepsilon))$, etc. We seek the velocity c in the form

$$c = c_0 + \varepsilon c_1 + \varepsilon^2 c_2 + \cdots, \tag{4.4.7}$$

but, as the analysis of consistency for the problems for higher-order terms shows, we should refrain from expanding z_0 and T as power series. Since we will only discuss

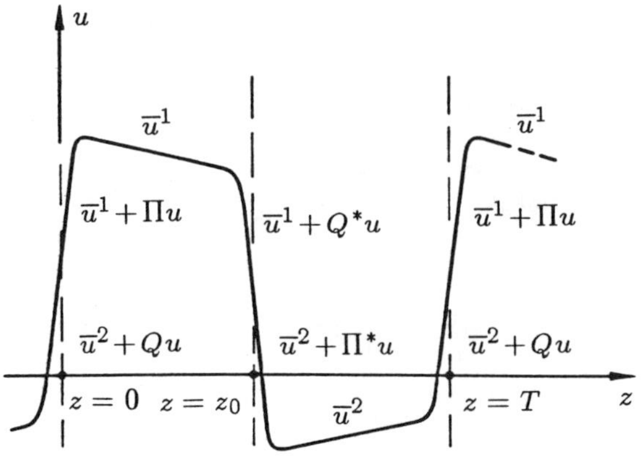

Figure 4.7: *Disposition of regular and boundary functions* (4.4.5)*for relaxation wave solution on the example of u-function.*

here the zeroth-order approximation, to simplify the notation, let us write $\overline{u}^i$, Πu, Qu, etc., instead of $\overline{u}_0^i$, $\Pi_0 u$, $Q_0 u$, etc. We will also denote c_0 by c.

Substituting (4.4.6),(4.4.7), (4.4.5) and similar expressions for $v(z,\varepsilon)$ into (4.4.3), (4.4.4), we obtain in the standard way problems for the terms of the asymptotic solution. For the regular part of the asymptotics at the zeroth order, we have

(4.4.8)
$$f(\overline{u}^i) - \overline{v}^i - \mu = 0,$$
$$c\frac{d\overline{v}^i}{dt} = b\overline{u}^i - \gamma\overline{v}^i, \qquad i = 1, 2.$$

Here superscripts 1 and 2 correspond to functions defined on subintervals $[0, z_0]$ and $[z_0, T]$, respectively. Since z_0 is associated with the leading edge transition layer, on the subinterval $[0, z_0]$ we choose the solution

(4.4.9)
$$\overline{u}^1 = g_1(\overline{v}^1, \mu)$$

of the algebraic equation in (4.4.8), corresponding to the right stable branch of the null cline $F(u, v, -\mu) = 0$ (see Fig. 4.4). In the differential equation for $\overline{v}^1$, the zeroth-order approximation c to the velocity is still unknown. To find the initial conditions for $\overline{v}^1$ and $\overline{v}^2$, we must consider additional conditions for the v-functions at the zeroth order, which can be formulated as follows.

The "jump-switching" condition

(4.4.10)
$$\overline{v}^1(0+) + \Pi v(0) = 0.$$

(This condition can be easily derived from a phase plane analysis (see Fig. 4.6). When the solution, represented by a point in the (u, v)-plane for a given z, arrives at point A on the left stable branch of $F(u, v, -\mu) = 0$, the derivative v_t is negative, so v must decrease. But $v = 0$ at A, so it cannot decrease further due to the constraint $0 \le v \le 1$. Therefore $v = 0$ switches on the "jump up" transition layer which, by our assumption, occurs at $z = 0$. Hence, $v(0) = 0$, and in the zeroth-order approximation we obtain (4.4.10).)

The matching condition

$$\overline{v}^1(z_0-) + Q^*v(0) = \overline{v}^2(z_0+) + \Pi^*v(0), \tag{4.4.11}$$

which states that we are looking for a v, continuous at the zeroth order at z_0.

Later we will also use the periodicity condition

$$\overline{v}^2(T-) + Qv(0) = 0 \tag{4.4.12}$$

to define the period of the relaxation wave solution. The notation $\overline{v}^1(0+)$, $\overline{v}^1(z_0-)$, etc., means that we use the right and left limits of functions at corresponding points. Since the equation

$$\frac{d\Pi v}{d\xi} = 0$$

with the decay condition $\Pi v(\infty) = 0$ has only the trivial solution $\Pi v(\xi) = 0$, and analogously, $Q^*v(\xi^*) = 0$, $\Pi^*v(\xi^*) = 0$ and $Qv(\xi) = 0$, we can rewrite (4.4.10), (4.4.11), (4.4.12) in the following form:

$$\overline{v}^1(0+) = 0, \tag{4.4.13}$$

$$\overline{v}^2(z_0+) = \overline{v}^1(z_0-), \tag{4.4.14}$$

$$\overline{v}^2(T-) = 0. \tag{4.4.15}$$

To define c, we consider the problem for $\Pi u(\xi)$ and $Qu(\xi)$, which describes the leading edge transition layer at the zeroth order. It is convenient to introduce the function

$$U(\xi) = \begin{cases} \overline{u}^1(0+) + \Pi u(\xi), & \xi \geq 0, \\ \overline{u}^2(T-) + Qu(\xi), & \xi \leq 0. \end{cases} \tag{4.4.16}$$

(Note that this is nothing but a relation between the "inner" solution U that can be obtained using the *matching technique* in the vicinity of the point $z = 0$ (or the point $z = T$ due to periodicity) and corresponding boundary functions. Functions $\overline{u}^1$ and $\overline{u}^2$ can be treated as outer solutions in the subintervals $(0, z_0)$ and (z_0, T), respectively.) Here $\overline{u}^1(0+)$ and $\overline{u}^2(T-)$ are the maximal and the minimal roots of the algebraic equation $f(w) - \mu = 0$. The FHN system will still be of excitable type if $\mu < \mu^*$, where μ^* is such that the roots $w_1^* < w_2^* < w_3^*$ of the equation $f(w^*) - \mu^* = 0$ satisfy the equality $2w_2^* = w_1^* + w_3^*$ (see, e.g., Keener [75] and Murray [99]). Function $U(\xi)$ will satisfy

$$\frac{d^2U}{d\xi^2} - c\frac{dU}{d\xi} + f(U) - \mu = 0, \qquad U(-\infty) = \overline{u}^2(T-), \qquad U(\infty) = \overline{u}^1(0+). \tag{4.4.17}$$

In our case $\mu \ll 1$ is a small parameter. Thus, we can expect that $\mu < \mu^*$, and then there exists a unique c for which a solution of (4.4.17) exists and is unique.

Let us obtain explicit asymptotic formulae for c and U. The roots of the equation $f(w) - \mu = w(a-w)(w-1) - \mu = 0$ can be represented as power series expansions in μ (see, e.g., Kevorkian [76]):

$$
\begin{aligned}
w_1(\mu) &= -\frac{\mu}{a} + O(\mu^2), \\
w_2(\mu) &= a + \frac{\mu}{a-a^2} + O(\mu^2), \qquad (4.4.18)\\
w_3(\mu) &= 1 - \frac{\mu}{1-a^2} + O(\mu^2).
\end{aligned}
$$

Higher-order terms can be easily written out. Since $0 < a < \frac{1}{2}$, we have $v_1 < v_2 < v_3$. Taking $\overline{u}^2(T-) = v_1(\mu)$ and $\overline{u}^1(0+) = v_3(\mu)$, we can write the expressions for U and c as

$$
U(\xi) = \frac{w_3(\mu) + w_1(\mu)\exp\left\{\frac{1}{\sqrt{2}}(w_1(\mu) - w_3(\mu))\xi\right\}}{1 + \exp\left\{\frac{1}{\sqrt{2}}(w_1(\mu) - w_3(\mu))\xi\right\}}, \qquad (4.4.19)
$$

$$
c(\mu) = \frac{1}{\sqrt{2}}(w_1(\mu) - 2w_2(\mu) + w_3(\mu)) > 0, \qquad (4.4.20)
$$

(cf. Murray [99, pp. 304–305]). It follows from (4.4.19) and (4.4.16) that the boundary function Πu decays exponentially as $\xi \to \infty$, and Qu decays exponentially as $\xi \to -\infty$ (this property of boundary functions allows us to construct higher-order terms of the asymptotic solution).

Taking into account that the zeroth-order approximation c to the velocity and the initial condition (4.4.13) for $\overline{v}^1$ are known, and substituting the expression (4.4.9) for $\overline{u}^1$ into the equation for $\overline{v}^1$, we obtain

$$
\begin{aligned}
c\frac{d\overline{v}^1}{dz} &= bg_1(\overline{v}^1, \mu) - \gamma\overline{v}^1 > 0, \qquad (4.4.21)\\
\overline{v}^1(0+) &= 0.
\end{aligned}
$$

In the equation for $\overline{v}^1$, the right-hand side is strictly positive. Thus, the solution $\overline{v}^1$ of (4.4.21) is monotonically increasing and it can be written in quadratures as

$$
z = \int_0^{\overline{v}^1(z)} \frac{c\,d\eta}{bg_1(\eta, \mu) - \gamma\eta} \equiv R_1(\overline{v}^1). \qquad (4.4.22)
$$

For the known $\overline{v}^1(z)$, the function $\overline{u}^1(z)$ is given by (4.4.9).

Before solving for $\overline{v}^2(z)$, we have to find the point z_0 at which the initial condition (4.4.14) is prescribed. By virtue of the monotonicity of $\overline{v}^1(z)$, this point corresponds to that where $\overline{v}^1$ has its maximal value $\overline{v}_{\max}$, i.e.,

$$
z_0 = R_1(\overline{v}_{\max}). \qquad (4.4.23)
$$

The value $\overline{v}_{\max}$ can be found while solving for the boundary functions Π^*u and Q^*u that help describe the trailing edge transition layer. As with Πu and Qu, let us introduce

$$U^*(\xi^*) = \begin{cases} \overline{u}^2(z_0+) + \Pi^*u(\xi^*), & \xi^* \geq 0, \\ \overline{u}^1(z_0-) + Q^*u(\xi^*), & \xi^* \leq 0, \end{cases} \tag{4.4.24}$$

which must satisfy the following problem:

$$\frac{d^2U^*}{d\xi^{*2}} - c\frac{dU^*}{d\xi^*} + f(U^*) - \overline{v}_{\max} - \mu = 0, \tag{4.4.25}$$
$$U^*(-\infty) = \overline{u}^1(z_0-), \qquad U^*(\infty) = \overline{u}^2(z_0+).$$

We are looking for a solution, periodic in z, for which the trailing edge moves with the same velocity as the leading edge. Therefore, the c in (4.4.25) is defined by (4.4.20). Then there exists a unique $\overline{v}_{\max}$ for which the solution of (4.4.25) exists and is unique (see Murray [99]). For the cubic polynomial $f(w)$, the value of $\overline{v}_{\max}$ can be found explicitly:

$$\overline{v}_{\max} = f(w_{\max}) + f(w_{\min}) - 2\mu > 0, \tag{4.4.26}$$

where

$$w_{\max,\min} = \frac{1}{3}\{w_1 + w_2 + w_3 \pm \sqrt{(w_1 + w_2 + w_3)^2 - 3(w_1w_2 + w_2w_3 + w_3w_1)}\}.$$

Here $w_1 < w_2 < w_3$ are given by (4.4.18), $f(w_{\max}) > 0$, and $f(w_{\min}) < 0$. For a known $\overline{v}_{\max}$, the solutions $\tilde{w}_1 < \tilde{w}_2 < \tilde{w}_3$ of the cubic equation

$$f(w) - \overline{v}_{\max} - \mu = 0$$

can be easily found (the explicit expressions for these solution are presented, e.g., in Murray [99, p. 705]). Then the solution U^* of (4.4.25) can be written in the following form, similar to (4.4.19):

$$U^*(\xi^*) = \frac{\tilde{w}_3(\mu) + \tilde{w}_1(\mu)\exp\left\{\frac{1}{\sqrt{2}}(\tilde{w}_3(\mu) - \tilde{w}_1(\mu))\xi^*\right\}}{1 + \exp\left\{\frac{1}{\sqrt{2}}(\tilde{w}_3(\mu) - \tilde{w}_1(\mu))\xi^*\right\}}. \tag{4.4.27}$$

Here we choose $\overline{u}^1(z_0-) = \tilde{w}_3 > 0$ and $\overline{u}^2(z_0+) = \tilde{w}_1 < 0$. The value of c is as in (4.4.20). From (4.4.27), (4.4.24) the exponential decay of Π^*u and Q^*u can be easily established.

Substituting $\overline{v}_{\max}$ defined by (4.4.26) into (4.4.23), we obtain $z_0(\mu)$, the point where the initial condition (4.4.14) for $\overline{v}^2(z)$ is prescribed. On the subinterval $[z_0, T]$ (with the period of the structure still unknown), we choose the solution

$$\overline{u}^2 = g_3(\overline{v}^2, \mu) \tag{4.4.28}$$

of the algebraic equation in (4.4.8) that corresponds to the left stable branch of the null cline $F(u, v, -\mu) = 0$ (see Fig. 4.4). By virtue of (4.4.8), (4.4.28), and (4.4.14), we obtain

$$c\frac{d\overline{v}^2}{dz} = bg_3(\overline{v}^2, \mu) - \gamma\overline{v}^2 < 0, \qquad (4.4.29)$$

$$\overline{v}^2(z_0+) = \overline{v}^1(z_0-) = \overline{v}_{\max}.$$

Since the right-hand side of (4.4.29) is strictly negative, $\overline{v}^2(z)$ is monotonically decreasing. The implicit solution of (4.4.29) can be written in the form

$$z - z_0 = \int_{\overline{v}_{\max}}^{\overline{v}^2(z)} \frac{c\,d\eta}{bg_3(\eta, \mu) - \gamma\eta} \equiv R_2(\overline{v}^2), \qquad (4.4.30)$$

similar to (4.4.22). For known $\overline{v}^2(z)$, function $\overline{u}^2(z)$ is given by (4.4.28). Using the periodicity condition (4.4.15), we finally obtain

$$T(\mu) = R_2(0) + z_0(\mu). \qquad (4.4.31)$$

This completes the construction of the asymptotic solution to the zeroth order. Note that along with the terms of the asymptotic expansion (4.4.5), we obtained the zeroth-order approximation c to the velocity of the structure, as well as the parameters z_0 and T. Since the derivatives of the regular functions $\overline{u}^1(z)$, $\overline{u}^2(z)$ and $\overline{v}^1(z)$, $\overline{v}^2(z)$ in the zeroth order are not matched at the points $z = 0$ (equivalently, $z = T$) and $z = z_0$, we must construct the boundary functions at the first order to compensate for these jumps. In the first- and higher-order approximations, we obtain linear problems for the regular and boundary functions that can be solved successively. The consistency conditions for higher-order terms imply that, of three parameters entering the problem (c, z_0 and T), only the velocity c of the structure should be sought as a power series expansion in ε. A detailed analysis is presented in Kalachev [69].

4.4.3 Concluding remarks

The FitzHugh-Nagumo equations are widely used for modeling the nerve axon as well as moving waves in muscle tissue, the brain cortex, etc. It is important to mention that similar analyses can be carried out for some general function $f(u)$ exhibiting properties close to those of cubic polynomials and for values of current I that are not necessarily small, whereas for small $I = -\mu$ and for cubic polynomials all the results can be obtained quite explicitly. Our approach, based on the boundary function method, can also be applied to investigating a moving front and moving pulse solutions of more general *reaction-diffusion* type systems. For example, the asymptotic moving pulse solution for one model of calcium-induced calcium release within a cell (formulated in terms of two reaction-diffusion equations) was discussed in Kalachev [68].

Exercise

Using the boundary function method, construct the zeroth-order approximation for the moving pulse solution of

$$\varepsilon\frac{\partial u}{\partial t} - \varepsilon^2\frac{\partial^2 u}{\partial x^2} = f(u) - v,$$

$$\frac{\partial v}{\partial t} = u - v$$

in the infinite spatial domain. (This is the system (4.4.1) with $I = 0$.)

Hint. Look for the solution depending on the variable $z = x + ct$ (where c is the velocity, and $z = 0$ corresponds to the leading edge of the moving pulse). Assume $u(x,t) = u(z)$ and $v(x,t) = v(z)$, and use the asymptotic representation

$$u_0 = \begin{cases} \bar{u}_0^1(z) + Q_0 z(\xi), & z \le 0 \quad (\xi = z/\varepsilon \le 0), \\ \bar{u}_0^2(z) + \Pi_0 z(\xi) + Q_0^* z(\xi^*), & 0 \le z \le z_l \quad (\xi = z/\varepsilon \ge 0, \quad \xi^* = (z - z_l)/\varepsilon \le 0), \\ \bar{u}_0^3(z) + \Pi_0^* z(\xi^*), & z \ge z_l \quad (\xi^* = (z - z_l)/\varepsilon \ge 0), \end{cases}$$

with similar representation for v_0 (compare with (4.4.5)). The zeroth-order approximation for the velocity c and the quantity z_l will be defined during the asymptotic process.

See also Casten, Cohen, and Lagerstrom [40] for alternative approach to construct approximation for the moving pulse solution.

4.5 On some other applied problems

We discussed in detail a variety of applied problems that led to the singularly perturbed equations, and whose solutions were characterized by the presence of boundary and/or internal transition layers. Let us briefly mention some other applied problems that were solved using the boundary function method and related approaches.

(1) Substantial attention for a number of years has been given to problems of optimal control, formulated in terms of singularly perturbed equations. A characteristic statement of a problem of this kind (without inequality constraints $|u(t)| \le u_0$ on the control function u) involves minimizing some functional, e.g.,

$$I_\varepsilon = \varphi(y(T), z(T)) + \int_0^T F(y, z, u, t)dt$$

along the trajectories of a system

$$\frac{dy}{dt} = f(y, z, u, t, \varepsilon), \qquad \varepsilon\frac{dz}{dt} = g(y, z, u, t, \varepsilon),$$

$$y(0, \varepsilon) = y^0, \quad z(0, \varepsilon) = z^0, \quad y \in R_m, \quad z \in R_M.$$

Introducing the Hamiltonian and the adjoint variables, we may transform this problem to a boundary value problem in the *conditionally stable* case. This and many other singularly perturbed optimal control problems are discussed in a survey by Vasil'eva and Dmitriev [152], where numerous references are also presented. In Belokopitov and Dmitriev [3] the solution of the optimal control problem is constructed using the boundary layer expansion directly, without using the adjoint variables (the so-called *direct scheme* of the solution). For other approaches to solving singularly perturbed optimal control problems see Kokotović, Khalil, and O'Reilly [81]. Singular arcs in the "cheap control" problems are discussed, e.g., in O'Malley and Jameson [115].

(2) The boundary function method is also effective in solving problems of rigid body dynamics under electro- and hydrodynamic influences (see Kobrin [78], Kobrin, Martynenko, and Novozhilov [79], and Kobrin and Martynenko [80]).

(3) In § 4.3, the problem of a (p-n)-junction in the theory of semiconductor devices was discussed. Some interesting problems also appear during the study of semiconductor films; to such problems the boundary function method can be applied as well (see Belyanin and Vasil'eva [8]).

(4) Asymptotic solutions for the semilinear Poisson equation in a two- and three-dimensional semiconductor structures were constructed in Kalachev and Obukhov [70], [71]. Numerical computations have shown that, when the asymptotic solutions are used as initial iterates, the convergence of the numerical process for the drift-diffusion model is speeded up by a factor of 5–10. In Kalachev, Kruchkov, and Obukhov [72] the case of gate contact is considered and the estimation of the remainder is presented. Asymptotic solution of the stationary drift-diffusion model in the case of large generation-recombination terms in a two-dimensional domain modeling a semiconductor device is discussed in Belyanin, Kalachev, and Mamontov [7].

(5) In Belyanin et al. [9], a one-dimensional problem for a diode (or thyristor structure) was posed as the optimal design problem. For a given voltage-current characteristic, the synthesis of the device with such characteristic is discussed, when a doping level N $(-1 \leq N \leq 1)$ is considered as a control function. This problem also involves singular perturbation analyses.

(6) The boundary function method was successfully used to solve problems of molecular aerodynamics, which were described by a simplified version of the Boltzmann equation (see Nagnibeda [100] and Vasil'eva and Butuzov [149]).

(7) In the theory of alloys (cf. Voroshin and Husid [165]), we may encounter kinetic systems possessing the following feature: the eigenvalues $\lambda_i(t)$ of the matrix $\overline{F}_z$ (see § 2.1.2) are zero at the initial moment of time $t = 0$, and for $t > 0$ the usual condition $\mathrm{Re}\lambda_i(t) < 0$ holds. It turns out that the boundary function method is applicable in these problems as well.

(8) The asymptotic solution of the singularly perturbed integral equations from the theory of slow neutrons and theory of epidemics were investigated by the boundary function method in Nefedov [105], [106]. Problems concerning transport of neutrons with short mean free paths that can be transformed to integro-differential equations were investigated in Latishev [84] and Latishev and Tupchiev [85].

(9) A singularly perturbed problem modeling heat and mass transfer in a two-component medium was considered in Nefedov [107].

(10). The asymptotic solution of the problem on acoustic oscillations in a medium with small viscosity with and without resonance is constructed by Butuzov and his colleagues in [27], [28]. A similar problem from the dynamics of viscous stratified rotating fluids is considered in Nefedov [108].

Bibliography

[1] U. ASCHER, R. MATTHEIJ, AND R. RUSSELL, *Numerical Solution of Boundary Value Problems for Ordinary Differential Equations*, Prentice Hall, Englewood Cliffs, NJ, 1988.

[2] N. BAKHVALOV, *The optimization of methods of solving boundary value problems with a boundary layer*, Comput. Math. Math. Phys., 9 (1969), pp. 139–166.

[3] S. BELOKOPITOV AND M. DMITRIEV, *Direct method of solution of optimal control problems with fast and slow motions*, Izv. Akad. Nauk SSSR, Ser. Tekhn. Kibern., (1985), No. 3, pp. 147–152. (In Russian.)

[4] M. BELYANIN, *On numerical-asymptotic solution of the nonstationary singularly perturbed problem from the theory of semiconductor devices*, Differentsial'nye Uravneniya, 21 (1985), pp. 1436–1440. (In Russian.)

[5] ——, *On asymptotic solution for one model of $p-n$-junction*, Zh. Vychisl. Mat. i Mat. Fiz., 26 (1986), No. 8, pp. 306–311. (In Russian.)

[6] ——, *On the asymptotics in one-dimensional model of some semiconductor devices*, Zh. Vychisl. Mat. i Mat. Fiz., 28 (1988), No. 1, pp. 34–51. (In Russian.)

[7] M. BELYANIN, L. KALACHEV, AND E. MAMONTOV, *Application of the boundary function method for the simulation of some semiconductor devices*, Mat. Model., 3 (1991), No. 8, pp. 63–71. (In Russian.)

[8] M. BELYANIN AND A. VASIL'EVA, *On an inner transition layer in a problem of the theory of semiconductor films*, Zh. Vychisl. Mat. i Mat. Fiz., 28 (1988), No. 2, pp. 224–236. (In Russian.)

[9] M. BELYANIN, A. VASIL'EVA, A. VORONOV, AND A. TIKHONRAVOV, *An asymptotic approach to the problem of designing a semiconductor device*, Mat. Model., 1 (1989), No. 9, pp. 43–63. (In Russian.)

[10] L. BOBISUD AND C. CHRISTENSON, *A singular singularly perturbed system of nonlinear parabolic equations from chemical kinetics*, J. Math. Anal. Appl., 74 (1980), pp. 296–310.

[11] Y. BOGLAEV, *On numerical methods for solving singularly perturbed problems*, Differentsial'nye Uravneniya, 21 (1985), pp. 1804–1806. (In Russian.)

[12] N. BOGOLIUBOV AND Y. A. MITROPOLSKII, *Asymptotic methods in the theory of nonlinear oscillations*, Hindustan Publ. Corp., Delhi, 1961.

[13] A. Y. BOMBA, *Asymptotic method for approximately solving a mass transport problem for flow in a porous medium*, Ukrainian Math. J., 34 (1982), pp. 400–403.

[14] A. BUSH, *Perturbations Methods for Engineers and Scientists*, CRC Press, Boca Raton, 1992.

[15] V. BUTUZOV, *Asymptotic properties of the solution of a finite difference equation with small steps in a rectangular region*, Comput. Math. Math. Phys., 12 (1972),

No.3, pp. 14–34.

[16] ——, *The angular boundary layer in mixed singularly perturbed problems for hyperbolic equations*, Math. USSR-Sb., 33 (1977), pp. 403–425.

[17] ——, *Corner boundary layers in singularly perturbed problems for partial differential equations*, Differential Equations, 15 (1979), pp. 1318–1328.

[18] V. Butuzov and L. Kalachev, *Asymptotic solution of the boundary value problem for a singularly perturbed system of parabolic equations with different powers of small parameters*, Technical Report, deposited in VINITI, 1984.

[19] ——, *Asymptotic approximation of the solution of a boundary value problem for a singularly perturbed parabolic equation in the critical case*, Math. Notes, 38 (1986), pp. 819–831.

[20] ——, *Using the boundary function method for simulation of a fast nonisothermal reaction with diffusion and thermoconduction*, Teoret. Osnovi Khimicheskoi Tekh., (1987), pp. 837–841. (In Russian.)

[21] ——, *Asymptotic solution of the combustion problem in the case of autocatalytic reaction*, Comput. Math. Math. Phys., 28 (1988), No. 4, pp. 39–47.

[22] ——, *Asymptotic derivation of the ambipolar diffusion equation in physics of semiconductors*, Mat. Model., 4 (1992), No. 8, pp. 66–74. (In Russian.)

[23] V. Butuzov and V. Mamonov, *Singularly perturbed elliptic boundary value problem in the critical case*, Differential Equations, 18 (1982), pp. 754–758.

[24] ——, *On a singularly perturbed quasilinear parabolic problem with unsmooth corner boundary functions*, Comput. Math. Math. Phys., 27 (1987), No. 4, pp. 33–39.

[25] ——, *Smoothing in a singularly disturbed quasilinear parabolic problem*, Comput. Math. Math. Phys., 27 (1987), No. 2, pp. 45–51.

[26] V. Butuzov and N. Nefedov, *A problem in singular perturbation theory*, Differential Equations, 12 (1976), pp. 1219–1227.

[27] V. Butuzov, N. Nefedov, and E. Fedotova, *Asymptotic solution of the linearized problem of propagation of sound in a bounded medium with low viscosity*, Comput. Math. Math. Phys., 27 (1987), No. 1, pp. 146–153.

[28] V. Butuzov, N. Nefedov, and E. Polezhaeva, *The asymptotic solution of linearized problem on the natural and forced resonance oscillations of a medium of low viscosity*, Comput. Math. Math. Phys., 29 (1989), No. 1, pp. 41–49.

[29] V. Butuzov and A. Nesterov, *On some singularly perturbed problems with nonsmooth boundary functions*, Soviet Math. Dokl., 25 (1982), pp. 420–423.

[30] ——, *Singularly perturbed elliptic boundary value problems with nonsmooth boundary functions*, Differential Equations, 21 (1985), pp. 1183–1187.

[31] ——, *Some singularly perturbed hyperbolic problems with transition layers*, Differential Equations, 22 (1986), pp. 1189–1194.

[32] V. Butuzov and A. Nikitin, *A system of singularly perturbed elliptic equations*, Differential Equations, 17 (1981), pp. 1104–1115.

[33] ——, *Singularly disturbed elliptic boundary value problem in a rectangle*, Comput. Math. Math. Phys., 24 (1984), No. 5, pp. 25–31.

[34] ——, *Uniform asymptotic form of the solution of a singularly perturbed elliptic system in a rectangle*, Moscow Univ. Comput. Math. Cybernet., (1985), No. 3, pp. 1–8.

[35] V. Butuzov and Y. Udodov, *Asymptotic solution of a system of singularly perturbed equations of elliptic type with an angular boundary layer*, Comput.

Math. Math. Phys., 21 (1981), No. 3, pp. 134–147.

[36] V. BUTUZOV AND T. URAZGIL'DINA, *A singularly perturbed system of elliptic equations in critical case*, Differential Equations, 22 (1986), pp. 1092–1101.

[37] ——, *Asymptotic solution of a quasistatic thermoelasticity problem for a slender rod*, J. Appl. Math. Mech., 51 (1987), pp. 761–769.

[38] V. BUTUZOV AND A. VASIL'EVA, *Asymptotics of a solution of a contrast-structure type*, Math. Notes, 42 (1987), pp. 956–961.

[39] ——, *The asymptotic theory of contrasting spatial structures*, Comput. Math. Math. Phys., 28 (1988), No. 2, pp. 26–36.

[40] R. CASTEN, H. COHEN, AND P. LAGERSTROM, *Perturbation analysis of an approximation to the Hodgkin-Huxley theory*, Quart. .Appl. Math., 32 (1975), pp. 365–402.

[41] K. CHANG AND F. HOWES, *Nonlinear Singular Perturbation Phenomena: Theory and Application*, Springer-Verlag, New York, 1984.

[42] E. CODDINGTON AND N. LEVINSON, *Theory of Ordinary Differential Equations*, McGraw-Hill, New York, 1955.

[43] J. COLE, *Perturbation Methods in Applied Mathematics*, Ginn-Blaisdell, Waltham, MA, 1968.

[44] E. DOOLAN, J. MILLER, AND W. SCHILDERS, *Uniform Numerical Methods for Problems with Initial and Boundary Layers*, Boole Press, Dublin, 1980.

[45] S. DVORYANINOV, *A periodic solution of a singularly perturbed autonomous parabolic system*, Differential Equations, 16 (1980), pp. 1040–1044.

[46] W. ECKHAUS, *Matched Asymptotic Expansions and Singular Perturbations*, North-Holland, Amsterdam, London, 1973.

[47] ——, *Asymptotic Analysis of Singular Perturbations*, North-Holland, Amsterdam, New York, Oxford, 1979.

[48] A. ERDÉLYI, *Asymptotic Expansions*, Dover Publications, New York, 1956.

[49] V. ESIPOVA, *Asymptotic properties of solutions of general boundary value problems for singularly perturbed conditionally stable systems of ordinary differential equations*, Differential Equations, 11 (1975), pp. 1457–1465.

[50] M. FEDORYUK, *Asymptotic Methods for Linear Ordinary Differential Equations*, Nauka, Moscow, 1983. (In Russian.)

[51] R. FITZHUGH, *Impulses and physiological states in theoretical models of nerve membrane*, Biophys. J., 1 (1961), pp. 445–466.

[52] V. FLYUD AND V. TSIMBAL, *Asymptotics of the solution of a mixed problem for a singularly perturbed weakly constrained hyperbolic system*, Ukrainian Math. J., 37 (1985), pp. 385–390.

[53] D. FRANK-KAMENETSKII, *Diffusion and Heat Transfer in Chemical Kinetics*, Plenum Press, New York, 1969.

[54] I. GEL'FAND, *Some problems of the theory of quasi-linear equations*, Uspekhi Mat. Nauk, 14 (1959), pp. 87–158. (In Russian.)

[55] S. K. GODUNOV, J. MILLER, AND V. NOVIKOV, eds., *BAIL 4. Proceedings of the Fourth International Conference on Boundary and Interior Layers — Computational and Asymptotic Methods, Novosibirsk 1986*, Boole Press, Dublin, 1986.

[56] J. GRASMAN, *Asymptotic Methods for Relaxation Oscillations and Applications*, Springer-Verlag, Berlin, Heidelberg, New York, 1987.

[57] H. HAKEN, *Synergetics, an Introduction: Nonequilibrium Phase Transitions and*

Self-organization in Physics, Chemistry and Biology, Springer Series in Synergetics, Springer-Verlag, Berlin, New York, 1983.

[58] J. HALE AND K. SAKAMOTO, *Existence and stability of transition layers*, Japan J. Appl. Math., 5 (1988), pp. 367–405.

[59] H. HAN AND R. KELLOGG, *Differentiability properties of solutions of the equation* $-\varepsilon^2 \triangle u + ru = f(x,y)$ *in a square*, SIAM J. Math. Anal., 21 (1990), pp. 394–408.

[60] P. HEMKER AND J. MILLER, eds., *Numerical Analysis of Singular Perturbation Problems*, Academic Press, New York, 1979.

[61] D. HENRY, *Geometric Theory of Semilinear Parabolic Equations*, Springer-Verlag, Berlin, Heidelberg, New York, 1981.

[62] E. HINCH, *Perturbation Methods*, Cambridge University Press, Cambridge, 1991.

[63] A. HODGKIN AND A. HUXLEY, *A quantitative description of membrane current and its application to conduction and excitation in nerve*, J. Physiol. (London), 117 (1952), pp. 500–544.

[64] F. HOPPENSTEADT, *Analysis and Simulation of Chaotic Systems*, Springer-Verlag, Berlin, Heidelberg, New York, 1993.

[65] V. IAKUBOVICH AND V. STARZHINSKII, *Linear Differential Equations with Periodic Coefficients*, John Wiley, New York, 1975.

[66] A. IL'IN, *Differencing scheme for a differential equation with a small parameter affecting the highest derivative*, Math. Notes, 6 (1969), pp. 596–602.

[67] ——, *Matching of Asymptotic Expansions of Solutions of Boundary Value Problems*, American Mathematical Society, Providence, RI, 1992.

[68] L. KALACHEV, *Asymptotic solution for one model of calcium-induced calcium release*, Technical Report, University of Washington, 1993.

[69] ——, *A relaxation wave solution of the Fitzhugh-Nagumo equations*, J. of Math. Biol., 31 (1993), pp. 133–147.

[70] L. KALACHEV AND I. OBUKHOV, *Approximate solution of the Poisson equation in a model for a two-dimensional semiconductor structure*, Moscow Univ. Phys. Bull., 30 (1989), No. 3, pp. 69–74.

[71] ——, *Asymptotic solution of the Poisson equation in a three-dimensional semiconductor structure*, Zh. Vychisl. Mat. i Mat. Fiz., 32 (1992), pp. 1509–1514. (In Russian.)

[72] L. KALACHEV, S. KRUCHKOV AND I. OBUKHOV, *Asymptotic analysis of the Poisson equation solutions in semiconductors*, Mat. Model., 1 (1989), No. 9, pp. 129–140. (Russian).

[73] A. KAPILA, *Asymptotic Treatment of Chemically Reacting Systems*, Pitman, Boston, London, Melbourne, 1983.

[74] K. KASIMOV AND B. KADYKENOV, *A Cauchy problem, with an initial discontinuity, for singularly perturbed linear hyperbolic equations degenerating into first order equations*, Differential Equations, 19 (1983), pp. 1549–1557.

[75] J. KEENER, *Waves in excitable media*, SIAM J. Appl. Math., 39 (1980), pp. 528–548.

[76] J. KEVORKIAN, *Partial Differential Equations: Analytical Solution Techniques*, Wadsworth and Brooks/Cole, Pacific Grove, CA, 1990.

[77] J. KEVORKIAN AND J. D. COLE, *Singular Perturbation Methods in Applied Mathematics*, Springer-Verlag, Berlin, Heidelberg, New York, 1981.

[78] A. KOBRIN, *On the motion of the hollow body filled with viscous liquid about its*

center of mass in a potential body-force field, J. Appl. Math. Mech., 33 (1969), pp. 418–427.

[79] A. KOBRIN, I. MARTYNENKO, AND I. NOVOZHILOV, *On the precession equations of gyroscopic system*, J. Appl. Math. Mech., 40 (1976), pp. 208–215.

[80] A. KOBRIN AND Y. MARTYNENKO, *Motion of a conducting solid body about a center of mass in a slowly varying magnetic field*, Soviet Phys. Dokl., 26 (1981), pp. 1134–1136.

[81] P. KOKOTOVIĆ, H. KHALIL, AND J. O'REILLY, *Singular Perturbation Methods in Control: Analysis and Design*, Academic Press, London, New York, Sydney, 1986.

[82] J. S. KOLESOV, *Periodic solutions of quasilinear parabolic equations of second order*, Trans. Moscow Math. Soc., Vol. 21 (1970), pp. 114–146.

[83] P. LAGERSTROM, *Matched Asymptotic Expansions*, Springer-Verlag, Berlin, Heidelberg, New York, 1988.

[84] V. LATYSHEV, *Asymptotic behavior of a solution of the neutron-transfer equation for a cylindrical region in the case of a short mean free path*, Differential Equations, 20 (1984), pp. 1156–1161.

[85] V. LATYSHEV AND V. TUPCHIEV, *Asymptotic expansion of a solution of the neutron-transfer equation in the case of a short mean free path*, Differential Equations, 19 (1983), pp. 1405–1409.

[86] C. LIN AND L. SEGEL, *Mathematics Applied to Deterministic Problems in the Natural Sciences*, Macmillan, New York, 1974.

[87] S. LOMOV, *Introduction to the General Theory of Singular Perturbations*, American Mathematical Society, Providence, RI, 1992.

[88] P. MARKOWICH, *The Stationary Semiconductor Device Equations*, Springer-Verlag, Wien, New York, 1986.

[89] P. MARKOWICH, C. A. RINGHOFER, AND C. SCHMEISER, *Semiconductor Equations*, Springer-Verlag, Wien, New York, 1990.

[90] V. MASLOV, *Complex WKB-method in nonlinear equations* , Nauka, Moscow, 1977. (In Russian.)

[91] E. MEDEUOV, *Asymptotic solution of singularly perturbed Cauchy problem for one system of differential equations of hyperbolic type*, Izv. Akad. Nauk KazSSR, Seriya Fiz.-Mat., (1981), No. 5, pp. 69–71. (In Russian.)

[92] Z. O. MEL'NIK AND V. TSYMBAL, *A mixed problem for a first-order hyperbolic system with a small parameter multiplying the derivatives*, Differential Equations, 12 (1976), pp. 633–637.

[93] R. MEYER AND S. PARTER, eds., *Singular Perturbations and Asymptotics*, Academic Press, New York, 1980.

[94] E. MISHCHENKO AND N. ROSOV, *Differential equations with small parameter and relaxation oscillations*, Plenum Press, New York, 1980.

[95] I. MITROPOL'SKII, *The Averaging Method in Nonlinear Mechanics*, Naukova Dumka, Kiev, 1971. (In Russian.)

[96] N. MOISEEV, *Asymptotic Methods of Nonlinear Mechanics*, Nauka, Moscow, 1981.

[97] J. MURDOCK, *Perturbations: Theory and Methods*, John Wiley, Inc., New York, 1991.

[98] J. MURRAY, *Asymptotic Analysis*, 2nd ed., Springer-Verlag, Berlin, Heidelberg, New York, 1984.

[99] ——, *Mathematical Biology*, Springer-Verlag, Berlin, Heidelberg, New York, 1989.

[100] E. Nagnibeda, *Solution of equations of non-equilibrium gas*, Vestnik LGU, Ser. Mat., Mekh., Astr., 2 (1969), pp. 97–111. (In Russian.)

[101] J. Nagumo, S. Arimoto, and S. Yoshizawa, *An active pulse transmission line simulating nerve axon*, Proc. IRE, 50 (1962), pp. 2061–2071.

[102] A. Nayfeh, *Perturbation Methods*, John Wiley, New York, 1973.

[103] ——, *Introduction to Perturbation Techniques*, John Wiley, New York, 1981.

[104] ——, *Problems in Perturbation*, John Wiley, New York, 1985.

[105] N. Nefedov, *Asymptotic solution of a singularly perturbed integral equation with domains of integration of different degrees of smallness*, Vestnik MGU, Ser. Vichisl. Mat. i Kibern. (1978), No. 1, pp. 28–35. (Russian).

[106] ——, *On a class of singularly disturbed equations*, Comput. Math. Math. Phys., 18 (1978), pp. 89–101.

[107] ——, *Asymptotic solution of the problem modeling the heat exchange in mutually penetrating media*, Differentsial'nye Uravneniya, 21 (1985), pp. 1819–1821. (In Russian.)

[108] ——, *On some singularly perturbed problems for viscous stratified fluids*, J. Math. Anal. Appl., 131 (1988), pp. 118–126.

[109] A. Nesterov, *Asymptotic solution with a transition layer of a singularly perturbed hyperbolic system of equations*, Soviet Phys. Dokl., 34 (1989), pp. 325–327.

[110] G. Nicolis and I. Prigogine, *Self-organization in Nonequilibrium Systems: from Dissipative Structures to Order through Fluctuation*, John Wiley, New York, 1977.

[111] W. Nowacki, *Dynamic Problems of Thermoelasticity*, PWN - Polish Scientific Publishers, Warsaw, 1975.

[112] F. Olver, *Introduction to Asymptotics and Special Functions*, Academic Press, New York, London, 1974.

[113] R. O'Malley, *Introduction to Singular Perturbations*, Academic Press, New York, 1974.

[114] ——, *Singular Perturbations Methods for Ordinary Differential Equations*, Springer-Verlag, Berlin, Heidelberg, New York, 1991.

[115] R. O'Malley and A. Jameson, *Singular perturbations and singular arcs – Parts 1 and 2*, IEEE Trans. Aut. Contr., AC-20 (1975), pp. 218–226; AC-22 (1977), pp. 328–337.

[116] R. O'Malley and L. Kalachev, *Regularization of nonlinear differential-algebraic equations*, SIAM J. Math. Anal., 25 (1994), p. 615–629.

[117] L. Polak and A. Mikhailov, *Self-organization in the nonequilibrium physico-chemical systems*, Nauka, Moscow, 1983. (In Russian.)

[118] B. Pol'skii, *Numerical Modeling of Semiconductor Devices*, Zinatne, Riga, 1986. (In Russian.)

[119] L. Pontriagin, *Ordinary Differential Equations*, Addison-Wesley, Reading, MA, 1962.

[120] M. Protter and H. Weinberger, *Maximum Principles in Differential Equations*, Springer-Verlag, New York, Berlin, Heidelberg, Tokyo, 1984.

[121] J. Rinzel, *Models in neurobiology*, in Nonlinear Phenomena in Physics and Biology, R. Enus, B. Jones, R. Miura, and S. Rangnekar, eds., Plenum Press, New York, 1981, pp. 345–367.

[122] Y. ROMANOVSKII, H. STEPANOV, AND D. CHERNAVSKII, *Mathematical Biophysics*, Nauka, Moscow, 1984.

[123] J. SANDERS AND F. VERHULST, *Averaging Methods in Nonlinear Dynamical Systems*, Springer-Verlag, Berlin, Heidelberg, New York, 1985.

[124] D. SATTINGER, *A nonlinear parabolic system in the theory of combustion*, Quart. Appl. Math., 33 (1975), pp. 47–61.

[125] L. SEGEL AND M. SLEMROD, *The quasi-steady state assumption: a case study in perturbations*, SIAM Review, 3 (1989), pp. 466–477.

[126] S.-D. SHIH AND R. KELLOGG, *Asymptotic analysis of a singular perturbation problem*, SIAM J. Math. Anal., 18 (1987), pp. 1467–1511.

[127] T. SISOEVA, *Asymptotic solution of initial-value problem for the system of first order partial differential equations*, Technical Report, deposited in VINITI, 1981.

[128] D. SMITH, *Singular-Perturbation Theory*, Cambridge University Press, Cambridge, 1985.

[129] A. TIKHONOV, *On the dependence of the solutions of differential equations on a small parameter*, Mat. Sbornik, 22 (1948), pp. 193–204. (In Russian.)

[130] ——, *On a system of differential equations containing parameters*, Mat. Sbornik, 27 (1950), pp. 147–156. (In Russian.)

[131] ——, *Systems of differential equations containing small parameters in the derivatives*, Mat. Sbornik, 31 (1952), pp. 575–586. (In Russian.)

[132] A. TIKHONOV AND A. SAMARSKII, *Equations of Mathematical Physics*, Pergamon Press, Oxford, England, 1963.

[133] A. TIKHONOV, A. VASIL'EVA, AND A. SVESHNIKOV, *Differential Equations*, Springer-Verlag, Berlin, Heidelberg, New York, Tokyo, 1985.

[134] V. TRENOGIN, *Development and application of the asymptotic Lyusternik-Vishik method*, Russian Math. Surveys, 25 (1970), pp. 119–156.

[135] V. TSYMBAL, *Cauchy problem for the hyperbolic system of the first order with several small parameters*, Dokl. Akad. Nauk Ukr. SSR, Ser. Fiz., Mat. i Tekhn. Nauki, (1983), pp. 27–29. (Russian).

[136] J. TYSON AND J. KEENER, *Singular perturbation theory of traveling waves in excitable media (a review)*, Physica D, 32 (1988), pp. 327–361.

[137] T. URAZGIL'DINA, *Asymptotic expansion of the solution of the problem on heat transfer in thin plates*, Technical Report, deposited in VINITI, 1986.

[138] ——, *Asymptotics for one quasistatic problem of thermoelasticity in thin disc*, Technical Report, deposited in VINITI, (1987).

[139] M. VAN DYKE, *Perturbation Methods in Fluid Mechanics*, Academic Press, New York, 1964.

[140] V. VASIL'EV, A. VOL'PERT, AND S. KHUDYAEV, *A method of quasi-stationary concentrations for the equations of chemical kinetics*, Comput. Math. Math. Phys., 13 (1972), pp. 187–206.

[141] A. VASIL'EVA, *Almost discontinuous solutions of a conditionally stable system with a small parameter multiplying the derivatives*, Differential Equations, 8 (1972), pp. 1204–1209.

[142] ——, *Periodic solutions of parabolic equations with small parameter*, Differential Equations, 19 (1983), pp. 1510–1515.

[143] ——, *Transition layer in a solution of a first-order system of partial differential equations*, Differential Equations, 21 (1985), pp. 1043–1049.

[144] ——, *On stability of solutions of contrast structure type*, Mat. Model., 2 (1990),

No. 1, pp. 119–125. (In Russian.)

[145] ——, *On stability of contrast structures*, Mat. Model., 3 (1991), No. 4, pp. 114–123. (In Russian.)

[146] ——, *Contrast structures with two transition layers of threshold type and their stability*, Zh. Vychisl. Mat. i Mat. Fiz., 32 (1992), pp. 1582–1593. (In Russian.)

[147] ——, *On the development of singular perturbation theory at Moscow State University and elsewhere*, SIAM Review, 36 (1994), pp. 440–452.

[148] A. VASIL'EVA AND V. BUTUZOV, *Asymptotic Expansions of the Solutions of Singularly Perturbed Equations*, Nauka, Moscow, 1973. (In Russian.)

[149] ——, *Singularly Perturbed Equations in Critical Case*, Tech. Report MRC–TSR 2039, University of Wisconsin, Madison, WI, 1980.

[150] ——, *Singularly perturbed differential equations of parabolic type*, in Asymptotic Analysis II, Lecture Notes in Math., , Springer-Verlag, Berlin, 1983, pp. 38–75.

[151] ——, *Asymptotic Methods in the Theory of Singular Perturbation*, Vischaja Shkola, Moscow, 1990. (In Russian.)

[152] A. VASIL'EVA AND M. DMITRIEV, *Singular Perturbations in Optimal Control Problems*, Itogi Nauki i Tekhniki, Vol. 20, VINITI, Moscow, 1973, pp. 3–77. (In Russian.)

[153] A. VASIL'EVA, S. KASHCHENKO, Y. KOLESOV, AND N. ROSOV, *Bifurcation of self-oscillations of non-linear parabolic equations with small diffusion*, Math. USSR-Sb., 58 (1987), pp. 491–503.

[154] A. VASIL'EVA AND E. KUCHIK, *Singularly perturbed systems of two first-order equations with partial derivatives of the first order in the critical case*, Differentsial'nye Uravneniya, 23 (1987), pp. 344–345. (In Russian.)

[155] A. VASIL'EVA AND M. SAJDAMATOV, *On periodic solutions of a singularly perturbed equation of hyperbolic type*, Izv. Akad. Nauk. UzSSR, Ser. Fiz.-Mat. Nauk, (1986), pp. 9–14. (In Russian.)

[156] A. VASIL'EVA AND T. SISOEVA, *Application of the boundary function method to singularly perturbed partial differential equation of the first order*, in Approximation Methods For Differential Equations and Their Applications, Kuibishev, 1981, pp. 32–44. (In Russian.)

[157] A. VASIL'EVA AND V. STEL'MAKH, *Singularly disturbed systems of the theory of semiconductor devices*, Comput. Math. Math. Phys., 17 (1977), No. 2, pp. 48–58.

[158] A. VASIL'EVA AND V. TUPCHIEV, *Periodic nearly-discontinuous solutions of systems of differential equations with small parameter in the derivatives*, Soviet Math. Dokl., 9 (1968), pp. 179–183.

[159] A. VASIL'EVA, V. TUPCHIEV, AND A. YARKIN, *Periodic solutions of systems of differential equations with small parameter before derivatives, close to discontinuous solutions*, in Proceedings of the Fifth International Conference on Nonlinear Oscillations, Vol. 1, Kiev, 1970, Inst. of Math. Acad. Sci. Ukr.SSR, pp. 149–157. (In Russian.)

[160] A. VASIL'EVA AND V. VOLKOV, *Periodic solutions of singularly perturbed parabolic equations*, Differential Equations, 21 (1985), pp. 1187–1191.

[161] ——, *Periodic solutions of some singularly-perturbed equations of parabolic type*, Comput. Math. Math. Phys., 25 (1985), pp. 609–614.

[162] M. VISHIK, *Asymptotic Behavior of Solutions of Evolutionary Equations*, Cambridge University Press, Cambridge, London, New York, 1992.

[163] M. VISHIK AND L. LYUSTERNIK, *Regular degeneration and boundary layer for*

linear differential equations with small parameter multiplying the highest derivatives, Usp. Mat. Nauk , 12 (1957), pp. 3–122. (In Russian.) Amer. Math. Soc. Transl., Ser.2, Vol. 20, 1962, pp. 239-364.

[164] A. VOL'PERT AND S. KHUDYAEV, *Analysis in the Class of Discontinuous Functions and the Equations of Mathematical Physics*, Nauka, Moscow, 1973. (In Russian.)

[165] L. VOROSHNIN AND B. HUSID, *Diffusive Mass Transfer in Multi-component systems*, Nauka, Minsk, 1979. (In Russian.)

[166] W. WASOW, *Asymptotic Expansions for Ordinary Differential Equations*, Interscience, New York, 1965.

[167] V. ZYKOV, *Modeling of Wave Processes in Excitable Media*, Manchester University Press, Manchester, 1988.

Index